Windows on the Wild®

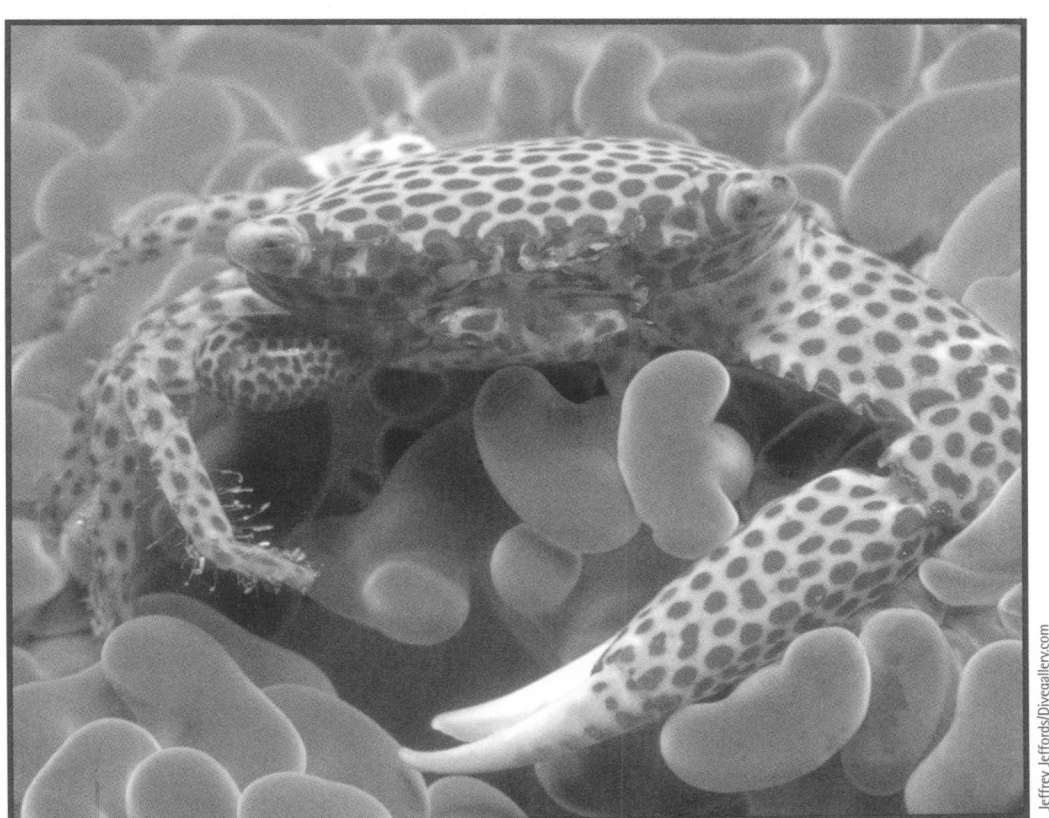

Jeffrey Jeffords/Divegallery.com

Oceans of Life
An Educator's Guide
to Exploring Marine Biodiversity

D1511394

Dear Educator,

The diversity of life in our seas is staggering. It ranges from microscopic plankton that release life-giving oxygen into the atmosphere to bustling, colorful coral reefs that support more than a third of all fish species in the world. In fact, our oceans provide habitats for more groups of organisms than do all our terrestrial habitats combined. More than three billion people depend on marine resources for food, jobs, medicines, recreation, and other services. And more than 16 percent of all animal protein consumed worldwide comes from oceans. Yet we know so little about our oceans that we don't fully understand just how important marine biodiversity is to our health, economy, and survival.

Unfortunately, what we do know is that marine biodiversity is severely threatened. In 2003, the science journal *Nature* published a study indicating that 90 percent of the oceans' large fish species have been depleted by commercial fishing. But fish aren't the only animals in trouble: Corals, whales, sea turtles, and many other ocean creatures are being adversely affected by human activities—from pollution to overfishing. And as the human population increases, the pressure on marine species and habitats will also increase—unless we turn the tide.

The encouraging news is that people around the world are taking action. They're doing everything from implementing new fishing techniques that harvest fish sustainably to creating marine protected areas that establish "safe spaces" for threatened or endangered species. But one of the most important things we can do to protect our oceans is to educate young people about the amazing diversity of marine life. And that's what this module is all about.

Oceans of Life is designed to provide ideas about how to integrate marine biodiversity into your teaching. We've included activities and background information that will work at many levels in both formal and nonformal settings. While the target audience for this module is middle school (grades six through nine), many of the activities can be adapted for use with older and younger students.

And, as you'll see, marine biodiversity is far from being a topic that's just appropriate in science classes. The activities included touch on social studies, language arts, mathematics, art, and drama, making this topic an ideal focus of interdisciplinary studies.

Oceans of Life is not intended to be used as a textbook. Rather, we hope that you will use it as a source of information and activities for building a unit or a course on this interesting and important topic. We encourage you to combine material from Oceans of Life with that found in a variety of other resources, including the other Windows on the Wild modules (Biodiversity Basics, Wildlife for Sale, and Building Better Communities). You can also supplement it with current newspaper, magazine, and Internet articles, and add a local flavor by using community-based case studies and stories gathered from your town, state, or region. By drawing from a range of sources, you'll be able to provide a diversity of viewpoints while adding depth and breadth to your students' marine studies.

Let us know what you think of the module by filling out the feedback form on pages 379-380 or on the Web at **www.worldwildlife.org/windows**. The WOW site provides a variety of supporting materials, including a "bonus" case study on salmon, the frameworks, correlations to science and social studies standards, workshop dates, online activities, links to the Biodiversity 911 traveling exhibition site, and other resources.

We hope you find this module to be inspiring and thought-provoking as you dive into marine biodiversity issues.

Judy Braus

Judy Braus, Director of Education

WWF CONSERVATION PROGRAM
Executive Staff

Kathryn S. Fuller
President and Chief Executive Officer

David Sandalow
Executive Vice President

Ginette Hemley
*Managing Vice President and
Vice President, Species Conservation*

Bruce Bunting
Vice President, Center for Conservation Finance

Guillermo Castilleja
*Vice President and Regional Director,
Latin American and Caribbean Secretariat*

Jason Clay
*Vice President,
Center for Conservation Innovation*

Eric Dinerstein
*Vice President,
Conservation Science and Chief Scientist*

William M. Eichbaum
Vice President, Endangered Spaces

Richard N. Mott
Vice President, International Policy

Brooks Yeager
Vice President, Global Threats

DEVELOPMENT TEAM

Director of Education
Judy Braus

Manager of Education and Outreach
Betty Olivolo

Managing Editors
Betty Olivolo, Claire Miller

Conservation Education Coordinator
Jeffrey England

Environmental Education Specialists
Ethan Taylor, Robyn Mofsowitz

Administrative Assistants
Susan Kevin, Nadia Cureton

Writers and Editors
Nicole Ardoin, Rita Bell, Gerry Bishop,
Dan Bogan, Judy Braus, Anne Canright,
Jenny Carless, Nora Deans, Joe Heimlich,
Claire Miller, Betty Olivolo, Sara St. Antoine,
Cindy Van Cleef, Christy Vollbracht, Luise Woelflein

Research Assistants
Lauren Arvidson, James Choe, Ryan Cree,
Christine Dell'Amore, Noah Domont, Anne Gessler,
Tina Leonard, Vanessa Jones, Florence Miller, Keri Parker,
Carlie Rodriguez, Kimberly Scott, Megan Thaler,
Dawn Turney

Module Design
Cutting Edge Design, DMP Design

Cover Design
Jim Nuttle, Inc.

Illustrators
Kirsten Carlson, Cutting Edge Design, Greg Davies,
Meryl Lee Hall, Jim Haynes, Joel Hickerson, DMP Design,
Cody Strathe (see Credits, inside back cover)

Marketing and Outreach Consultant
Barb Pitman

Project Evaluator
Lou Iozzi

ACKNOWLE

WINDOWS ON THE WILD Advisory Board and Council

Janet Ady
Chief, Division of Education Outreach, National Conservation Training Center, U.S. Fish and Wildlife Service

Julian Agyeman
Assistant Professor, Department of Urban and Environmental Policy, Tufts University

Gerry Bishop
Editor, Ranger Rick Magazine, National Wildlife Federation

Rich Block
Chief Executive Officer/Director, Santa Barbara Zoological Gardens

Dan Bogan
Environmental Scientist, Environment and Natural Resources Institute, University of Alaska Anchorage

Bruce Carr
Director, Conservation Education, American Zoo and Aquarium Association

Randy Champeau
Director, Wisconsin Center for Environmental Education, University of Wisconsin, Stevens Point

Dwight Crandell
Executive Director, St. Louis Science Center

Nora Deans
Environmental Education Consultant

Carol Fialkowski
Conservation Education Director, Field Museum of Natural History

Paul Grayson
Vice President, Indianapolis Zoo

Steve Hage
Environmental Science Educator, School of Environmental Studies, Minnesota

Joe Heimlich
Associate Professor, School of Natural Resources, The Ohio State University

Lou Iozzi
Professor of Natural Resources and Education, Rutgers University

Thane Maynard
Vice President, Conservation Foundation, Cincinnati Zoo & Botanical Garden

Kathy McGlauflin
Vice President of Education, Project Learning Tree, American Forest Foundation

Annie Miller
Educator, Jefferson Junior High School, Washington, D.C.

Terry O'Connor
Manager of Conservation Education, Woodland Park Zoo

Mary Schleppegrell
Assistant Professor of Linguistics and Director of ESL Program, University of California, Davis

Danie Schreuder
Professor of Environmental Education, University of Stellenbosch, South Africa

Samuel Scudder
Educator, Hart Junior High School, Washington, D.C.

Talbert Spence
Consultant, Talbert B. Spence Consulting

Cynthia Vernon
Vice President for Education and Conservation Programs, Monterey Bay Aquarium

Cherie Williams
Public Education Program Specialist, The Seattle Aquarium

Keith Winsten
Curator of Education, Brookfield Zoo

MARINE ADVISORY BOARD and EXPERT REVIEWERS

*denotes Marine Advisory Board member

*C. Michael Bailey
Gulf Coast Coordinator for Marine Recreational Fisheries, National Marine Fisheries Service, NOAA

*Kate Barba
National Education Coordinator, Estuarine Reserves Division, NOAA

Lillian Becker
Permit Specialist, Office of Protected Resources, NOAA

*Gerry Bishop
Editor, Ranger Rick Magazine, National Wildlife Federation

Herb Broda
Associate Professor of Education, Ashland University

*Scott Burns
Director, Fisheries Conservation, World Wildlife Fund

Christine Calardo
Education Specialist, Oregon Coast Aquarium

Acknowledgments

Holly Casman
*Aquarium Manager,
Albuquerque Aquarium*

Randy Champeau
*Director,
Wisconsin Center for
Environmental Education,
University of Wisconsin,
Stevens Point*

*Julie Childers
*Education Specialist,
Mote Marine Laboratory*

*Manuel Cira
*Cultural Manager,
Nausicaa, National Sea Centre,
France*

*Debi Clark
*Education Manager,
The Living Seas,
Walt Disney World*

*Vicki Clark
*Marine and Seafood
Education Specialist,
Virginia Institute of
Marine Science*

Jason Clay
*Senior Fellow, Center for
Conservation Innovation,
World Wildlife Fund*

*Becky Clayton
*Curator of Education,
The Florida Aquarium*

Jim Cliff
*Education Specialist,
Oregon Coast Aquarium*

*Gary Cook
*Director, Project WET,
Montana State University*

Grant Craig
*Marine Educator,
Discovery Hall Program,
Dauphin Island Sea Laboratory*

Ryan Cree
*Web and Publications Associate,
League of Conservation Voters*

Lu Eldredge
*Invertebrate Zoologist and Executive
Secretary, Pacific Science
Association, Bishop Museum*

*Stacia Fletcher
*School Programs Manager,
South Carolina Aquarium*

Eileen Flory
*Editor and Exhibits Specialist,
Oregon Coast Aquarium*

*Rosanne W. Fortner
*Professor of Natural Resources,
The Ohio State University*

*Nina Garfield
*National Estuarine Research
Reserves System,
National Ocean Service*

*Thomas Grasso
*Director of Marine Conservaton
Policy, World Wildlife Fund*

Amy Haddow
*Education Director,
Alaska Sealife Center*

*Martin Hall
*Chief Scientist,
Tuna-Dolphin Program,
Inter-American Tropical Tuna
Commission*

Lara Hansen
*Senior Program Officer,
Global Threats,
World Wildlife Fund*

Anne Harris
*Visitor Programs Educator,
Aquarium of the Americas*

Bill Hastie
*Outreach Coordinator,
Oregon Watershed Enhancement
Board*

*Josetta Hawthorne
*Executive Director,
Council of Environmental Education*

Paul Holthus
*Executive Director,
Marine Aquarium Council*

*Catherine Hubbard
*BioPark Education Curator,
Albuquerque Aquarium*

Elizabeth Rose Iglasias
*Educator,
South Park Middle School,
Corpus Christi, Texas*

*Julie Johnson
*Executive Vice President & Chief
Operating Officer,
New Jersey State Aquarium*

Ken Kassem
*Conservation and GIS Analyst,
Conservation Science,
World Wildlife Fund*

Marty Kodis
*Legislative Specialist,
U.S. Fish and Wildlife Service*

*Shelly Lakly
*Director of Education,
Zoo Atlanta*

*Mercedes Lee
*Vice President,
Blue Ocean Institute*

*Ghislaine Llewellyn
*Marine Conservation Scientist,
Conservation Science,
World Wildlife Fund*

Ron Lukens
*Assistant Director,
Gulf States Marine Fisheries
Commission*

*Tom Martin
*President,
Earth Force*

ACKNOWLE

Cathleen McConnell
Education Specialist,
Oregon Coast Aquarium

*Whit McMillan
Conservation Education Manager,
South Carolina Aquarium

Amber Neilson
Education Consultant,
Olympia, Washington

Andy Oliver
Education Consultant,
Washington, DC

Steve Olson
Director of Government Affairs,
American Zoo and Aquarium
Association

Mark Oswell
National Media & Constituent
Affairs, National Marine Fisheries
Service, NOAA

Libby Palmer
Community Science Specialist,
Port Townsend Marine Science
Center

Sue Perin
Residential Education Faculty,
Teton Science School

*Jim Pfeiffenberger
Exhibits Manager,
Alaska SeaLife Center

*Brady Phillips
Program Specialist,
National Marine
Sanctuaries Division, NOAA

Carl F. Rebstock
Senior Interpreter,
Monterey Bay Aquarium

Rachel Reinhart
Program Coordinator,
Washington Wildlife and
Recreation Coalition

Pete Salmansohn
Education Coordinator, Seabird
Restoration Program,
National Audubon Society

Chris Schmitz
Director of Public Programs and
Volunteer Services,
Oregon Coast Aquarium

Dan Schrag
Professor of Earth and
Planetary Sciences,
Harvard University

*Tara Schultz
Director of Education,
Texas State Aquarium

Nancy Sefton
Photographer/Video Production,
Triton Video Project

*Ursula Sexton
Senior Research Associate,
WestEd

*Anne Smrcina
Education Coordinator,
NOAA Stellwagen Bank

*Billy Spitzer
Vice President for Programs
and Exhibits,
New England Aquarium

*Vikki Spruill
Executive Director,
SeaWeb

*Kim Standish
Assistant Director of Education,
Mystic Aquarium

*Sara St. Antoine
WWF Consultant/Writer

Joe Starinchak
Outreach Coordinator,
U.S. Fish and Wildlife Service

Karen Stickman
Water Quality and Aquatic
Environment Monitoring Project,
Native American Fish and Wildlife
Society

*Pam Stryker
Teacher and Secretary,
The National Marine Educators
Association

Tim Tynan
Sustainable Fisheries Division,
National Marine Fisheries Service,
NOAA

*Cynthia Vernon
Vice President for Education and
Conservation Programs,
Monterey Bay Aquarium

*Bert Vescolani
Vice President of Education and
Programs,
Shedd Aquarium

*Christy Vollbracht
WWF Consultant/Writer

*Sharon H. Walker
Associate Dean for Outreach,
College of Marine Sciences,
The University of Southern
Mississippi

Steven Webster
Senior Marine Biologist,
Monterey Bay Aquarium

Kelley Wharity
Visitor Programs Educator,
Aquarium of the Americas

Jim Wharton
Education Specialist,
Oregon Coast Aquarium

*Cherie Williams
Public Education Program
Specialist, The Seattle Aquarium

DGMENTS

Morgan Witman
Educator,
Smithfield High School
Smithfield, Rhode Island

Luise Woelflein
Educational Program Coordinator,
Bureau of Land Management,
Alaska State Office

PILOT EDUCATORS

Emily Becker
Tri-Valley School,
Healy, Alaska

Michelle Belt
St. Mary's Ryken High School,
Leonardtown, Maryland

Eileen Berteling
Montgomery Middle School,
Rockville, Maryland

Marcia Bisnett
Miami Norland Senior High School,
Miami, Florida

Catherine W. Carter
Georgia Perimeter College,
Decatur, Georgia

Terri Carter
K-Beach Elementary School,
Soldotna, Alaska

Janet Charnley
The Evergreen School,
Shoreline, Washington

Cindy Connolly
Benjamin Banneker Middle School,
Burtonsville, Maryland

Candace Curtis
Lake Weir High School,
Ocala, Florida

Phyllis D'Angio
Long Wood Middle School,
Middle Island, New York

Jay Drag
Eagle Ridge Science and Technology
Magnet School,
Coral Springs, Florida

Kelly Drinnen
Moody Gardens,
Galveston, Texas

Kathy Ehrlick
Montgomery Village Middle School,
Rockville, Maryland

Laura Elkins
Hyde Leadership Public Charter
School,
Washington, D.C.

Jay Freundlich
Langley Middle School,
Langley, Washington

Leslie Gates
Chaffee Zoological Gardens,
Fresno, California

Dave Grant
Brookdale Community College,
Lincroft, New Jersey

Natalie Hansen
E. Rudd Intermediate School,
Van Vleck, Texas

Laurie Hart
St. Joseph's Academy,
St. Louis, Missouri

Patricia Hewitt
The University of Tennessee at
Martin,
Martin, Tennessee

Rindy Higgins
Maritime Aquarium/CT Sea Grant,
Norwalk, Connecticut

Elizabeth Rose Iglesias
South Park Middle School,
Corpus Christi, Texas

Anne James
Riverside Zoo,
Scottsbluff, Nebraska

Andrew Kirk
Eagle Ridge Science and Technology
Magnet School,
Coral Springs, Florida

Pamela Kucsan
Forest Oak Middle School,
Gaithersburg, Maryland

Ardi Kveven
Snohomish High School,
Snohomish, Washington

Annette Matzner
Norwood School,
Bethesda, Maryland

Stephen Messinger
Southhampton Intermediate School,
Southhampton, New York

Paula Miles
Edwardsburg Middle School,
Edwardsburg, Michigan

Delyth Morgan
Royal Oak Middle School,
Victoria, British Columbia

Kathy Mullin
Ocean Quest/Sturgis Charter School,
East Dennis, Massachusetts

Lanis Petrik
Brookfield Zoo,
Brookfield, Illinois

Kim Raccio
Maritime Aquarium/CT Sea Grant,
Norwalk, Connecticut

Joseph Ruak
Dept. of Lands and Natural
Resources, Division of Fish and
Wildlife,
Saipan, Mariana Islands

ACKNOWLE

Jo Schiebel
Calvary Middle School,
Silver Spring, Maryland

Marge Selfridge
Tuckerton Elementary School,
Tuckerton, New Jersey

Terry Slaven
Sherrod Elementary School,
Palmer, Alaska

Hans Swygert
Furman Middle School,
Sumter, South Carolina

Sacheen Tavares
Florida Sea Grant,
Davie, Florida

Bill Weinsheimer
Black Hills High School,
Olympia, Washington

Morgan Whitman
Smithfield High School,
Smithfield, Rhode Island

Nora Wittmore
Benjamin Banneker Middle School,
Burtonsville, Maryland

DGMENTS

TABLE OF CONTENTS

1 Biodiversity Break-Down 64
Use the "Joy to the Fishes and the Deep Blue Sea" poster to introduce students to the three levels of marine biodiversity and to some of the richest marine ecoregions around the world. Have students create their own posters.

2 Sea for Yourself ☼. 72
Take a trip to the seashore; visit a natural history museum, science center, zoo, or aquarium; or go on a "virtual tour" of oceans to get more closely acquainted with marine life and the three levels of marine biodiversity.

3 Ocean Explorers ☼. 82
Conduct an outdoor simulation to demonstrate why it's difficult to discover new species in the ocean.

4 Services on Stage 86
Act out four short skits that demonstrate some of the many services marine biodiversity provides, then design a print or video ad that educates people about the importance of marine biodiversity.

5 Going Under 94
Explore different perspectives of what it's like to be under the sea.

CHAPTER 2: Coral Reefs. 100
Background Information 102

1 Build-a-Reef 110
Draw or build models of a coral colony and display them in a classroom "coral reef."

2 Postcards from the Reef. 116
Gather clues provided in a series of fictional postcard messages to determine the location of coral reefs around the world. Learn more about the benefits of healthy coral reefs to people and wildlife, as well as some of the threats coral reefs face.

3 Coral's Web 128
Use information cards to create a coral reef food web, and then explore some of the ways that natural and human forces affect this web of life.

4 Coral Bleaching: A Drama in Four Acts 💡 140
Perform skits that show the interdependence of corals and zooxanthellae, as well as the devastating effects of rising ocean temperatures and coral bleaching.

☼ = Outside Activity   💡 = Challenging Activity

© WWF/Fritz Pölking

"The ocean represents everything that allows us to live on Earth. Making this a personal connection for people everywhere is one of the most important jobs anyone can do."

–Francesca Cava,
National Geographic Society's Sustainable Seas Expedition

Welcome to Windows on the Wild and Oceans of Life

"For each of us, then, the challenge and opportunity is to cherish all life as the gift it is, envision it whole, seek to know it truly, and undertake with our minds, hearts, and hands to restore its abundance. It is said that where there's life there's hope, and so no place can inspire us with more hopefulness than the great, lifemaking sea, that singular, wondrous ocean covering the blue planet."

–Carl Safina, marine ecologist

World Wildlife Fund

Windows on the Wild: Oceans of Life

Welcome to Windows on the Wild®

"Biodiversity represents the very foundation of human existence. Besides profound ethical and aesthetic implications, it is clear that the loss of biodiversity has 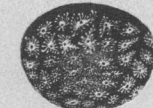 serious economic and social costs. The genes, species, ecosystems, and human knowledge that are being lost represent a living library of options available for adapting to local and global change. Biodiversity is part of our daily lives and livelihood and constitutes the resources upon which families, communities, nations, and future generations depend."

—Global Biodiversity Assessment,
Summary for Policy Makers
United Nations Environment Programme

Biodiversity! Although the term may seem intimidating to some, you couldn't choose a more engaging and stimulating topic—or one as all encompassing and important for our future.

Biodiversity is the variety of life on Earth. It's everything from the tiniest microbes to the tallest trees, from creatures that spend their entire lives deep in the ocean to those that soar high above the Earth's surface. It's also the word used to describe the wealth of habitats that house all life forms and the interconnections that tie us together. All of Earth's ecosystems and the living things that have evolved within them—including the fantastic range and expression of human cultures—are part of our planet's biodiversity.

Oceans of Life is part of World Wildlife Fund's environmental education program called *Windows on the Wild*, or *WOW*. *WOW* uses the topic of biodiversity as a "window" to help learners of all ages explore the incredible web of life. *WOW* also explores the complexity of biodiversity—and looks at it within scientific, social, political, cultural, and economic contexts.

We believe that biodiversity is an important and powerful topic that draws in learners. As a theme, it cuts across many disciplines and provides real-world contexts and issues that promote critical and creative thinking, citizenship skills, and informed decision making. Biodiversity also illustrates the complexity of environmental issues and makes clear that there are many perspectives as well as much uncertainty.

The diversity of life on Earth shapes and nourishes every facet of our existence. But because those connections are seldom obvious, we have often pursued short-term interests with limited regard for the well-being of other species and the places they live. At the same time, social and economic inequities have forced some people to overexploit resources to meet their basic needs. As a result, biodiversity is rapidly declining. If we want to ensure the long-term health of the planet, we need to develop an informed and motivated citizenry that understands what biodiversity is and why it's important, as well as the many factors that affect it—from habitat loss to climate change. And we need citizens who have the skills and confidence to rise to the challenge of protecting biodiversity and who feel empowered to do so. Education, we believe, is one of the best tools we have for achieving this goal.

An Overview of Oceans of Life

Oceans of Life is the third educator's guide in the *Windows on the Wild* middle school program. As you read through the table of contents, you'll see that this module focuses on the many issues associated with marine biodiversity. From the threat of invasive alien species to the wonders of coral reefs, marine biodiversity provides rich opportunities to engage students in discovering a part of the planet that is largely unexplored but increasingly threatened.

Oceans of Life uses case studies focused on familiar species and engaging topics to help learners of all ages explore life in the oceans. The module takes students from mountain streams and coastal wetlands to the deepest, darkest depths of the ocean—and to a variety of ecosystems in between—on its tour of Earth's marine diversity. The module's case studies challenge students to consider the complexities of conserving marine biodiversity in ways that support the future of wild species and habitats, as well as the vitality of human communities.

NOAA Coral Kingdom

This module explores a variety of complex and controversial issues, including trawling and aquaculture, sustainability of current fishing practices, and different cultural perspectives on marine species and habitats. Through this broad-based approach to marine topics, the module aims to help learners understand how diverse points of view are essential to the successful conservation of marine biodiversity. See the box on the right for specifics on the key concepts emphasized throughout the module.

Oceans of Life can help bring students closer to the rich tapestry of life in the oceans. Not only will the module give students a better appreciation for the kinds of life thriving in the oceans, but it will also introduce them to the tough issues we'll all face in conserving marine biodiversity in the future.

"Life . . . lies deep in burrows and tubes and passageways. It tunnels into solid rock and bores. It encrusts and spreads over a rock surface or wharf piling, keeping alive the sense of continuing creation and of the relentless drive of life."

–Rachel Carson, ecologist

Marine Biodiversity Key Concepts and Ideas

- The world's oceans contain an incredible variety of life. Marine biodiversity not only includes a rich diversity of species, but it also includes a diversity of ecosystems where species live, as well as the genetic diversity within species. In addition, marine biodiversity is often intricately connected to life on land.

- Marine biodiversity provides many benefits to people. From maintaining the balance of gases in the air to feeding billions of people every day, the seas help keep people healthy and economies running smoothly. Marine species and habitats also hold cultural significance for many people around the world, and provide recreational, aesthetic, and spiritual benefits.

- A variety of human activities are threatening the future of marine biodiversity. Unsustainable fishing operations, habitat destruction, pollution, and invasive alien species are just a few of the threats that are causing major disruptions in marine ecosystems. Many of these issues are directly related to increasing human populations, the rising demand for marine products, and economic development along the world's coastlines.

- There are many things people can do to help conserve marine biodiversity. Even people living far from the ocean can have a positive effect on marine ecosystems by reducing water pollution in their local waterways, buying sustainably caught seafood at local stores and restaurants, and speaking out for the future of our oceans.

How Oceans of Life Is Organized

Oceans of Life includes background information on marine biodiversity, followed by five introductory activities to set the stage for your students. The remainder of the module is organized into five case studies—coral reefs, shrimp, sharks, alien species, and salmon*—which touch on a variety of engaging topics related to marine conservation. Each case study includes background information on the topic, accompanied by hands-on activities to help you teach the important concepts. We have also developed mini case studies, which include short descriptions of several other fascinating marine topics along with brief activity ideas.

In the appendices, you'll find a unit-plan section that provides some themes and ideas on how to link the activities to create broader units. We've also developed a biodiversity education framework, which includes concepts and skills related to marine issues and biodiversity. (The framework is located on the *WOW* Web site at **www.worldwildlife.org/ windows**).

The module contains a variety of resources to help you teach about marine biodiversity, including:

- a glossary of terms for students and educators;
- a list of important legislation regulating marine species and habitats, including U.S. and international laws and treaties; and
- a resource section with suggestions for where to find additional information and activities.

Oceans of Life is designed for use with students in sixth through ninth grades, but we've found that many of the activities work well with younger and older age groups, too. We've designed the activities for use in schools as well as nonformal settings, such as museums, zoos, aquariums, nature centers, and other community education institutions. Additionally, a number of the activities can be adapted for use in home teaching.

Please take a few minutes to fill out the feedback form (on page 379) so that we can incorporate your ideas into the next revision. We depend on suggestions from educators, students, scientists, and others to make sure our educational materials are the best they can be. Each activity has been field-tested in formal and nonformal settings and revised based on comments from educators and students. Scientists and other marine experts also reviewed several drafts of this module as it was being developed.

* *You can find the Salmon Case Study at* **www.worldwildlife.org/windows/marine.**

BIOFACT

The volume of the world's oceans is estimated to be approximately 328 million cubic miles, comprising 85 percent of the total water on the Earth's surface.

The Building Blocks of Windows on the Wild

The goal of *Windows on the Wild* is not to teach your students what to think about biodiversity, but rather to introduce them to fascinating topics, raise challenging questions, and guide them to explore, analyze, evaluate, and discuss those issues from an informed position. We're aware that biodiversity issues can be both controversial and complicated. Your students may bring many diverse perspectives to the biodiversity-related issues you introduce. But with careful guidance, these different points of view can contribute to a dynamic learning environment in which students clarify their own thinking, learn how to listen to others, and gain new insights about intriguing and relevant issues.

We hope that the *WOW* series opens your students' minds to the wondrous diversity of life around them. We also hope it engages them in thoughtful dialogue about their place on the planet—and about the future of the world we all share.

WOW is built on a set of underlying principles about education. As you read through the activities in this module, you'll see many familiar strategies and approaches—from constructivist education, which values prior experiences and knowledge, to innovative assessment strategies, group learning, problem solving, interdisciplinary teaching, and experiential learning. (For more about the teaching strategies we've used in this module and others, see "Putting the Pieces Together" on pages 322-325.)

Education should challenge students to think critically and creatively about their world—to question how and why we do things, and how we might do them differently. It should promote positive change (both personally and within communities), help students envision a better society, increase respect and tolerance for others, and build effective citizenship skills and stewardship. In the *WOW* program, we emphasize four overlapping themes that we believe can help create a more sustainable society: futures education; community action and service learning; education for sustainability; and creating a sense of wonder. We've touched on each in the following subsections.

Futures Education: Looking Ahead

When kids watch movies about the future, they often see a world gone awry. In fact, many writers and filmmakers center their fictional future breakdown on environmental disasters. So how can we teach about biodiversity loss and other environmental issues without making students cynical, or even terrified, about the future they'll inherit? The answer may lie in futures education—education that encourages students to envision a positive future and the role they can play to make such a future happen.

Thinking constructively about the future may be more important now than ever before. Increasingly, the way we choose to live is affecting global natural systems, from the atmosphere to the oceans. Rather than letting your students feel like victims of an inevitable future disaster, you can encourage them to see themselves as active participants in creating a more livable future.

> *"Now the leaders of the world must demonstrate the political will and leadership that is so desperately needed if we are to save the oceans and the circle of life that supports us all."*
>
> **–Dawn M. Martin,
> Executive Director, Oceana**

In the *Windows on the Wild* modules, you'll find a number of activities designed to help your students look forward. These activities encourage them to imagine futures that will contribute to their own quality of life, as well as to the well-being of both local and global communities. (For more information on futures education, contact the World Future Society at 7910 Woodmont Avenue, Suite 450, Bethesda, MD 20814. **www.wfs.org**.)

Service Learning— A Closer Look

More and more schools across the country are embracing service learning as a way to engage students in community activities that solve real-life problems, apply academic skills, and help others in the process. The National and Community Service Act of 1990 defines service learning as "thoughtfully organized service experiences that meet actual community needs . . . and that help students learn and develop through active participation." Many educators predict that service learning will continue to grow and that, by encouraging students to take part in projects that focus on the environment, health, the arts, the elderly, politics, and other important community issues, we will produce a nation with more caring, committed, and skilled citizens.

Learning from the Community

In schools and communities around the world, educators are finding that one of the best ways to prepare students for their future role as active, voting citizens is to get them involved in local issues. By addressing a real community need, students can learn about the political process, environmental issues, careers, project planning, and what it means to be a responsible citizen.

Biodiversity lends itself to service learning and community investigations. Every community faces environmental challenges that affect the well-being of both people and wild species: pollution, rapid development, transportation problems, park planning, and so on. By getting involved in a biodiversity-related project, your students will invest energy in their community and see that they can help to improve its condition.

At their best, service learning and community action are learner-centered and teacher-facilitated, with constant opportunities for both students and facilitators to reflect on the problem, approach, and achievements of the project. This cycle of action, reflection, and revised action is known as action research. Many educators already rely on action research to improve their teaching. And educators report increased motivation and maturity among their students as the students become more involved in and thoughtful about their learning process. (For more about service learning, see *Enriching the Curriculum Through Service Learning*, edited by Carol W. Kinsely and Kate McPherson, Association for Supervision and Curriculum Development, 1995. For related insights, read *Environmental Education for Empowerment: Action Research and Community Problem Solving* by William B. Stapp, Arjen E. J. Wals, and Sheri L. Stankorb, Kendall/Hunt Publishing, 1996. For tips on getting your students involved in local issues, see "Getting Involved!" on pages 358–377 in *Biodiversity Basics*.)

The White Sea in northwestern Russia is named for the ice that covers the sea for more than 200 days each year.

Education for Sustainability

Many of the activities in *Oceans of Life* explore pathways to sustainability—meeting the needs of the present without compromising our ability to meet the needs of the future. In the process, the activities examine the relationships among ecological integrity, economic prosperity, and social equity. And, because these three goals often come into conflict, the module also works to develop students' ability to listen, analyze, negotiate, persuade, and compromise.

Thinking in terms of sustainability—and finding ways to balance economic issues, social equity, and ecological integrity—also requires thinking beyond our immediate needs and interests. The activities in this module encourage students to consider the perspective of other individuals, communities, and cultures, and to look forward to assess the way actions today will affect the lives of people and other species in the United States and around the globe in the future. These activities also challenge students' thinking about fairness, individual and community responsibility, and other concerns that are critical to our understanding of sustainability. At the same time, the activities within this module, and many of those listed in the resources section, encourage students to work on developing a set of personal ethics—a framework by which they can make decisions.

Oceans of Life, and other modules in the *Windows on the Wild* series, introduce the theme of sustainability to help students, educators, and the public create a more positive vision of the future. To facilitate this process, the activities encourage creative thinking and problem-solving skills—both of which are vital to taking action toward a more sustainable and equitable future.

Environmental Education and Agenda 21

In 1977, at the first meeting of its kind, representatives from more than 70 countries gathered in Tbilisi, Georgia (in the former U.S.S.R.), to discuss education and the environment. The goal was to "create new patterns of behavior of individuals, groups, and society as a whole towards the environment." Since then, environmental education has been evolving to meet the challenges facing the environment and society. Today, many countries have national environmental education programs and are trying to implement Agenda 21—the detailed plan resulting from the 1992 Earth Summit, which was sponsored by the United Nations Conference on Environment and Development (UNCED). Held in Rio de Janeiro, Brazil, the UNCED Summit embraced education as a key to our efforts to build a more sustainable society. At the 2002 World Summit on Sustainable Development, held in Johannesburg, South Africa, participating countries agreed that education is a critical tool in achieving sustainable development. During this ten-year follow-up to Rio, numerous partnerships were created to work toward specific targets for improving basic education, as well as environmental education, over the next five years.

BIOFACT

Water temperatures at the surface of the Persian Gulf often exceed 90° F in the summertime.

One View of Sustainable Development

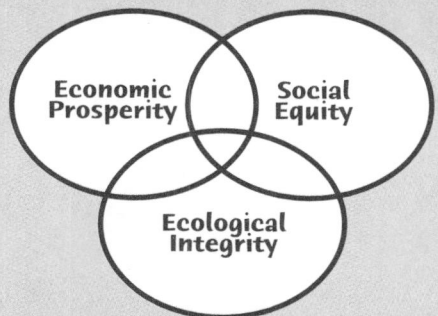

Economic Prosperity

Social Equity

Ecological Integrity

"Sustainable development meets the needs of the present without compromising the ability of future generations to meet their own needs. Choosing to be sustainable in businesses, schools, government institutions, and our individual lives demands a national commitment to the nation's economic prosperity, ecological integrity, and social equity." (President's Council on Sustainable Development, April 1995)

Creating a Sense of Wonder

Increasingly, the wonders of the natural world are lost on many of us. Each day the saga of human affairs dominates the media and demands our attention, leaving little room for any awareness of what's happening right outside our doors—in the soil beneath our feet, the trees lining the block, and the sky overhead.

At WWF, we believe that it's important to nurture a perspective that includes an understanding of natural processes and rhythms. After all, how can people be expected to value or protect something that they've had little or no exposure to and have little or no understanding of?

When people become aware of the fantastic phenomena that routinely take place in the natural world, most experience a sense of awe and appreciation. These phenomena range from the incredible journeys that salmon make from the sea back to the streams where they hatched and the vast and varied habitats created by tiny coral polyps, to the amazing diversity of sharks. Some have argued that we're "hardwired" to respond to these things with a sense of wonder, or at least with curiosity.

In the *Windows on the Wild* program, we provide many opportunities for educators to draw out students' own natural curiosity and sense of wonder. We believe that studying biodiversity is one of the best ways to stimulate these natural predispositions— building on students' inherent sense of wonder will help them understand and appreciate biodiversity. But there are also important educational advantages to presenting the wonder and "gee whiz" of biodiversity. Provoking curiosity leads students to ask questions and think creatively, to explore, and to challenge previous knowledge.

To help your students learn more about their local environment, we encourage you to spend as much time outside with them as you can—and to make use of the many natural areas and outdoor educational institutions that exist in your community, from local nature centers to zoos to city and rural parks.

Environmental Justice for All

The Environmental Justice movement is an active effort to make sure people of all races, cultures, and incomes have equal opportunities to live and work in healthy environments. By encouraging students to examine issues from all sides and to develop an ethical framework for making decisions and taking action, you can help them understand the role that social equity plays in creating a more sustainable world.

(For more information, visit the U.S. Environmental Protection Agency's Environmental Justice Web site at www.epa.gov/compliance/environmentaljustice/index.html or the Center for Community Action and Environmental Justice site at www.ccaej.org.)

Exploring

Oceans cover nearly three-quarters of our planet, and yet few of us have explored beyond the nearshore waters. Fewer yet understand much about the creatures that live in or around the sea, or their many kinds of habitats. In this section, you'll find background information on marine biodiversity, including the ecosystems that make up the marine environment, the services marine systems provide, the reasons for the general decline of marine biodiversity, and what we can do to protect our oceans.

Oceans of Life

An Introduction to Marine Biodiversity

If you were to travel across the ocean on a ship,  you might spend days without seeing many signs of life. You could search the horizon for movement or hang over the rail to look down into the deep blue of the water, but most likely you wouldn't see very much except for the flat, unbroken surface of the sea. If you were lucky, you might catch a glimpse of a  whale's tail as it dives, or you might spot a giant ocean sunfish basking on the surface. But these chance encounters would give you only the tiniest hint of how much life exists in the sea.

A s you search the ocean's surface, you wouldn't see the huge tuna that migrate hundreds or thousands of miles to mate or hunt prey. You wouldn't hear the complex songs of whales communicating over vast distances. You wouldn't notice if giant barracudas were swimming just beneath your ship—the only shady place for miles. You wouldn't see the billions of microscopic plants absorbing carbon dioxide and giving off oxygen. And you'd never know that some of the biggest, fastest, and oldest forms of life on the planet are living and dying in the water around you. Your trip across the ocean would reveal, at best, only a hint of the incredible variety of life at sea, known as *marine biodiversity.*

Most of us, including marine scientists, don't know a lot about marine biodiversity. That's because,  according to some estimates, up to 95 percent of the ocean remains unexplored.[1] In the last century alone, important new species have been discovered in the ocean. In the 1930s, for example, fishers off the coast of South Africa hauled in a bizarre-looking blue and white fish with bulging eyes, unusual fins, and heavy scales.

coelacanth

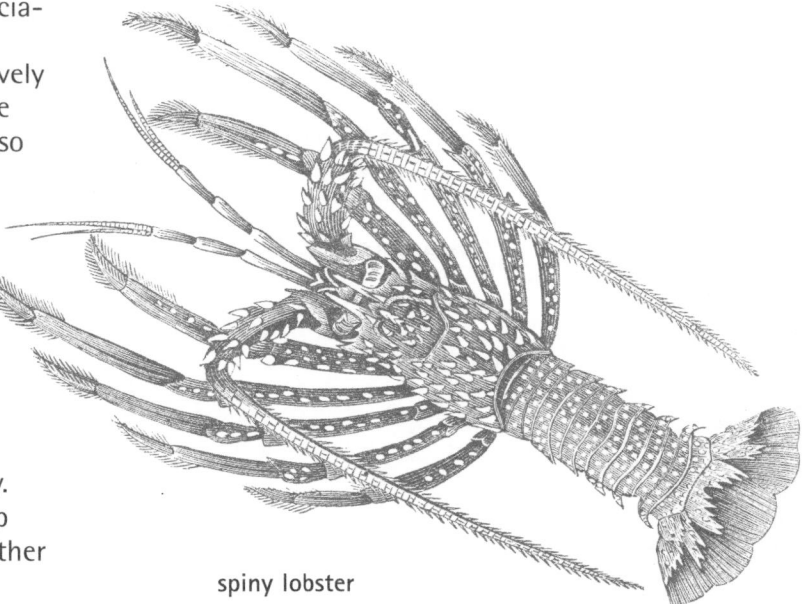
spiny lobster

Unable to identify the fish, the fishers took their find to scientists, who were shocked to discover that this was a coelacanth (SEE-la-kanth)—a fish that was thought to have gone extinct with the dinosaurs 65 million years ago. According to some scientists, the deep sea could contain up to 10 million species that have yet to be discovered.[2]

Because we know so little about the ocean and the life it contains, our understanding and appreciation of it lags behind our understanding and appreciation of life on land. Most of us see relatively few creatures from the ocean. The ones we do see are often lying on a bed of ice in a supermarket, so we don't have the same feelings about fish and other sea animals that we do about more familiar animals such as rabbits and robins. And because we don't pass through the ocean every day as we might pass through a community park, we often don't have the same concerns about damaging the ocean that we do about our local lands.

We also don't often realize how important marine biodiversity is to our health and economy. Many people are unaware that the sea helps keep the Earth's atmosphere livable for humans and other animals; provides jobs, goods, and services to millions of people; is a major source of protein around the world; and could be an important new source for cancer- and infection-fighting medicines. Without realizing how much we can be harmed by the loss of marine biodiversity, we may not worry when it's threat-ened. This lack of understanding and appreciation has led to serious problems for the world's oceans. But there are actions we can take to help protect marine biodiversity. In fact, many important and effective efforts are already underway.

On the following pages, we will explore the meaning and importance of biodiversity, the threats to the variety of marine life, and the actions we can take to help protect our amazing oceans.

BIOFACT

Under the ocean's surface exist mountains higher than Mount Everest, canyons deeper than the Grand Canyon, and the largest living structure on Earth, Australia's Great Barrier Reef.

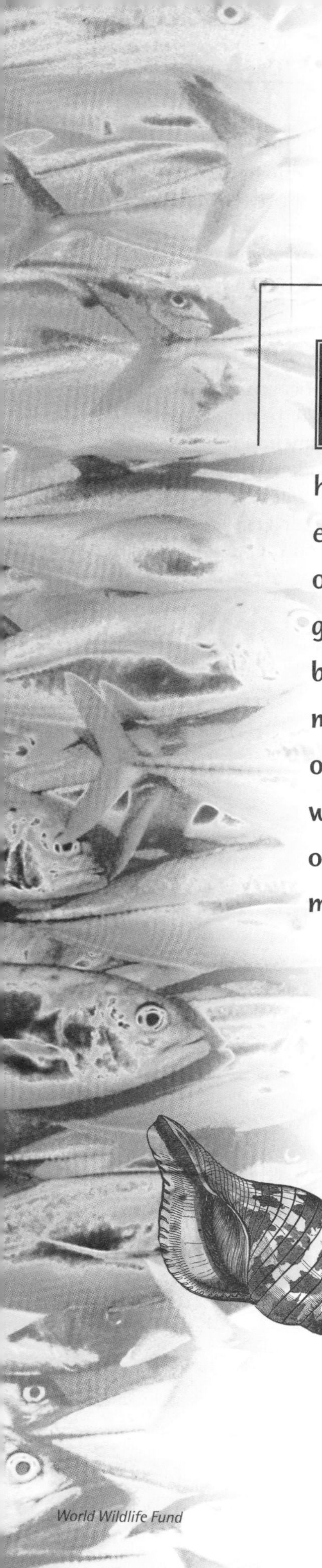

From deep, dark canyons to colorful coral reefs, an incredible diversity of habitats provides homes for an equally incredible diversity of ocean life. Almost every major group of organisms on Earth can be found in the sea, but, with much of the ocean's vastness out of sight and beyond our reach, we've so far managed to discover only a tiny fraction of the world's marine biodiversity.

"More has been learned about the nature of oceans in the past 25 years than during all preceding history. However, what we know about the oceans is still outweighed by what we do not know."

—**Independent World Commission on the Oceans**

Diving In!

WWF-Canon/Roger LeGuen

What Is Marine Biodiversity?

Biological diversity, or *biodiversity* for short, is the variety of life on Earth. Most people think of biodiversity as the amazing variety of *species* on the planet, from humpback whales to tiger salamanders. But biodiversity also encompasses the array of *genes* that species contain and the variety of *ecosystems* the species are found in. An ecosystem is a community of plants, animals, and other living things that are linked by energy and nutrient flows and that interact with each other and with the physical environment. An ecosystem can be as large as an ocean or rain forest or as small as a tide pool or rotting log.

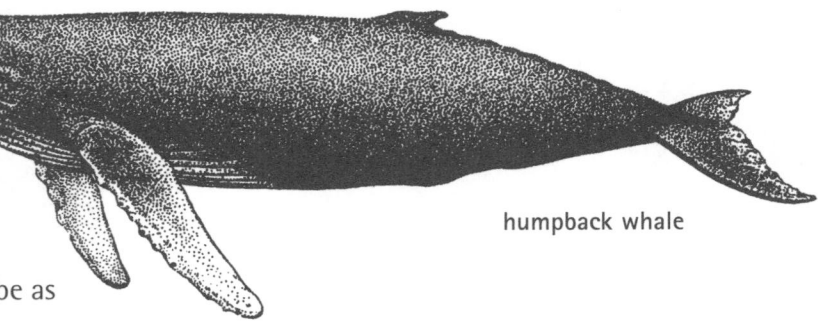

humpback whale

Marine biodiversity is the diversity of species and genes found in and around the Earth's saltwater ecosystems, including oceans, seas, bays, sounds, and all the areas where fresh water and salt water meet.

The Meaning of 'Marine'

The term "marine" means different things to different people. For most, the term is used to describe things found in or around the ocean. Some people use the term more broadly to describe anything found under water—in freshwater systems such as lakes, rivers, and streams, as well as the ocean—but this isn't the usual use of the word.

While the diversity of life found in freshwater habitats is vitally important and often connected to life in the ocean in complex ways, most of this guide focuses on the diversity of life connected to saltwater ecosystems. If you'd like to learn more about freshwater diversity, or would like resources for teaching about fresh water, check the Resources section on pages 360-369.

On Land and at Sea

How does marine biodiversity compare to biodiversity on land? Many scientists think there may be more species on land, but the sea contains many more animal phyla, or major groups of animals.[3] All of the 33 animal phyla that have been named, are represented in marine and coastal ecosystems, while not all of those phyla are represented in land ecosystems.[4] According to many scientists, this is an important measure of the sea's diversity.

Marine biodiversity isn't found just *in* bodies of salt water; it's also found *around* these bodies of water. Shorebirds, marsh plants, and any species that directly depend upon oceans for their survival are also considered part of the world's marine biodiversity.

Why So Diverse?

Some biologists think one reason for the high diversity of species in the ocean is that life may have originated and evolved there billions of years before appearing on land.[5] Biologists also think that the complexity and vastness of the oceans creates a wide diversity of habitats.[6] This, in turn, has led to the evolution of a high diversity of species. The ocean floor, for example, is incredibly varied, with huge open plains, chains of volcanic mountains, deep trenches, and rich coastal zones. In addition to all of this habitat complexity, the ocean is extremely deep: Life in the sea, unlike life on land, is not confined to the areas near the ground. Marine creatures inhabit the ocean bottom, the entire water column, the sea surface, and even the air above it—taking advantage of much more living space than there is on land. Add factors such as currents, light availability, and water chemistry, and you have the makings of a complex seascape that can support a high diversity of life.[7]

More to Explore

There's still a lot that we don't know about the ocean and the diversity it contains. For example, until recently, many scientists thought the bottom of the deep sea had little or no life. After all, what could possibly survive in a place with extreme pressure, near-freezing temperatures, and no light? But with new technology, we're starting to discover how many different and bizarre creatures have adapted to this dark, cold realm. And the discovery of deep-sea *hydrothermal* (hot water) *vents* has changed the way we think about life in the deep sea. The vents occur where extremely hot water (sometimes over 600° F) pours from openings in the ocean floor. Some of these springs, called *black smokers*, get their name because the water coming from them reacts with minerals in the rocks and turns black, making the vents look as if they're spewing out black smoke. Scientists were surprised

to find entire ecosystems built around these vents, with an amazing diversity of species including worms, clams, crabs, and mussels.

One group of organisms found at these vents could prove to be especially important to science. The organisms are tiny microbes that convert chemical energy into energy they can use through a process called *chemosynthesis*. Scientists recently tested the DNA of one of these microbe species and found that more than half of its genes were completely unlike any previously known genes.

WWF-Canon/Catherine Holloway

Because the organisms are genetically unique, some scientists think they should be placed in their own kingdom, the broadest level of taxonomic classification. And some scientists speculate that these creatures could be direct descendants of the first organisms to have evolved on Earth.[8]

What's more, scientists studying the deep sea floor recently found 898 species in 225 square feet of ocean bottom—an area smaller than the size of most backyard swimming pools.[9] Of those species, nearly half were new to science. If finds such as these occurred so recently, and if 95 percent of the ocean remains unexplored, what other new forms of life await discovery? Some scientists believe that there may be millions of unknown species still to be discovered in the ocean.

The inch-long conch fish, *Astraopogon stellatus*, hides in the shell of a queen conch during the day and comes out at night to feed.

Ocean Ecosystems

The oceans are made up of a variety of ecosystems, each of which harbors different species and plays a different role in ocean ecology. This quick guide covers some of the common ecosystems that best illustrate the variety of ocean life.

Coral Reefs

Coral reefs are among the oldest and largest ecosystems on the planet. Rivaling tropical rain forests in their biodiversity, coral reefs are home to about one-third of all known fish species. Reefs are made up of colonies of tiny animals called **polyps**. Many coral polyps secrete hard, calcium carbonate "skeletons," and over hundreds or even thousands of years, these structures accumulate to become huge reefs. Coral reefs are known for their bright colors, varied textures, and the diversity of fish, sponges, clams, sea stars, anemones, and many other forms of life that they support. (For more information, see the Coral Reef Case Study on pages 100-149.)

pufferfish

beach grass

Rocky Shores

In the United States, rocky shores are found mainly along the West Coast and the Northeast Coast. The rocky shores of both coasts provide habitat for a wide variety of organisms that cling to the hard substrate or find shelter in and around rocky crevices and hollows. Along portions of the West Coast, for example, huge strands of algae called **kelp** anchor themselves to the rocks offshore. The strands can grow to be more than 100 feet long, creating kelp forests that provide habitat to hundreds of species of animals. Along these rocky shores, the rising and falling tides help to create a variety of habitats—from constantly wet to mostly dry—that in turn give rise to an amazing diversity of life uniquely adapted to the changing conditions.

Sandy Shores

Sandy shores harbor a wide diversity of worms, clams, snails, and other creatures that live in sand or mud. In turn, those species provide food to large numbers of shorebirds—making some sandy shores important stopover points for migrating birds. Sandy shores are most common on the East Coast and Gulf Coast of the United States. The sand dunes that often form inland of the sandy shores provide habitat for an additional array of plants, insects, reptiles, birds, mammals, and other organisms.

Ocean Ecosystems (Cont'd)

Coastal Wetlands

As the name implies, coastal wetlands are areas along the coast that are usually wet. The term **estuary** is used to describe an area where fresh water and salt water meet, usually where fresh water spreads out at the mouth of a river. **Salt marshes** are often found within estuaries and contain grasses that can withstand the salt water that floods the marsh with every incoming tide. Tropical **mangrove forests** are another type of coastal wetland with plants adapted to living in salt water. Most coastal wetlands are important nursery areas for fish and other marine species.

Open Ocean

Many parts of the open ocean contain few fish or other large animals and few nutrients. Large areas of the open ocean are deeper than 300 feet, the depth below which there isn't enough sunlight for most plants to photosynthesize. And where there are no plants—the foundation of most food chains—there may be little other life. (Major exceptions are the food chains that are based on chemical energy captured by bacteria around deep ocean vents [see the "deep ocean" description].) In some parts of the open ocean, however, currents bring nutrients from decaying sea creatures to the surface. Large schools of fish often use those areas as feeding grounds, and, of course, people are attracted to their abundance of fish.

Deep Ocean

The deepest parts of the open ocean contain a number of distinctly different ecosystems, including hot vents, cold seeps, mud bottoms, and rocky bottoms. Although there's no sunlight, many creatures have adapted to life in the deep sea. Microbes use the chemical energy of vents in the ocean floor to make their own food and thus form the foundation of different food chains. Many organisms in the deep ocean have lights on their bodies that attract prey. For example, the anglerfish has a light that dangles in front of its large jaws and lures prey close enough to be snatched up.

harbor seal

Polar Oceans

Even in the coldest parts of the world's oceans, an astounding array of marine biodiversity thrives. A variety of creatures ranging from microscopic plants and animals to large mammals such as seals and whales live on, under, or near the polar ice. While the Arctic is home to polar bears, Antarctica is famous for its penguins. Whales are found in all oceans, including the polar oceans, and one sea bird, the Arctic tern, makes an incredible migration every year between the Arctic and Antarctic regions. In both polar oceans, plankton—tiny floating organisms—make up the base of food chains.

Every day the ocean spews tons of oxygen into the atmosphere, as tiny plankton photosynthesize. That's just one of the many "sea services" we take for granted. All life on Earth depends on the healthy functioning of the world's oceans. Without the oceans and the biodiversity they contain, we wouldn't have the life-supporting atmosphere we depend on, we would have to look for a major new source of protein to feed the world's growing human population, we would be without important medicines to help fight infections and cancers, and the quality of our lives would be greatly reduced. Marine biodiversity helps support the world economy, both directly—by providing goods, services, and employment to millions of people—and indirectly—by providing critical ecosystem services that help keep the planet working.

Clearly, marine biodiversity isn't just important to us— it's critical for our survival.[10]

> *"Since the beginning of life as we know it, the sea has supported our planet. From it has come the air we breathe, the food we eat and the water we drink. In short, the sea has given us life. . . . Perhaps it's time we return the favor."*
>
> **—Bob Talbot, marine photographer**

World Wildlife Fund

Sea Services

The Ocean and the Air

Oceans help control the Earth's weather, temperature, and balance of gases in the atmosphere. No matter how far from the ocean you live, everything from the air you breathe to the water you drink was produced by the ocean. That's because most of the rain, snow, and other precipitation that falls to Earth comes from the ocean. Every year, over 100,000 cubic miles of water evaporate from the oceans, and much of this water falls to the Earth as precipitation, providing the land with the fresh water it needs to maintain healthy plant and animal communities.[11]

Water Cycle

Since the Earth formed over four billion years ago, very little "new" water has been created. Instead, the same water has cycled through the oceans, atmosphere, and land again and again.

Oceans also affect global temperatures. Currents carry warm tropical waters toward the poles and bring cooler waters toward the equator. Because these currents move such large amounts of water, they have profound effects on the Earth's atmosphere, and many of these effects can make the climate milder. For example, the northern reaches of Europe are warmed by the Gulf Stream, a current that brings warm tropical waters north across the Atlantic. Thanks to the Gulf Stream, palm trees grow in southern Scotland[12]—a region lying at about the same latitude as frigid Moscow! And while ports at the same latitudes in other parts of the world are clogged with ice, ports warmed by the Gulf Stream are open for business.[13] But these powerful currents can also cause huge problems for people. Changes in winds and currents in the Pacific Ocean sometimes develop into an El Niño weather pattern, bringing heavy rains, flooding, droughts, fires, and deep freezes to many parts of the world.[14]

Of course, the oceans' biodiversity has no effect on temperature and precipitation: Currents and evaporation would continue even if the

BIOFACT

The vast majority of water on the planet—about 97 percent—is contained in the oceans. Only about 3 percent of the Earth's water is fresh, and two-thirds of that is found in ice caps and glaciers.[15]

oceans were emptied of their life. However, marine biodiversity does play a key role in maintaining the balance of gases in the atmosphere, especially carbon dioxide and oxygen. *Phytoplankton* are the important players here. They are very tiny (mostly microscopic) plants and plantlike organisms that drift in large quantities throughout much of the world's oceans. Like the plants on land, phytoplankton take in carbon dioxide and release oxygen in the process of photosynthesis. To make food, the phytoplankton convert sunlight and carbon dioxide to sugars they use for energy. The phytoplankton live in the surface waters of the ocean where sunlight is abundant and where carbon dioxide is absorbed from the air. These tiny organisms release oxygen as a byproduct of photosynthesis. It's estimated that 50 to 75 percent of the atmosphere's oxygen comes from ocean photosynthesis.[16] With growing concern about the connection between increased carbon dioxide levels and global warming, the ocean's role in taking up carbon dioxide from the atmosphere could become more important than ever.[17]

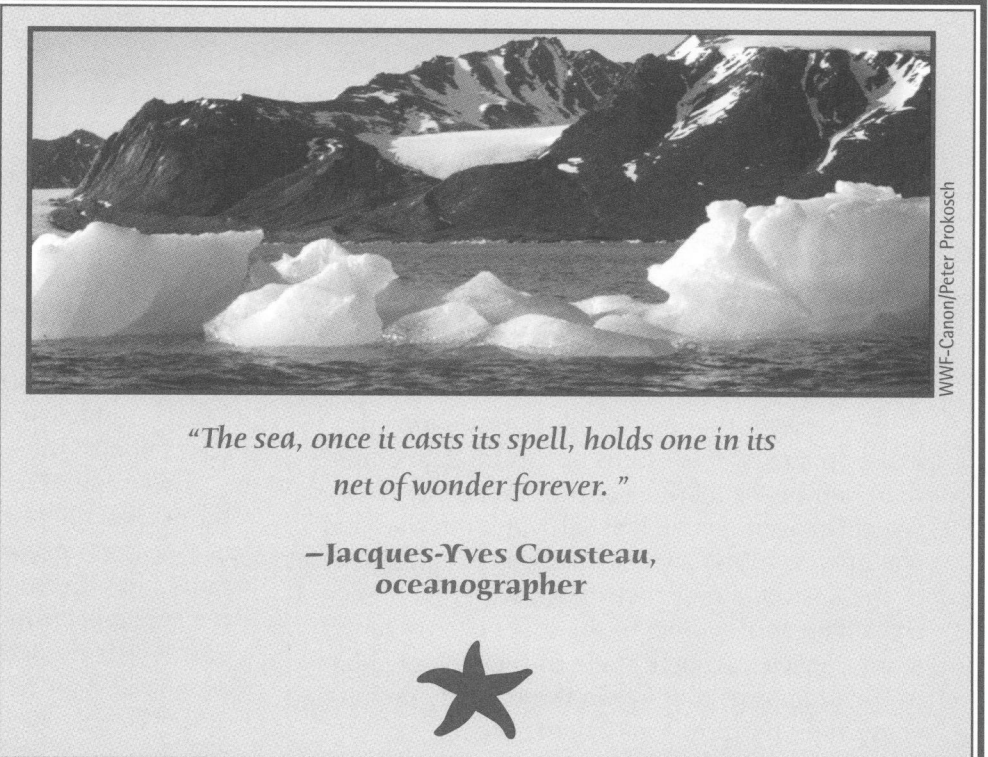

"*The sea, once it casts its spell, holds one in its net of wonder forever.*"

—Jacques-Yves Cousteau, oceanographer

Cost-Effective Coastlines

Anyone who has shopped for a house in a coastal area knows that land near the ocean is very valuable. Many people are starting to argue that coastal land may be even more valuable than we think. That's because, right now, we don't place a full monetary value on the services marine biodiversity provides for us.

In the past, we haven't often added the value of biodiversity into prices because we didn't know how much biodiversity was worth. Recently, economists have begun trying to put dollar figures on the services provided by biodiversity, and while these estimates are difficult to compute, they help give a rough idea of how much biodiversity contributes to the global economy. According to one estimate, the total value of the ecosystem services that biodiversity provides—including controlling the balance of gases in the air, recycling waste and nutrients, and controlling pollution—rings up at $33 trillion per year (about three times the total value of all goods and services produced annually in the United States).[18]

mangrove

The services that marine biodiversity provides make up almost two-thirds of this total, and more than half of the marine contribution comes from coastal ecosystems such as saltwater marshes and mangroves.[19] Why are coastal areas so valuable? Part of the answer is that they're rich with life. In fact, coastal areas are home to the vast majority of marine species at some stage of their lives.[20] Many of the organisms that live in these highly productive ecosystems perform a variety of services for people.

Mangrove forests, which grow along tropical shorelines, help protect coastlines from the effects of waves and storms that would otherwise pose much greater threats to coastal communities and farmlands. Also, the roots of mangrove trees and the shoots of marsh plants help slow a large portion of the more than 25 billion tons of sediments that are washed from the land into the oceans each year. When these sediments encounter mangrove roots and marsh grasses, a large portion of the sediments settle to the

bottom. As they fill in the wetlands and land plants move in, nutrient-rich coastal lands are formed—for free.[21]

Coastal wetlands also play a key role in filtering pollutants that have been washed from the land.[22] As pollutants move through these systems, some may settle into the sediments and plants absorb others. Although the chemicals are not completely eliminated when they are absorbed by plants, they're present in the water in much smaller concentrations and are released into the system much more slowly.

These and other services all add up to big savings for people. We benefit from these services every day, but we don't pay a cent for the removal of pollutants from the water, and we're never charged for the erosion control that mangroves provide. Many economists think we should start to consider the economic value of marine biodiversity. But whether or not we pay for it, we still benefit from it every day. And when it's lost, we don't factor that loss into our economic bottom line.

Food from the Sea

Marine biodiversity accounts for a major part of the world's food supply. From fish to clams to seaweed, the wild plants and animals of the sea feed billions of people around the world. On average, seafood provides almost 20 percent of the world's animal protein intake for humans,[23] making it the largest single source.[24] And in some parts of the world, such as Southeast Asia and the South Pacific,

tree coral

BIOFACT

Corals produce a natural sunscreen that chemists in Australia are purifying for human use.

seafood accounts for up to 90 percent of animal protein consumed.[25]

Every year, Americans spend more than $50 billion on fish and shellfish,[26] and each American eats, on average, more than 14 pounds of seafood. Since the 1960s, the amount of seafood eaten by Americans has been increasing, probably because of increased awareness of seafood's health effects. Low in saturated fat and high in protein, vitamins, minerals, and healthful oils, seafood has become a popular meal choice for many health-conscious Americans. And that's meant more jobs for people associated with the seafood industry. Hundreds of thousands of people work as fishers, processors, importers, and exporters of seafood products. These workers help to catch, process, and prepare the almost 10 billion pounds of seafood caught by American fishers each year.[27]

But seafood may be too nutritious for its own good. According to many scientists, the high demand for it has caused people who fish to put too much pressure on fisheries and to take more fish and shellfish than populations can replace. (For more about this problem, see Chapter 3.)

Marine Medicine

If you've recently taken a medication, chances are good that it had its origins in biodiversity. In fact, over 25 percent of medicines contain compounds derived from or modeled on chemicals found in the natural world.[28] Plants and fungi have proven to be important to drug-hunting scientists because they contain the most

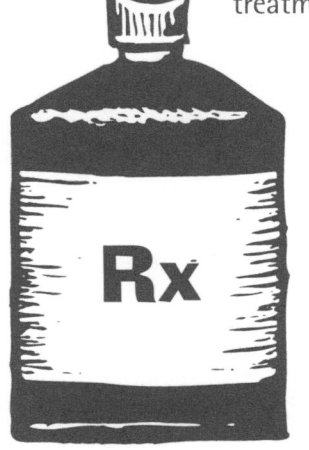

complex chemicals. These chemicals help defend the plants and fungi against animals that might otherwise find the immobile organisms an easy meal.

The sea not only has its share of plants and fungi, but it also has an incredible variety of immobile animals, making the potential for finding useful chemicals in the sea very high.[29] So far, hundreds of marine species have been identified as having medicinal properties,[30] and drug companies in several countries are searching for more.[31] Many of these organisms can be found in coral reefs, where defensive chemicals offer a great advantage. Drugs from sea life that have shown promise in lab tests or are already being used include antibiotics, tumor inhibitors, pain suppressants, anti-inflammatory drugs, skin care products, sunscreens, and medicines for treating heart and nerve conditions.[32]

In addition to the medicines that can be derived from sea life, some marine organisms are now being used in new medical treatments. Coral, for example, is being used to help reconstruct damaged bones.[33] Scientists are hopeful that, as we discover more marine biodiversity, we may uncover more potential drugs and medical treatments for use in the future.

People and the Ocean

Every year, millions of people crowd onto beaches all over the world. They spend millions of dollars on masks, snorkels, and other equipment that will help them get a good look at life in the sea, they write songs and paint pictures about the ocean and its life, they hire captains to take them to productive fishing spots, they walk along shorelines collecting shells to display in their homes, they shop for

of the Bering Sea, for instance, derive a large part of their diet, clothing, and mythology from the sea.

Other Americans have close ties to the sea. Every year 180 million people visit our coastlines.[34] In fact, the fastest growing segment of the tourism industry is marine based,[35] although a large majority of these people may not be interested in nature and wildlife. However, 34 million go on whale-watching trips, while nearly 36 million people visit aquariums annually.[36]

According to some psychologists, the connections we form with the living world around us are important parts of the way we develop as humans.[37] They suggest that all the time and money people spend getting to the ocean (and other natural areas) and inter-acting with the life there is an indication of how much we need these connections. If we don't get opportunities to see gulls, explore seashores, look at dolphins, and learn about penguins, we will miss out on some important parts of what makes us human.

WWF-Canon/Susan Wells

seafood to prepare and enjoy with friends, and they visit aquariums to see marine life. It's clear that people value marine biodiversity not just because it helps maintain the atmosphere or provides a huge part of the food supply, but also because it's important for aesthetic, spiritual, and cultural reasons, among others.

Throughout history, marine biodiversity has played an important role in societies around the globe. The ancient Greeks used images of marine creatures to decorate their mosaics and pottery, and Polynesians told stories of the origins of the Earth in a giant clam. Today, many cultures are still closely tied to marine biodiversity. The Inuits

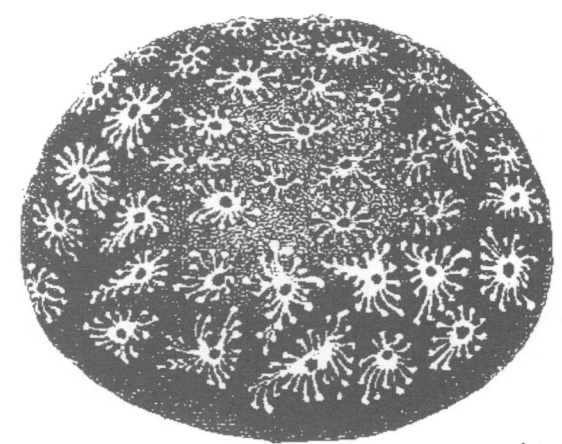

sea squirt

BIOFACT

Chemicals harvested from Caribbean sea squirts may play a role in halting the spread of cancer. A chemical produced by these brightly colored animals stops cancer cells from resisting anti-cancer drugs, making those drugs more effective in fighting the disease.

Marine Biodiversity and the "Web of Life"

Almost all living things on Earth are closely interconnected. In innumerable ways, species depend directly on each other for their survival. Many ecologists use the analogy of a spider web to describe the ways that living things are connected to one another. Like a web, what happens in one part of a system can have important effects on other parts of the system. Remove any one species, and there could be serious repercussions.

In the California kelp forest, it's easy to see how interconnected species can be: Sea otters are top predators in the kelp forest and consumers of sea urchins. When otters are eliminated (as they have been in many areas), they no longer serve as a check on sea urchin numbers. The urchin population then rises quickly, and urchins consume huge amounts of kelp, forming large "barrens" in their path. Loss of the kelp leads to losses of other species that depend on it.[38] Many other marine systems display this same vulnerability to disruptions in the wake of species losses.

Not only is every species important in marine systems, but species and activities on the land also affect life in the sea, creating a web that extends over the entire planet. Not all marine species spend all of their lives in the ocean; in fact, many marine fish are hatched hundreds or thousands of miles from the coast in mountain streams. Called *anadromous* fish, these species live their adult lives in the ocean but spawn (lay their eggs and fertilize them) in rivers. Salmon are some of the most famous anadromous fish, and they're dependent on healthy forests for their survival. If forest and stream conditions aren't just right, the fish can't spawn and their numbers fall dramatically.[39] That's just one of the many examples of how species depend on each other for survival throughout the ocean and on land.

THE FAR SIDE® BY GARY LARSON

"Tell it again, Gramps! The one about being caught in the shark frenzy off the Great Barrier Reef!"

sea otters

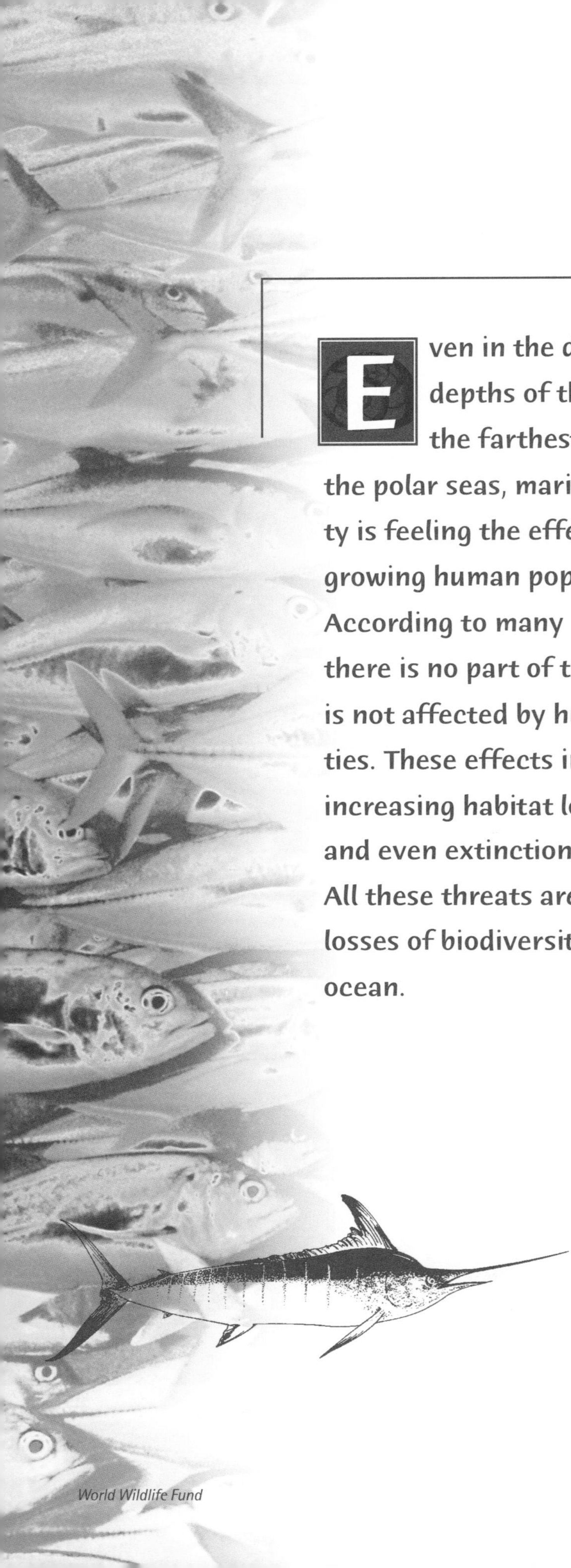

Even in the deepest depths of the ocean and the farthest reaches of the polar seas, marine biodiversity is feeling the effects of a growing human population. According to many scientists, there is no part of the ocean that is not affected by human activities. These effects include increasing habitat loss, pollution, and even extinction of species. All these threats are adding up to losses of biodiversity in the ocean.

"It is a curious situation that the sea, from which life first arose, should now be threatened by the activities of one form of that life. But the sea, though changed in a sinister way, will continue to exist; the threat is rather to life itself."

—Rachel Carson, ecologist

Troubled Waters

section 2, chapter 3

Large-Scale Losses

While there are still many fishers who work on a small scale in local waters, there are a growing number of "factory," or industrial-scale, fishing boats that catch fish on a large scale. A typical industrial fishing boat dragging a net on the ocean bottom can bring in 300,000 to 1 million pounds of fish in a single haul of the net.[40]

These huge catches are made possible by recent advances in fish-finding equipment and fishing gear. It is not unusual, for example, for bluefin tuna fishers to use spotter planes to help them find tuna from the air, and satellites have been used to find fish from space.[41] New technologies greatly shift the advantage to large-scale fishers and make it increasingly easy for them to exploit populations of fish.

In addition to improvements in finding fish, fishers also have increased their ability to catch fish. With the use of driftnets and longlines, fish can be caught on a larger scale than ever before. New materials have allowed some boats to set up to 40 miles of drift nets every day. Fishers in the North Atlantic use up to 2 million miles of nets every year—enough to circle the Earth 88 times.[42] Long-lines are another large-scale fishing method with an enormous capacity to catch fish. A longline may stretch up to 40 miles with a baited hook placed every few feet.[43]

All of these technologies have produced huge global catches. U.S. fishers alone catch about four million tons of fish a year[44] (equivalent to the weight of 10,000 Boeing 747 airplanes). Worldwide, an estimated 85 million tons of seafood are caught per year[45] (that's what 190,000 Boeing 747s would weigh). No other wild

species on land or in the ocean is caught on as large a scale as fish and shellfish.

These large-scale catches come with a price though. Research is showing that many populations can't withstand the heavy fishing pressure. Starting in the 1950s, catches increased steadily and dramatically for decades, leading many to think that the ocean was a limitless resource. But recently, global catches have been leveling off, despite increases in the number of boats and improvements in the fishing gear being used. In other words, large-scale fishers are trying harder and catching less.

Many fisheries haven't been able to withstand the fishing pressure and have collapsed. Areas such as the Grand Banks off New England that were once productive fishing grounds now have very few of the fish that once supported the communities on land. The local economies in those areas are suffering as livelihoods have been lost and fishers and others connected to the fishing industry scramble to find new jobs. The United Nations estimates that about 70 percent of the world's fisheries are either being fished at full capacity or are already overexploited or depleted.[46]

What's more, many scientists worry that we don't know enough about many commercially targeted marine species to accurately predict how their numbers will respond to fishing pressures.[47] Inaccurate or incomplete information about a species' life span, reproductive capabilities, or other biological factors could drastically change managers' recommendations for how many fish can be caught. Errors in information about

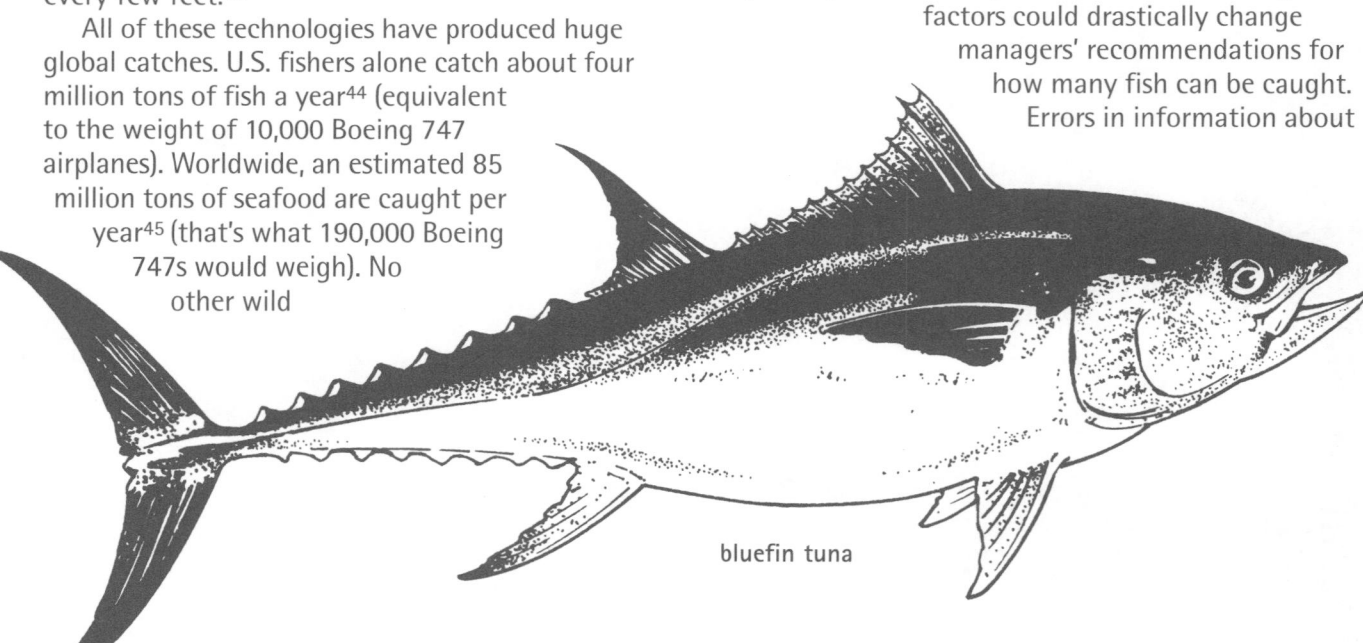

bluefin tuna

the life span of New Zealand's orange roughy fish led managers to recommend a maximum take that was six times too high. Consequently, the population declined steeply.[48] Scientists also worry that we can't accurately measure how many fish are in a population. The ocean is big, and it's hard to know how many fish are out there. If we don't know with any certainty how many fish there are, how can we know how many fish to take?

are not always closely monitored, it's hard to know how severe the impact is on marine biodiversity, but it's clear that valuable resources are being wasted and marine ecosystems are being disrupted. (See the Shrimp Case Study on pages 150-192.)

Habitat Loss

From scraping the ocean floor clean of life with giant trawl nets to clearing productive coastal ecosystems with heavy equipment to make way for coastal developments, a variety of activities threaten the habitats of marine biodiversity. Habitats are the places where species get all the nutrients, water, shelter, and space that they need to survive. And humans are the number-one threat to marine habitats around the world. For example, dynamite fishing in coral reefs is threatening one of the most species-rich marine habitats.[52] Fishers use

What's the Catch with Bycatch?

Not only are we catching too many of the species we want to catch, but we're also catching large numbers of unwanted fish. Unfortunately, most large-scale fishing gear isn't very specific about what it hooks, entangles, or scoops up. Because fishers can't control what bites their hooks or swims into their nets, large catches of non-target species, which are called *bycatch*, are common. The bycatch is simply tossed overboard, dead or dying. For example, in 2003, the International Whaling Commission reported that nearly 308,000 whales, dolphins, and porpoises are killed each year worldwide by entanglement in fishing equipment. Overall, it's estimated that up to one-quarter of the worldwide global catch is bycatch.[49]

Before the introduction of special nets that allow turtles to escape, shrimp trawlers in the Gulf of Mexico discarded as many as 50,000 dead sea turtles annually,[50] in addition to an estimated 10 billion unwanted fish.[51] Even with improvements in the nets, approximately four pounds of unwanted sea life are discarded for every pound of shrimp caught. Because the amounts and types of bycatch

Nets and other fishing gear result in 20 million tons of bycatch being thrown overboard every year.

BIOFACT

Between 1960 and 1998, the average size of a North Atlantic swordfish, *Xhiphias gladius*, dropped from 270 pounds to 90 pounds.

small bombs to help them catch fish, but the bombs kill everything within several feet of the explosion, destroying reefs that can take centuries to develop. In addition, half the world's coastal wetlands have been lost over the last 100 years.[53] These wetlands provide habitat for commercially important fish species, migrating birds, and mammals, and they provide us with valuable ecosystem services (see pages 33-34). Losing wetlands means almost certain disruption to coastal ecosystems.

It's important to keep in mind that habitat loss also occurs when habitats are fragmented (broken up into smaller pieces) or when they're degraded in other ways. Many coastal ecosystems, for example, are being fragmented by development. In some areas, sand dunes—important for a variety of coastal plant species—are now found only in tiny patches nestled between oceanfront high rises and coastal roads. And many rivers that are important spawning grounds for marine fish species have been degraded with dams that alter the habitat and block the movement of fish.[54]

Too Much of a Good Thing

Millions of Americans tune into TV news when a big oil spill occurs. And thousands of people take part in coastal cleanups every year because they've seen pictures of birds and seals strangled by six-pack rings and other marine pollution. But marine ecosystems face more insidious threats that have required a lot of scientific detective work to uncover.

Scientists are just now discovering the reasons behind large ocean areas devoid of life, called "dead zones," which develop at certain times of the year in certain places. According to some researchers, dead zones are the result of heavy loads of nutrients running off agricultural fields.[55] These nutrients can trigger huge blooms of algae. When the mats of algae die and decompose, they rob the water of oxygen. Without oxygen in the water, some of the local inhabitants leave, and those that can't leave die, creating vast areas that are like underwater wastelands. Even more troubling, the dead zone that regularly appears near the mouth of the Mississippi River has been growing.[56] In the

summer of 2001, an area almost the size of Massachusetts was found to be stripped of oxygen. And many scientists are beginning to suspect that the recent increases in red tides could also be tied to increasing amounts of nutrients in the coastal ocean.[57]

Aliens Invading!

The invasion of nonnative species, often called introduced or alien species, is a major threat to marine biodiversity. In science fiction, alien species arrive from outer space, but in the ocean, introduced species often arrive with ships delivering goods from across the ocean. These huge ships take water, called *ballast*, into large tanks to help balance their load. When the ships arrive in a new port, they discharge the water before reloading. But as this water is discharged, so are all the creatures that were taken in at the last port and that survived the trip. Up to 50 species have been found in the ballast water of a single ship.[58] And if conditions are right, these species can survive, and sometimes even flourish, in their new home.

San Francisco Bay has been especially plagued by species introductions: Resource managers there are battling invasions of Asian clams and green crabs that are threatening local species. In fact, there are about 250 alien species in the bay, which together account for as much as 99 percent of the biomass. The Asian clam is now one of the most common organisms found on the floor of the bay, and green crabs are quickly advancing up the coast, competing with native species along the way. And the problem isn't limited to U.S. waters—hundreds of millions of

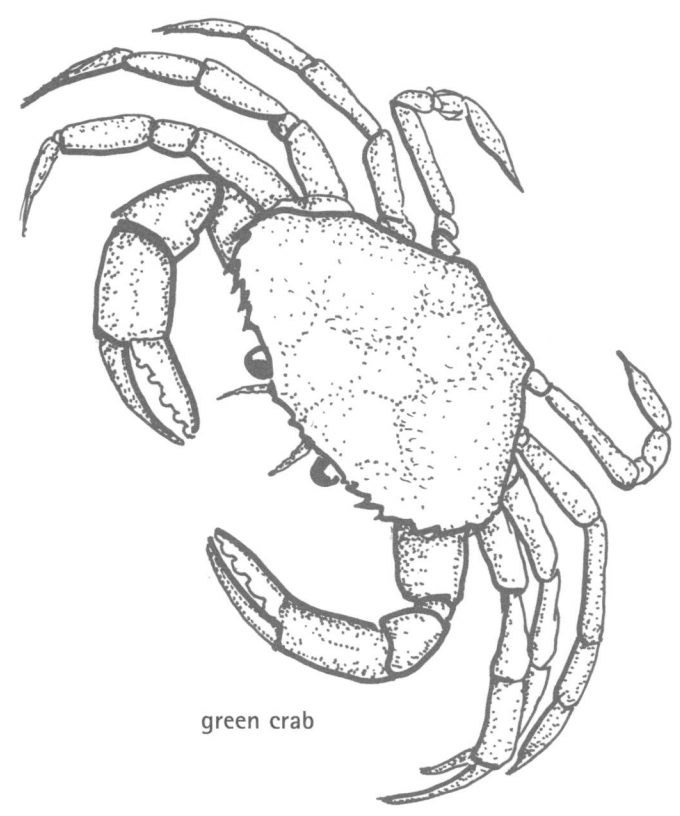

green crab

tons of comb jellies can now be found in the Black Sea, where these jellies are not native. Because of the jellies' fierce competitiveness and voracious appetite, an estimated 85 percent of the marine biodiversity that once thrived in the Black Sea is now gone.[59] In fact, alien species are found in and around almost every port in the world. (See the Alien Species Case Study on pages 248-291.)

Mississippi Delta

Historical and Projected Densities of Coastal and Noncoastal Counties of the Continental United States

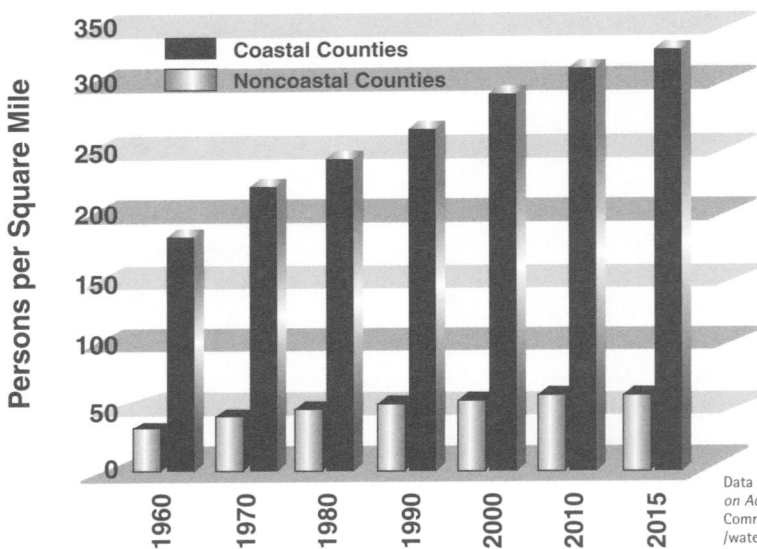

Data adapted from *Coastal Sprawl: The Effects of Urban Design on Aquatic Ecosystems in the United States* (Pew Oceans Commission). Available online at www.pewoceans.org/reports/water_pollution_sprawl.pdf.

Population Pressure

According to many analysts, the ultimate threat to marine biodiversity is the growing human population and the constant need for more food worldwide. Problems such as exploitation, pollution, habitat destruction, and species introduction tend to be highest in areas where people are concentrated— and people are concentrated near the coast. Worldwide, more than 2.5 billion people live within 100 miles of the coast, and coastal populations are growing at a higher rate than the general population. In the contiguous United States, coastal areas make up about 17 percent of the land area but are home to about 53 percent of the population.[60] And those same coastal areas gain more than 3,000 new residents every day.[61]

Estuaries are often eliminated near coastal cities to make way for housing, ports, and industries. Along with the habitat that's lost, much more is degraded by the chemicals, nutrients, and other wastes that so many people generate. And with heavy traffic of huge ships delivering products to the booming populations, the risk of invasive species traveling in the ballast water is high in these already degraded systems.[62]

Many argue that it's not simply a growing human population that's causing problems. They suggest that America's patterns of consumption also lead to losses of marine biodiversity. By comparing the consumption of energy in industrialized countries with that of developing countries, analysts say it's clear that some countries are using more than their share. The United States, for example, uses 12 times as much petroleum as India[63]—even though India's population is almost 4 times as large.[64] And the average Chinese household uses less than three percent of the energy consumed in an average American home.[65] The high demands Americans have for oil leads to an increased need for oil transport across the ocean and increased potential for oil-spill disasters. And American consumption of farmed shrimp and other seafood requiring fishmeal for nourishment further depletes fish populations. It takes about three pounds of wild-caught fish to feed one pound of farmed shrimp or salmon.[66] Growing numbers of people are demanding more species that are especially sensitive to exploitation, and restaurants around the world are giving their customers what they want. Asian consumers, for example, are creating a demand for large numbers of shark fins for soups and other dishes, and shark populations may not be able to recover from the heavy fishing pressure.

But despite the oceans of trouble being caused by the world's human population, there is some hopeful news, too. Many people around the world are realizing the importance of maintaining marine biodiversity, and they are finding ways to keep from depleting the oceans. The next chapter looks at some of the ways people are working to help protect the world's marine environment.

Jeff Foott

*"There needs to be a stronger commitment to stewardship of the sea,
to monitoring the activities of humans in that environment, and to
the protection of all the sea's creatures."*

—Les Watling, conservation biologist

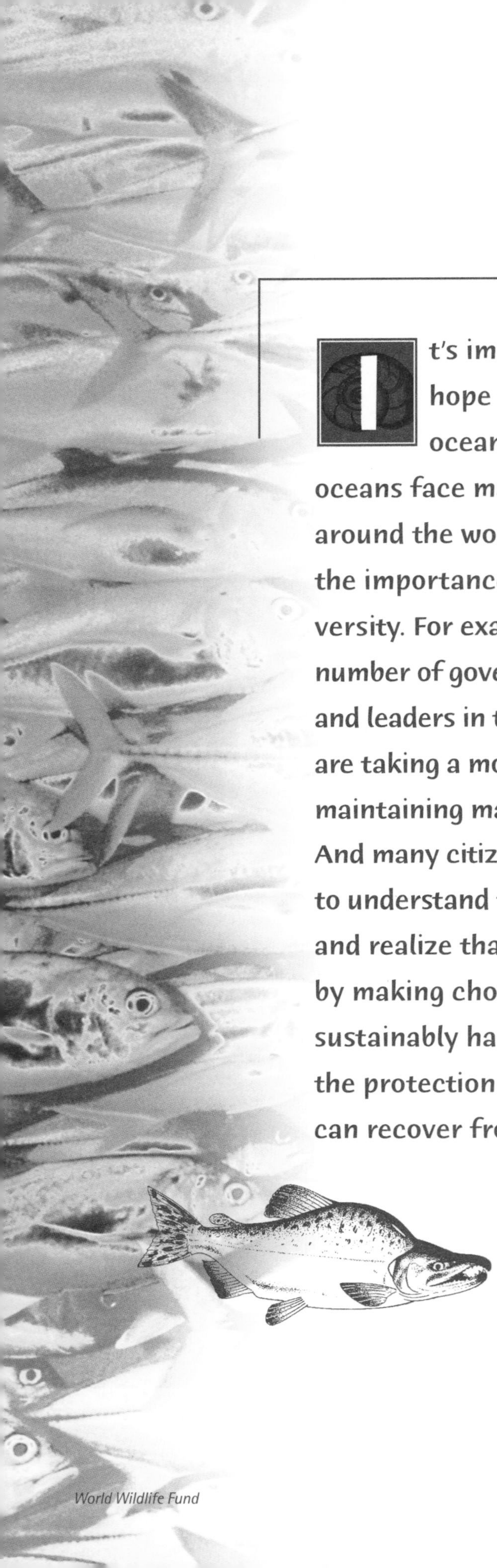

It's important not to lose hope for the world's oceans. Although the oceans face many threats, people around the world are realizing the importance of marine biodiversity. For example, a growing number of government officials and leaders in the seafood industry are taking a more active role in maintaining marine biodiversity. And many citizens are beginning to understand their power as consumers and realize that they can protect diversity by making choices that create demand for sustainably harvested seafood and urging the protection of areas where marine species can recover from human pressures.

> "We owe it to ourselves and to future generations to become committed and responsible ocean stewards—not just for today—but for the long-term health of the oceans."
>
> —Roger Rufe, President, The Ocean Conservancy

Creating a Sea Change

section 2, chapter 4

Smart Shopping

Every time you check out at the register of your local grocery store, you are making a statement about the kinds of products you support. If you are concerned about the harmful effects of over-packaged products, you can buy only products that have a minimal amount of packaging. This allows you to voice your preference to companies. If enough consumers make their preferences clear, companies will be forced to respond to consumer demand by providing the kinds of products that consumers want and limiting the amount of packaging they use.

Consumers can also use this power to help encourage smart fishing practices.

bottlenose dolphin

A good example is when shoppers across the country stopped buying tuna until companies proved that they did not harm dolphins with tuna nets. National campaigns to educate people about the problem led to public outrage about the destructive fishing practice and a consumer boycott of tuna. Congress responded by adopting standards to protect dolphins in certain tuna fisheries. Tuna companies realized that, to stay in business, they'd have to respond to the public's concern—and they did. Today, any tuna caught by a method that can ensnare dolphins must be accompanied by an official observer on the boat. These observers must ensure that no dolphins were seriously harmed or killed in the fishing process. Another "ecosystem-safe" method is catching tuna with a rod and reel. Tuna caught by this method is called *troll caught.*

But how can consumers know how different fish species were caught, or if the populations of those fish are big enough to support large-scale fishing? Scientists, policymakers, and conservationists are working together to create a labeling system. For example, the Marine Stewardship Council (MSC) has certified several fisheries that conform to sustainable fishing practices,[67] and they are working on certifying more. The MSC labeling system doesn't just protect the species certified; it is designed to protect entire marine ecosystems and regulate overall fishing practices within those ecosystems. Unfortunately, it takes a long time to certify that the fishers in an area are using sustainable fishing practices that don't harm the environment. But many hope that, as the MSC label becomes more recognizable and customer demand increases, more fisheries will want to be certified. By buying only fish that are caught in sustainable ways, consumers let fishers know that they can make profits without overfishing. By letting fishers know that they don't have to provide an endless supply of all kinds of fish, but rather only those fish whose populations are healthy, consumers can help reduce the problem of overfishing.

Is Aquaculture the Answer?

If you've bought shrimp, salmon, or trout in a restaurant, fish market, or grocery store, there's a good chance that it wasn't caught in the ocean or a river but instead was raised on a fish farm. Practiced in China for thousands of years, farm-raising fish, also called **aquaculture**, is now catching on all over the world. Since 1985, the amount of fish raised on farms has doubled. Worldwide, aquaculture supplies about one-third of the fish eaten by people.[68]

Many experts think that aquaculture could be the answer to at least some of our fishing problems. Farmed fish are raised in large numbers either to be released for fishers to catch or to be sent directly to stores. With the human population increasing and many fish species declining, supporters say that aquaculture could supply the world with fish that might otherwise be unavailable in the coming years. And by helping to reduce fishing for wild species, aquaculture could give those fish populations a chance to recover.[69]

But many others cite problems with fish farming. Establishing these farms often means disturbing or destroying natural fish habitats along coasts and in rivers—habitats that provide important ecosystem services (see pages 33-34). Some fish farms release large amounts of waste, causing water pollution.[70] And if farmed fish are released or escape into the wild, they can threaten the native species by spreading diseases to the native fish as well as competing with them for food and other resources.[71] What's more, fish farmers often must catch or buy large amounts of wild species to feed the fish they're raising. So the problem of overfishing isn't always avoided; instead, some fish farming today often helps one species at the expense of others.

Still people argue that aquaculture holds such promise in providing large amounts of fish that we should find ways to solve its

shrimp farms

WWF-Canon/Tantyo Bangun

current problems. We should find ways to raise fish without destroying or polluting habitats or threatening native fish with accidental releases of farmed fish. We must also remember that farmed fish are not a replacement for wild fish; we should protect wild species and their habitats even if we no longer depend on them for food. At its best, supporters argue, aquaculture can both produce and protect fish so that we can feed a growing population and still maintain the wild diversity of the sea for future generations. (See the section on shrimp farming on pages 156-157.)

There are additional ways of finding out which fish to buy. Several organizations have developed lists that tell consumers which fish are currently overfished and which ones are safe to buy. (For example, see seafood cards at **www.mbayaq.org** and **www.audubon.org**.) Some of the lists also inform consumers about which kinds of farmed fish are

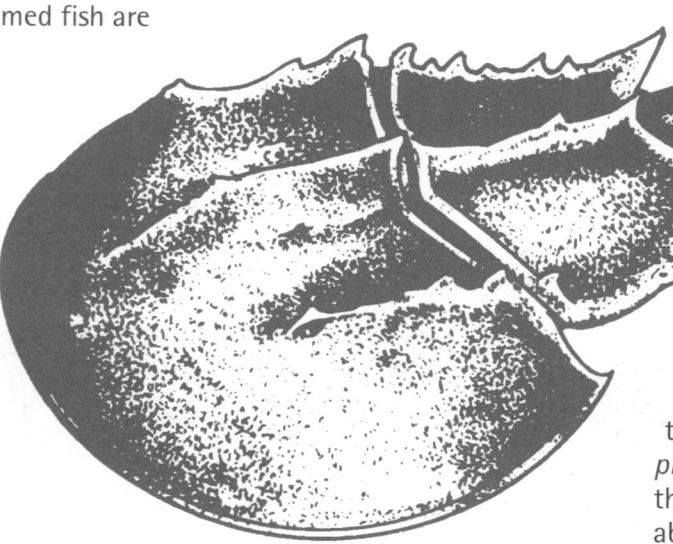

horseshoe crab

raised in ways that don't threaten other species. By being a well-informed consumer, you can cast your vote for marine biodiversity.

The Precautionary Principle

Imagine this scenario: Scientists gathering data on the population of horseshoe crabs in Delaware Bay find signs that their numbers are declining. But other studies in the same area have conflicting findings. If you lived along the Delaware Bay, what actions would you encourage local officials to take if they weren't certain that the horseshoes' numbers were declining, or if they thought the population was declining but didn't know by how much?

Keep in mind that horseshoe crabs provide a major food source for migrating shorebirds, whose migration from South America is timed to meet the horseshoe crabs as the crabs emerge from the sea to lay their eggs on the shores of the Delaware Bay. The birds feast on horseshoe crab eggs in one of the only major stops on their flight to the Arctic. Without this food source, the birds might not have the energy to finish the trip and their populations could become threatened within a few seasons.

Would you wait until more research was done that could confirm the horseshoes' decline? Or would you act right away, taking precautions to protect the horseshoe crabs while more research was being done?

Many conservationists think we should take the second approach, following what is called the *precautionary principle*. People who believe in using the precautionary principle argue that information about the populations of marine species is often difficult to obtain and to interpret. If we wait until we do more research and then find that the numbers are in fact declining, it may be too late to protect many species and help them recover. And, they argue, uncertainty is always a part of scientific research. Scientists are never able to make predictions or judgments with absolute certainty. So it's up to us to put into place regulations and practices that will protect marine biodiversity while we gather more information about species' biology, behavior, or population sizes. Although people who fish for or use the species may oppose restrictions, they will benefit from early action if the population numbers truly are declining. If future research reveals that populations are in trouble, the early action taken should lead to a more effective and efficient recovery effort. And if we're lucky, as we uncover more information, we'll find that populations are healthy and we can ease any precautions that may be more stringent than necessary.

The Global 200: Setting Conservation Priorities

The map below is an example of how scientists at WWF and its partners around the world have considered a variety of factors—from richness (the number of species) to ecological importance (flood control, water purification, and so on) to uniqueness (species and landscapes that are found nowhere else)—to determine conservation priorities. From the windswept tundra of Alaska's North Slope to the warm tropical waters of the Sulu-Sulawesi Sea in the western Pacific, this map highlights more than 200 of the richest, rarest, and most distinct natural areas on the planet. Together, these areas—called ecoregions—are part of a comprehensive assessment of the world's biodiversity that WWF is calling the Global 200.

The Global 200 represents a science-based approach to setting priorities for conservation. At its core is a simple concept: If we conserve the broadest variety of the world's habitats, we can conserve the broadest variety of the world's species, along with the ecological and evolutionary processes that maintain the web of life.

Many of these bio-rich areas are in trouble: Nearly half of the Global 200 terrestrial ecoregions are critically endangered and only a quarter are still relatively intact. In addition, many of the marine and freshwater ecoregions are also endangered and face serious threats.

Many scientists believe that the Global 200, which was developed with input from hundreds of experts worldwide, represents an important blueprint for long-term conservation action. WWF is now working with its partners to develop strategies for protecting these ecoregions, with a focus on the need to address biological, social, economic, cultural, and political factors.

Visit **www.worldwildlife.org/wildworld** to find out more.

Global 200

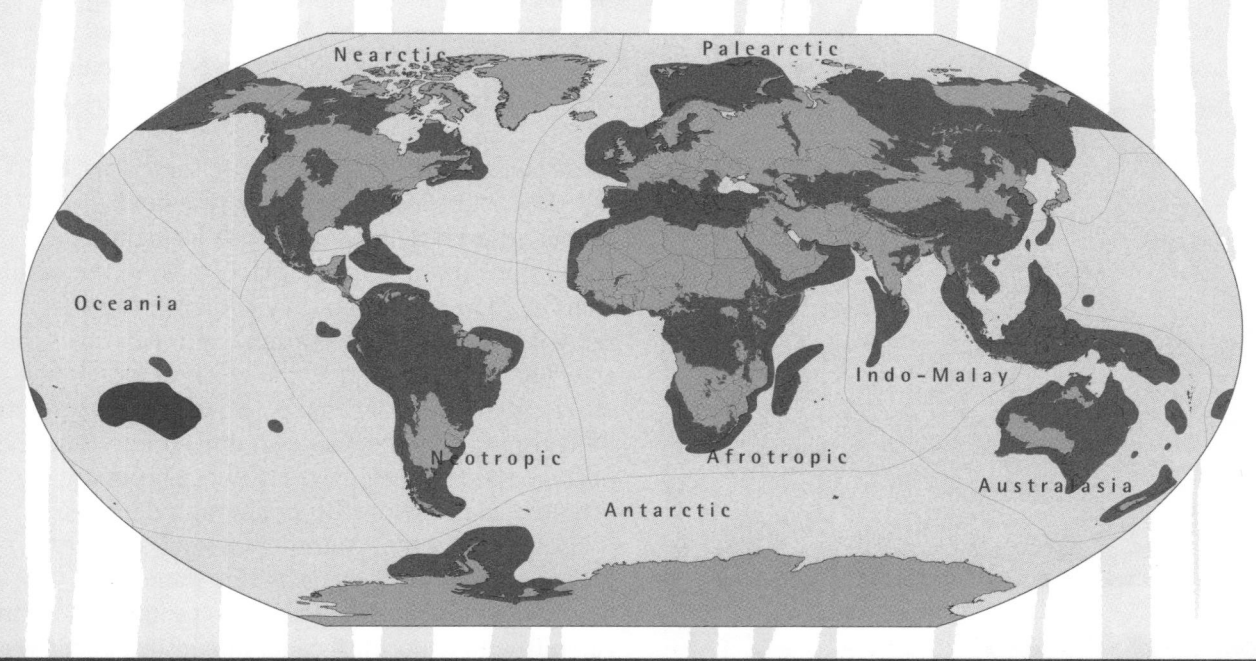

A Glittering Success

Called the "crown jewels" of the western Indian Ocean, the coral reefs of the East African Marine ecoregion are filled with a dazzling array of species. Dugongs swim among the mangroves, and all five of the Indian Ocean's sea turtle species live in the crystal-clear waters. In 2002, WWF and local communities worked together to set aside two areas in the region: The Bazaruto National Park in Mozambique was doubled in size to 542 square miles, and the Quirimbas Islands archipelago was declared a 579-square-mile national park. With these new additions, the East African Marine ecoregion now contains the three largest marine protected areas in the western Indian Ocean.

© WWF-Canon/Cat Holloway

Cooperation Across Borders

Marine life—especially migratory species such as whales, tuna, and sharks—doesn't stick to human-made boundaries. In 1994, the United Nations adopted the Law of the Sea, which gave coastal countries rights to govern the water within 200 nautical miles of their coasts.[72] But no matter how much protection one country affords marine species in its territory, as soon as the species leave that country's marine borders, they are at risk if neighboring countries don't provide the same protection. To protect wide-ranging species, countries must work to develop international agreements.

Several international bodies exist to help manage migratory species. The International Commission for the Conservation of Atlantic Tuna (ICCAT), for example, was formed to help countries agree on fishing limits for tuna and certain other highly migratory species in the Atlantic, including swordfish.[73] Many conservationists have been critical of this commission, however, because different countries' representatives have different ideas about how best to conserve these species. In the end, countries for which marine biodiversity conservation is a low priority can influence the regulations that are put in place, making the regulations less effective than they could be. Another related problem of international agreements is that they are only as strong as the weakest member makes them. If, for instance, five countries agree to limit fishing of swordfish, and one country does not enforce the regulations, then all countries are affected as the swordfish population declines.

Another way that countries interact to preserve marine biodiversity is by restricting trade with countries whose practices they don't support. Countries can refuse to import fish caught in unsustainable ways and refuse to allow ships into their ports that have not adopted double-hull technology to help prevent oil spills. But these efforts often create hardships for countries whose trade has been restricted, and they can create political unrest or bad feelings between countries. The use of trade restrictions has been challenged in international bodies that regulate trade, such as the World Trade Organization. Although trade restrictions may be an effective tool for encouraging biodiversity conservation, both people and the marine environment benefit when all countries can agree to support trade *and* conserve marine biodiversity.

Agreement across borders is also important within countries. In the United States, fisheries are often managed by different bodies in different states. Again, fish don't always stay in the waters regulated by one state. To best manage fish that range throughout U.S. waters, it's important for states to work together to enforce regulations that they all agree upon. The striped bass, for example, has made a dramatic recovery in the eastern United States because several states joined in a concerted effort to bring back the fish.

Although countries and states may not always agree on how best to protect marine biodiversity, international and interstate agreements are nevertheless a critical tool in the protection of marine biodiversity. These agreements encourage discussion about how best to preserve biodiversity and help us balance many competing needs.

Restoring Habitats

Across the country and around the world, individuals, governments, and organizations are leading efforts to restore marine habitats to conditions as close as possible to what they were before humans damaged them. This effort, called *habitat restoration*, has led to some important improvements in the habitats of many marine species.

On some rivers, dams have been removed to restore free-flowing waterways that are vital salmon spawning grounds. Where native trees and grasses were once choked out of coastal areas by invading species, people have organized to remove the invasive species to restore the natural coastal community. And where trash littered beaches and threatened birds and marine mammals along coasts, people have gathered to clean up the debris and have worked to limit the amount of trash that finds its way into the sea.

The National Marine Fisheries Service and the FishAmerica Foundation recently announced a $1-million partnership to help communities restore their coastal and marine habitats. This new initiative funds projects that include seagrass restoration in Virginia and Maryland, coral reef and mangrove restoration in Florida, marsh habitat restoration in Texas, kelp restoration off the California coast, and invasive plant removal in the state of Washington.[74] In each case, students and other citizen volunteers team up with local officials to help bring local marine habitats back to healthier conditions, demonstrating that restoration isn't just the work of scientists and managers but can be important work of communities interested in preserving the biodiversity in their own backyards.

kelp

WWF-Canon/Rob Webster

"Today, marine sanctuaries are places in the sea, as elusive as a sea breeze, as tangible as a singing whale. They are beautiful, or priceless, or rare bargains, or long-term assets, or fun, or all of these and more. Above all, sanctuaries are now and with care will continue to be 'special places.' Each of us can have pleasures of defining what that means."

–Sylvia Earle, marine biologist

Safe Spaces

We've learned from our experience on land that species whose numbers are declining need places where they can escape from human threats. For the past 125 years, we've been establishing national parks and wildlife refuges to help protect wildlife habitat and prevent exploitation on land. But we've been much slower to provide the same kinds of protection to life in the sea: Most marine reserves have been created in just the past 20 years.[75] And aside from areas that have been declared off-limits for whaling, marine reserves account for less than one percent of the ocean's area.[76]

Scientists and policymakers are encouraging governments to create more marine reserves because of their potential benefits to marine biodiversity. One of the main benefits of marine reserves is that they can provide important refuges for species that are heavily fished. Although creating large areas that prohibit fishing may seem detrimental to the liveli-

hood of fishers, many scientists argue that, over time, fishers will benefit. Because individual fish can grow large and reproduce within the protected areas, populations can grow and fish can disperse into areas where fishing is allowed—ultimately increasing catches for fishers in the areas surrounding the reserves.[77]

Marine reserves can also serve as important sites for research, monitoring, education, and recreation. Because reserves can be designed with any of these goals in mind, they take many forms. Some reserves' rules don't allow any fishing within the reserve boundary. Others have a patchwork design that allows a variety of activities in different areas: Fishing may occur in some and recreation in others, while still others are completely off-limits to people.[78]

Marine reserves also vary widely in their size. In the United States, the National Marine Sanctuary Program has set aside a 1.3-square-mile site that surrounds the wreckage of the *U.S.S. Monitor.* Off the coast of Hawaii, a new marine reserve has been created that is 113,800 square miles— as large as the combined areas of Florida and Georgia.

As citizens, there are a variety of actions we can take to help make marine reserves effective protectors of marine biodiversity. We can let our governmental representatives know that we support the creation of more marine reserves that are big enough and have strict enough rules to help marine species recover from human threats. If we visit marine reserves, we can learn about and follow the reserves' rules designed to protect marine biodiversity. Since most reserves don't have the funding they need to enforce all of the rules all of the time, individuals have to take responsibility for following the rules that make marine reserves effective. In the Florida Keys, for example, it's up to individual snorkelers and divers not to drop their anchors on the reefs or take pieces of coral home with them as souvenirs. By having a citizenry that supports the establishment of marine reserves and the enforcement of their rules, we can give marine biodiversity spaces where it can recover and thrive.

Developing a Sea Ethic

Values are very hard to define and sometimes even harder to stick to. It's not easy to articulate just how we feel about complicated issues, and it can be difficult to decide how to act. Still, it's important to think about what matters to us as community members, voters, consumers, or family members and act in ways that we think are right. If we value clean air, for example, we might organize a community carpool to help cut down on car emissions, support legislation that will help enforce clean air regulations, or teach our families habits that will help keep the air clean. Similarly, if we value marine biodiversity, we might organize efforts to restore marine habitats in our community, support legislation to help establish more marine reserves, buy fish that were sustainably caught, and take family vacations to places where we can appreciate marine biodiversity.

According to some policy makers, we've been much better at defining how we value biodiversity on the land than we have been about defining our values of marine biodiversity. In light of public sentiment and policy decisions, it's pretty clear that we value eagles, bison, and many other wild animals on land, but it's less clear how we feel about the wild animals of the sea, such as crabs or fish. And because we don't often think about how we value life in the sea, we might not realize how our decisions affect it. Writer Carl Safina has argued that many of us have a "land ethic" that guides our decisions about land use, but "we have yet to extend our sense of community below the high-tide line. Many still view the ocean as the blank space between continents. We need now a 'sea ethic.'"[79]

The first step to developing a sea ethic will be to define how we value marine biodiversity. And this doesn't simply mean how we value it economically. By realizing that healthy oceans help make possible all varieties of life on this planet and are crucial to the future of that life, we can learn to appropriately value the oceans. By recognizing that we benefit economically, ecologically, and personally from having oceans full of life, we can begin to make informed decisions about how and when to use (and not to use) the oceans.

Perhaps more than anything else, we need to develop a sense of wonder about the sea—its deep chasms and strong currents, its microscopic plankton and giant whales, its fascinating complexity and staggering diversity. Only by feeling excited about and inspired by the oceans and the life they contain will we be likely to develop an ethic that will protect marine biodiversity. And with as much as 95 percent of the oceans still unexplored, the potential for our curiosity, exploration, and wonder is almost as big as the ocean.

Endnotes for Background

1 Boyce Thorne Miller. 1999. *The Living Ocean*. Washington, DC: Island Press, p. xiv.

2 WWF/IUCN Marine Policy. 1998. *Creating a Sea Change*. Washington, DC: WWF International Marine Programme and IUCN, p.7.

3 Boyce Thorne Miller. 1999, p. xiii.

4 Denis Hayes. 2002. *Ocean Policies for the New Millennium*. Testimony presented to the U.S. Commission on Ocean Policy in Seattle, Washington, June 13, 2002. p. 2.

5 Rachel Carson. 1950. *The Sea Around Us*. New York: Oxford University Press, p. 7.

6 Boyce Thorne Miller. 1999, p. 40.

7 Anne Platt McGinn. 1999. Charting a New Course for the Oceans. In *State of the World 1999*. New York: W.W. Norton and Company, p. 81.

8 Independent World Commission on the Oceans. 1998. *The Ocean Our Future*. Cambridge: Cambridge University Press, p. 172.

9 Harold V. Thurman. 1997. *Introductory Oceanography*. Upper Saddle River: Prentice Hall, p. 445.

10 Boyce Thorne Miller. 1999, p. xiii.

11 Independent World Commission on the Oceans. 1998, p. 166.

12 Jim Fisher. 2002. Hardy Palms. **homepages.nildram.co.uk/~jimella/palm.htm**

13 Rachel Carson. 1950, pp. 168–169.

14 Anne Platt McGinn. 1999, p. 83.

15 National Geographic, Wonder of Learning Books. *Captain Conservation: All About Water*, 1993. **www.nationalgeographic.com/education/teacher_store/products/WE31026.html**

16 Boyce Thorne Miller. 1999, p. 11.

17 Boyce Thorne Miller. 1999, pp. 10–11.

18 Independent World Commission on the Oceans. 1998, p. 102.

19 Independent World Commission on the Oceans. 1998, p. 102.

20 Personal communication with Richard Barber, 21 April 2000.

21 Elliott Norse, ed. 1993. *Global Marine Biological Diversity*. Washington, DC.: Island Press, pp. 25-27.

22 Boyce Thorne Miller. 1999, p. 14.

23 WWF/IUCN Marine Policy. 1998, p. 13.

24 Elliott Norse, ed. 1993, p. 17.

25 WWF/IUCN Marine Policy. 1998, p. 13.

26 The National Fisheries Institute. 1999. About Our Industry/Harvesting. **www.nfi.org/about/retail.php**

27 The National Fisheries Institute. 1999. **www.nfi.org/?a=about&b=harvesting**

28 Edward O. Wilson. 1992. *The Diversity of Life*. New York: Harvard University Press, p. 283.

29 Elliott Norse, ed. 1993, p. 20.

30 WWF/IUCN Marine Policy. 1998, p. 15.

31 Elliott Norse, ed. 1993, p. 20.

32 Boyce Thorne Miller. 1999, p. 13.

33 Boyce Thorne Miller. 1999, p. 13.

34 National Oceanic and Atmospheric Association (NOAA). Year of the Ocean Web site. **www.yoto98.noaa.gov**

35 WWF/IUCN Marine Policy. 1998, p. 16.

36 American Zoo and Aquarium Association. 2002. Total Attendance, 2000. **www.aza.org/Newsroom/NewsroomStatistics**

37 Stephen R. Kellert. 1997. *Kinship to Mastery*. Washington, DC: Island Press, pp. 1-9.

38 Geoffrey Waller, ed. 1996. *SeaLife*. Washington, DC: Smithsonian Institution Press, p. 86.

39 Carl Safina. 1997. *Song for the Blue Ocean*. New York: Henry Holt and Company, pp. 159-161.

40 Boyce Thorne Miller. 1999, p. 17.

41 Carl Safina. 1997, p. 24.

42 Elliott Norse, ed. 1993, p. 93.

43 Boyce Thorne Miller. 1999, p. 17.

44 The National Fisheries Institute. 1999.
    www.nfi.org/?a=about&b=harvesting

45 NOAA. 1998. Ecological Effects of Fishing by S.K. Brown, P.J.
    Auster, L. Lauk, and M. Coyne. NOAA's *State of the Coast
    Report.*
    oceanservice.noaa.gov/websites/retiredsites/sotc_pdf/IEF.pdf

46 Food and Agriculture Organization of the United Nations
    (UN FAO). 2000. State of the World's Fisheries and
    Aquaculture.
    www.fao.org/DOCREP/003/X8002E/x8002e04.htm#P0_0

47 Elliott Norse, ed. 1993, pp. 90-93.

48 Elliott Norse, ed. 1993, p. 91.

49 Dayton L. Alverson. 1994. *A Global Assessment of
    Fisheries Bycatch and Discards.* Rome: UN FAO.

50 NOAA. 1998.
    state-of-coast.noaa.gov/bulletins/html/ief_03/ief.html
    See also WWF/IUCN Marine Policy. 1998, p. 21.

51 Elliott Norse, ed. 1993, p. 95.

52 WWF/IUCN Marine Policy. 1998, p. 22.

53 WWF/IUCN Marine Policy. 1998, p. 19.

54 Independent World Commission on the Oceans. 1998,
    p. 197.

55 Joby Warrick. 1999. Death in the Gulf of Mexico.
    In National Wildlife, June/July 1999, pp. 48-52.

56 Joby Warrick. 1999, p. 48.

57 Anne Platt McGinn. 1999, p. 86. See also
    Boyce Thorne Miller. 1999, pp.23-26.

58 Elliott Norse, ed. 1993, pp. 90-93.

59 Anne Platt McGinn. 1999, p. 87.

60 NOAA. 1998. Population Distribution, Density, and Growth
    by T.J. Culliton. NOAA's *State of the Coast Report.*
    oceanservice.noaa.gov/websites/retiredsites/sotc_pdf/
    POP.pdf

61 NOAA. 1998.
    oceanservice.noaa.gov/websites/retiredsites/sotc_pdf/
    POP.pdf

62 Elliott Norse, ed. 1993, p. 85.

63 World Wildlife Fund. 1999. *Windows on the Wild:
    Biodiversity Basics.* Washington, DC: World Wildlife Fund,
    p. 43.

64 World Wildlife Fund. 1999, p. 43.

65 World Wildlife Fund. 1999. p. 43.

66 www.sciencenews.org/sn_arc98/11_7_98/Food.htm

67 Marine Stewardship Council. www.msc.org

68 FAO. 2000. *The State of the World Fisheries and
    Aquaculture, Part 1: Fisheries Resources: Trends in
    Production, Utilization, and Trade.*
    www.fao.org/DOCREP/w9900e/w9900e02.htm#P0_0

69 Claude E. Boyd and Jason W. Clay. 1998. Shrimp
    Aquaculture and the Environment. In *Scientific American,*
    June 1998.

70 Claude E. Boyd and Jason W. Clay. 1998.

71 Boyce Thorne Miller. 1999, p. 20.

72 United Nations. 2002. Oceans and the Law of the Sea.
    www.un.org/Depts/los/convention_agreements/conven-
    tion_overview_convention.htm

73 International Commission for the Conservation of Atlantic
    Tuna. www.iccat.es

74 National Oceanic and Atmospheric Administration (NOAA)
    Restoration Center.
    www.nmfs.noaa.gov/habitat/restoration/community

75 World Resources Institute. *Tools for Protecting Marine
    Biodiversity.* www.wri.org/wri/wr-96-97/bi_txt6.html

76 Independent World Commission on the Oceans. 1998, p.
    199.

77 Gary W. Allison, Jane Lubchenco, and Mark H. Carr.
    1998. Marine Reserves are Necessary but not Sufficient
    for Marine Conservation. In *Ecological Applications* 8(1)
    Supplement. 1998, pp. 579-592.

78 Boyce Thorne Miller. 1999, p. 117.

79 Carl Safina. 1997, p. 439.

All Web sites were accessed in 2003.

Introducing the Activities and Case Studies

"The object of education

is to prepare the young

to educate themselves throughout their lives."

–Robert Hutchins,
educator and former president of the
University of Chicago

Oceans of Life Activities

Activities, Activities, and More Activities!

Welcome to the activities section. In "Chapter 1: Introducing Marine Biodiversity," you'll find introductory activities that will help kick off your marine biodiversity studies. In the following five chapters, which are organized as case studies, the activities relate to specific topics and allow your students to explore various facets of marine biodiversity and related issues. (The case study topics include coral reefs, shrimp, sharks, aliens, and salmon.) All of the introductory and case study activities are designed to teach the concepts and skills outlined in the "Biodiversity Education Framework" (found at **www.worldwildlife.org/windows**).

You'll see that the activities vary in length, depth, and approach. We encourage you to adapt these ideas, combine them with other resources, and devise organizing themes that will best meet your particular objectives. We also encourage you to explore the sample units that link the activities using different themes and approaches. (The unit plans and a description of our educational approach start on page 322.)

Every activity follows the format described in the sample below. For additional ideas, check out the "Resources" (pages 360-369), as well as the suggestions at the end of each case study, to find dozens of exemplary educational materials that can help you build the best possible biodiversity and marine education programs.

1 Subjects: Lists specific disciplines.

2 Skills: Lists the key skills that students will use in the activity. You can find complete lists of skills in the "*Skills Framework*." (On the Web at www.worldwildlife.org/windows.)

3 Framework Links: Shows specific connections to the "*Biodiversity Education Framework*," which can be found on the Web.

4 Vocabulary: Highlights important words used in the activity that students might not know. Words in **bold** are defined in the glossary on pages 338-340.

5 Time: Gives an idea of how much time the activity will take based on pilot testing and educator comments. We have estimated that each session is about 45 minutes.

6 Connections: Lists related activities found in this module, as well as other *WOW* modules, that could be used before or after the activity to create more effective lessons and units.

7 Outdoor and Challenging: This symbol ☼ indicates that parts of the activity take place outside. This symbol 💡 indicates a more challenging activity.

8 Objectives: Describes the purpose of the activity and the types of learning (skills and knowledge) that might occur. Objectives describe what students should be able to do after taking part in the activity.

9 Introduction: Starts off each activity and often includes related background information. (Sometimes this background information is contained in boxes outside the normal text. In other cases, it's integrated into the introduction of the activity.)

10 What to Do: Includes step-by-step directions about how to conduct the activity. The first part of each step is in boldfaced type. The last step in the directions brings closure to the activity.

11 Assessment: Suggests strategies for evaluation. The assessments includes examples of what excellent, satisfactory, and unsatisfactory results might include.

12 Writing Ideas: Encourages creative and technical writing skills. Writing ideas can be integrated into the activity or used as an extension.

13 Extensions: Provides additional activities that relate to the core activity and can be used to encourage more in-depth investigation or discovery. Some can also be used as assessment strategies.

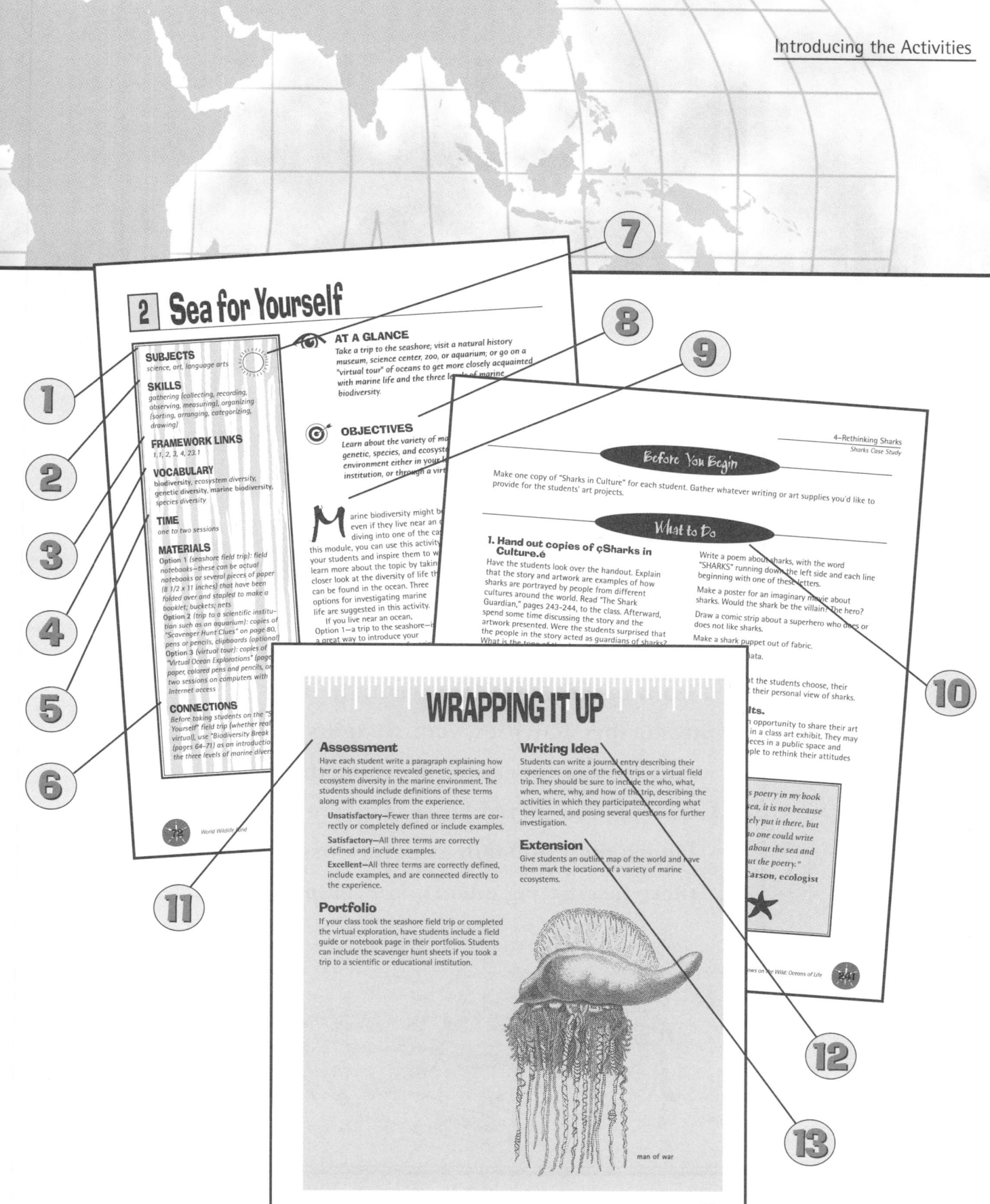

2 Sea for Yourself

SUBJECTS
science, art, language arts

SKILLS
gathering (collecting, recording, observing, measuring), organizing (sorting, arranging, categorizing, drawing)

FRAMEWORK LINKS
1.1, 2, 3, 4, 23.1

VOCABULARY
biodiversity, ecosystem diversity, genetic diversity, marine biodiversity, species diversity

TIME
one to two sessions

MATERIALS
Option 1 (seashore field trip): field notebooks—these can be actual notebooks or several pieces of paper (8 1/2 x 11 inches) that have been folded over and stapled to make a booklet; buckets; nets
Option 2 (trip to a scientific institution such as an aquarium): copies of "Scavenger Hunt Clues" on page 80, pens or pencils, clipboards (optional)
Option 3 (virtual tour): copies of "Virtual Ocean Explorations" (page__ paper, colored pens and pencils, on__ two sessions on computers with Internet access

CONNECTIONS
Before taking students on the "S__ Yourself" field trip (whether real__ virtual), use "Biodiversity Break__ (pages 64–71) as an introduction__ the three levels of marine divers__

AT A GLANCE
Take a trip to the seashore; visit a natural history museum, science center, zoo, or aquarium; or go on a "virtual tour" of oceans to get more closely acquainted with marine life and the three levels of marine biodiversity.

OBJECTIVES
Learn about the variety of ma__ genetic, species, and ecosyste__ environment either in your l__ institution, or through a virt__

Marine biodiversity might be__ even if they live near an o__ this module, you can use this activity__ your students and inspire them to w__ learn more about the topic by takin__ closer look at the diversity of life th__ can be found in the ocean. Three options for investigating marine life are suggested in this activity.

If you live near an ocean, Option 1—a trip to the seashore—i__ a great way to introduce your__

World Wildlife __nd

4–Rethinking Sharks
Sharks Case Study

Before You Begin

Make one copy of "Sharks in Culture" for each student. Gather whatever writing or art supplies you'd like to provide for the students' art projects.

What to Do

1. Hand out copies of çSharks in Culture.é
Have the students look over the handout. Explain that the story and artwork are examples of how sharks are portrayed by people from different cultures around the world. Read "The Shark Guardian," pages 243–244, to the class. Afterward, spend some time discussing the story and the artwork presented. Were the students surprised that the people in the story acted as guardians of sharks? What is the tone of th__

Write a poem about sharks, with the word "SHARKS" running down the left side and each line beginning with one of these letters.

Make a poster for an imaginary movie about sharks. Would the shark be the villain? The hero?

Draw a comic strip about a superhero who does or does not like sharks.

Make a shark puppet out of fabric.

__ata.

__t the students choose, their __ their personal view of sharks.

__lts.
__n opportunity to share their art __ in a class art exhibit. They may __eces in a public space and __ople to rethink their attitudes

__s poetry in my book
__ea, it is not because
__ely put it there, but
__o one could write
__about the sea and
__ut the poetry."

__arson, ecologist

__ws on the Wild: Oceans of Life   241

WRAPPING IT UP

Assessment
Have each student write a paragraph explaining how her or his experience revealed genetic, species, and ecosystem diversity in the marine environment. The students should include definitions of these terms along with examples from the experience.

Unsatisfactory—Fewer than three terms are correctly or completely defined or include examples.

Satisfactory—All three terms are correctly defined and include examples.

Excellent—All three terms are correctly defined, include examples, and are connected directly to the experience.

Portfolio
If your class took the seashore field trip or completed the virtual exploration, have students include a field guide or notebook page in their portfolios. Students can include the scavenger hunt sheets if you took a trip to a scientific or educational institution.

Writing Idea
Students can write a journal entry describing their experiences on one of the field trips or a virtual field trip. They should be sure to include the who, what, when, where, why, and how of the trip, describing the activities in which they participated, recording what they learned, and posing several questions for further investigation.

Extension
Give students an outline map of the world and have them mark the locations of a variety of marine ecosystems.

man of war

Windows on the Wild: Oceans of Life

Introducing Marine Biodiversity

The ocean's vast, blue surface belies the amazing and colorful world that lies below. What is it like to be under the sea, face to face with tropical fish, eels, and sharks? What do we know about the layers of ocean biodiversity and how that diversity supports our lives? Why do so many ocean mysteries remain? In the introductory activities, your students will get an ocean overview—learning the A, B, C's of our incredible seas before they launch their in-depth explorations of these fascinating underwater realms.

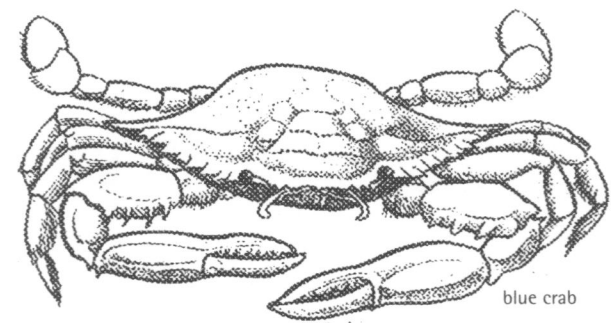

blue crab

WWF-Canon/Anton Vorauer

"If there is a key, it is education—education of the young scientists and engineers who will be responsible for the next generation of coastal-based research; education of specialists to work in marine-related fields; and education of our citizens who must have the scientific knowledge to make reasoned choices that will influence environmental decision-making."

—Maryland Sea Grant

SUBJECTS
science, art

SKILLS
gathering (researching), organizing (matching, arranging, categorizing, classifying)

FRAMEWORK LINKS
1, 1.1, 2, 3, 4, 21, 23.1, 25, 71

VOCABULARY
biodiversity, ecoregions, ecosystem, ecosystem diversity, genetic diversity, hotspots, marine biodiversity, species, species diversity

TIME
one to three sessions, depending on amount of time allotted for research

MATERIALS
"Joy to the Fishes and the Deep Blue Sea" marine biodiversity poster; black-and-white copies of marine biodiversity poster (page 68); copies of "Sea Bits" (page 69), "Describing Diversity" (page 70), and "Pick a Marine Ecoregion" (page 71); access to the Internet or reference books about marine ecosystems; scissors, string or yarn, tape, felt pens, colored pencils, paints (optional)

CONNECTIONS
To get students thinking about various aspects of biodiversity, try "What's Your Biodiversity IQ?" in Biodiversity Basics. *"Mapping Biodiversity," also in* Biodiversity Basics, *will provide a more thorough discussion of ecoregions—how they're defined, where they're located, and which ones are the most threatened. You can also access this activity on line at* **www.worldwildlife. org/windows.** *(Select "Activity Archive" under "WOW Online." Then select, "Mapping Biodiversity.")*

AT A GLANCE
Use the "Joy to the Fishes and the Deep Blue Sea" poster to learn about the three levels of marine biodiversity and some of the richest marine ecoregions around the world. Then create your own poster.

OBJECTIVES
Define marine biodiversity. Identify examples of genetic, species, and ecosystem diversity in the marine environment.

B iodiversity is big. It covers all life on the planet at all levels—from genes to species to entire ecosystems. Marine biodiversity—the diversity of life in the sea—is part of the big biodiversity picture.

To get the most out of the case studies in this module, it's important that students understand the big picture of marine biodiversity. This activity will help students appreciate what marine biodiversity is by exploring its three levels—genetic diversity, species diversity, and ecosystem diversity.

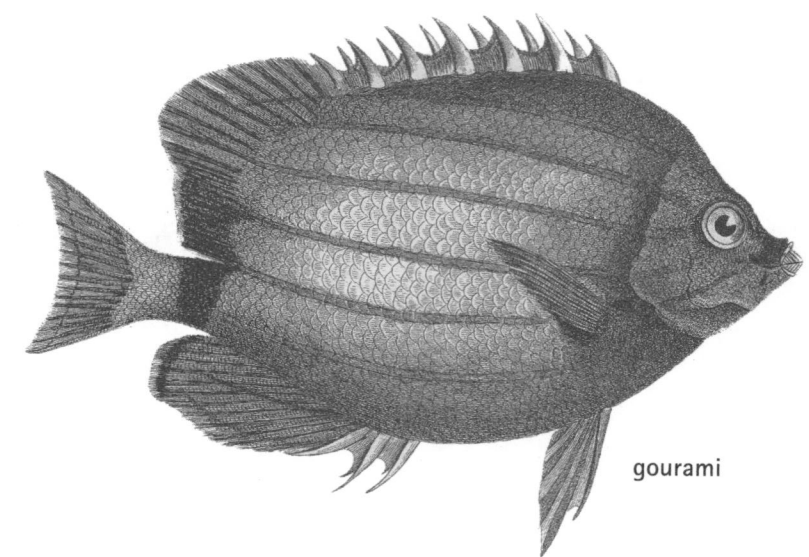

gourami

Before You Begin

If possible, put up a copy of the "Joy to the Fishes and the Deep Blue Sea" poster on a wall or bulletin board. For each student, make one copy of the small black-and-white "Joy to the Fishes" poster, "Sea Bits," "Describing Diversity," and "Pick a Marine Ecoregion " (pages 68-71). Arrange Internet access for students for one class session, or have on hand reference books that focus on marine biodiversity.

What to Do

1. Distribute copies of "Sea Bits" and the small reproductions of the marine biodiversity poster.

Tell the students that the "Joy to the Fishes" poster represents life in five different ocean ecosystems from five different parts of the world. Ask your students to describe what an ecosystem is. If they don't know the word, you can introduce it by talking about the ecosystems on the poster. (An ecosystem is a community of plants, animals, microorganisms, and other life forms that are linked by energy and nutrient flows and that interact with each other and with the physical environment.) The "Sea Bits" page contains information about these ecosystems and the species that live in them.

Explain to the students that their job is to label each of the five ecosystems depicted on the poster using the information provided on the "Sea Bits" sheet. Then they should match each marine fact with the appropriate species on the poster. If they want, students can first color the highlighted species. (Have the students refer to the full-size, full-color poster for a clearer view of the marine species.) Then have the students cut out the "Sea Bit" facts and affix them with tape or glue directly to the small poster or arrange them around the outside of the poster and connect the descriptions to the appropriate species with a piece of yarn or string.

2. Hand out the sheet with the columns labeled "Describing Diversity."

Now tell the students to read through the list of terms (listed on the "Describing Diversity" handout), answer the first two questions, and follow the instructions on how to label their posters. If you prefer, you can read and discuss the "Describing Diversity" page as a class and ask for volunteers to give examples of each level of biodiversity. Or you could assign this page as homework and have the students discuss it in class the following day. Either way, use the following answers to guide your discussion.

ANSWERS

- The entire poster represents biodiversity, with the exception of the dark night sky in the background.

- Except for the terrestrial areas on the globe, everything on the poster represents marine biodiversity as well. Remind the students that marine biodiversity refers not only to water-dwelling organisms, such as corals and dolphins, but also to those that live on the shore, such as storks, polar bears, and marsh grasses. Even terrestrial species, such as the monkey shown on the poster, can affect and be affected by marine biodiversity.

- Good examples of genetic diversity can be found in the "Sea Bits" information on fiddler crabs (genetic differences between the sexes) and polar bears (genetic variation in size). Why might it be important for species to be genetically diverse? *(Genes are what give a population the ability to adapt and survive in changing conditions over time. Individuals from a population that lacks genetic diversity—such as small populations of sea otters—may all be susceptible to the same diseases or environmental changes. In a more diverse population, it's more likely that some individuals will survive any given event, thereby carrying on the population.)*

■ The "Sea Bits" fact that most directly addresses species diversity is the one having to do with salmon. But all of the images of different plant and animal species are examples of species diversity.

■ The five ecosystems—estuaries, open ocean, kelp forests, coral reefs, and mangroves—are all examples of ecosystem diversity. How is ecosystem diversity different from species diversity? (Ecosystem diversity refers to whole communities of living things—a tropical rain forest, a coral reef, a salt marsh. Without a variety of ecosystems, the Earth couldn't sustain the wide range of species that enrich our planet.)

3. Discuss ecoregions.

Ask the students if they think all parts of the ocean have the same kinds of life. *(The students should now recognize that different ocean areas have different ecosystems and species.)* What accounts for the differences in ocean habitats? *(Water temperature, water depth, topographic features, and other factors.)*

Tell your students that because ocean regions differ so much, scientists have mapped them into distinct biogeographic areas called *ecoregions*. An ecoregion is a relatively large unit of land or water that is characterized by a distinctive climate, specific ecological features, and similar kinds of plant and animal communities. World Wildlife Fund (WWF) scientists and their collaborators are mapping the more than 1,000 terrestrial, marine, and freshwater ecoregions that exist around the world. Of these,

WWF has identified more than 200 as the richest, rarest, and most distinct natural areas that are priorities for conservation.

Ecoregions are usually larger than ecosystems and are always tied to a specific geographic area. For example, an individual coral reef anywhere in the world is an ecosystem, whereas the Mesoamerican Reef, found off the Caribbean coasts of Belize, Guatemala, Honduras, and Mexico, is an ecoregion made up of coral reef, mangrove, sea grass, pelagic, and estuarine ecosystems. There are other coral reef ecoregions around the world, such as the Great Barrier Reef off the eastern coast of Australia. Explain to the students that the ecosystems on the poster are critical areas for biodiversity and are found in the priority ecoregions of WWF's Global 200 (see page 51).

4. Have students look at the handout labeled "Pick a Marine Ecoregion."

Tell your students that now they will get a chance to make their own poster depicting life in one of the world's richest and rarest marine ecoregions. Have the students divide into teams of two or three and select one of the ecoregions listed under "Pick a Marine Ecoregion." Then have the students search for more information about the ecoregion. The posters the students create should show what several parts of this ecoregion looks like, identify some of the plant and animal species living there, and include a brief paragraph introducing the region. Ideally, students should provide "Sea Bits" information that highlights different aspects of biodiversity. Resources such as the

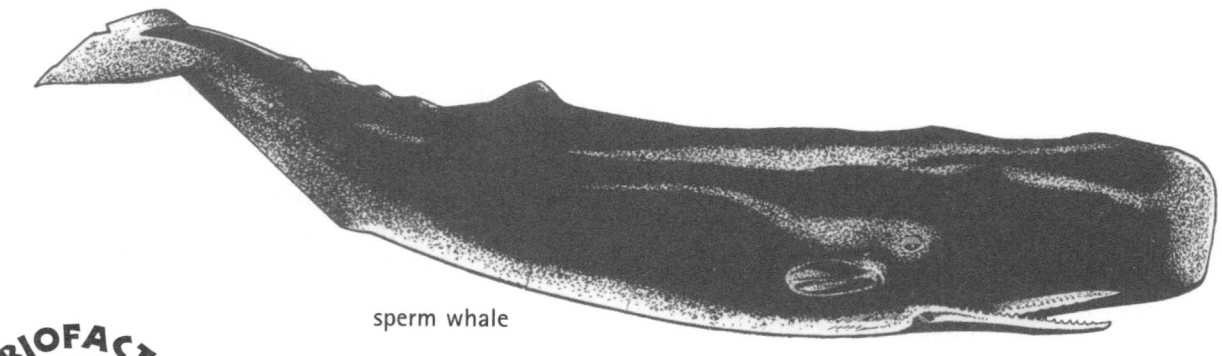

sperm whale

BIOFACT

Whales must come up for air every few minutes. However, some whales can stay under water for an incredibly long time—sperm whales can hold their breath for up to an hour!

Internet can be useful because they contain pictures of the ecoregions that will help students create their own paintings or drawings. Your students should be able to find information specifically devoted to these ecoregions on two Web sites: www.worldwildlife.org and www.worldwildlife.org/wildworld.

Note: Some of the locations listed in "Pick a Marine Ecoregion" are not independent countries (for example, St. Pierre and Miquelon are French territories).

5. Review completed posters.
Hang the completed posters around the room (or in the hallways) for others to see. Then allow time for each group to describe its poster to the rest of the class. Be sure each group points out its ecoregion on a world map before beginning its presentation.

WRAPPING IT UP

Assessment
On the board, create a chart with three columns labeled: (1) Marine Ecosystem, (2) Genetic Diversity, and (3) Species Diversity. Then divide the chart into three rows. Have students copy the chart and fill in the three columns and rows, identifying three marine ecosystems in the left-hand column. In each of the remaining two columns, the students should fill in types of genetic diversity (such as the differences in the sizes of polar bears and in the claws of fiddler crabs) and the species diversity they'd expect to see in each of these ecosystems.

Unsatisfactory—None of the rows and columns have correct responses.

Satisfactory—One correct response in the second and third column for at least two ecosystems.

Excellent—More than one correct response in each column for at least two ecosystems.

hawksbill turtle

Portfolio
Include the students' ecoregion posters (or copies of them) in their portfolios.

Writing Idea
Have students choose one of the marine ecoregions highlighted on the poster and create a travel brochure for it. In the brochure, students should describe what makes the region unique and what kinds of plants and animals live there.

Extension
Students can conduct more in-depth research on a chosen marine ecoregion, specifically focusing on conservation efforts. They should search for information on what communities and governments as well as local, national, and international conservation groups are doing to protect that particular ecoregion. They should brainstorm additional ideas for activities that may help conserve marine biodiversity in the ecoregion in the future.

Joy to the Fishes and the Deep Blue Sea

AND ALL THE AMAZING LIFE THAT DEPENDS ON OUR OCEANS

Protect Marine Biodiversity

This poster was developed by World Wildlife Fund, in partnership with the Monterey Bay Aquarium. It is part of Windows on the Wild, a World Wildlife Fund environmental education program supported by Eastman Kodak Company. To learn more, visit www.worldwildlife.org

Use the following information to label each of the five ecosystems depicted on the poster. Write the type of ecosystem at the bottom of each column on the poster.

The Mesoamerican **coral reefs** provide shelter and food for an amazing diversity of life, from tropical fish to sea turtles to manatees.

Six states and 64,000 square miles of land drain into the Chesapeake Bay **estuary**, which provides fertile habitat for crabs, fish, oysters, reptiles, wading birds, and more.

The cold **open waters** of the Bering Sea are inhabited by fish, seabirds, and large marine mammals such as Steller's sea lions, polar bears, and whales.

In Malaysia and other warm areas, **coastal mangroves** line the shores. Fish and crabs live in the water among their roots, and milky storks and proboscis monkeys perch in their branches.

Australia's **kelp forests** are like underwater jungles, with rock lobsters, weedy sea dragons, and sea slugs called *nudibranchs* living in and around the dense growth.

Now see if you can match these facts with the correct plant or animal on the poster:

Green moray eels are actually blue with a yellow mucus coating.

The tongue of a bowhead whale weighs almost a ton.

The grooves and channels of brain coral make it look like a petrified brain.

More than 45,000 diamondback terrapins were once harvested each year from the Chesapeake Bay so restaurants could make turtle soup. Now they are protected.

Male fiddler crabs use their one enormous front claw to intimidate rivals and attract mates. Females don't have these big claws.

Great blue herons build their nests in high trees, sometimes forming huge colonies. A steady rain of feces and occasionally regurgitated food discourages predators from advancing from below.

Marine scientists who study spotted dolphins identify individuals by their spot patterns and other markings.

Male polar bears can weigh from 800 to more than 1,300 pounds.

Sockeye, Chinook, Coho, Chum, and Pink may sound like people's nicknames, but they're actually Pacific salmon species. A sockeye is pictured on the "Joy to the Fishes" poster.

In Australia, male weedy sea dragons carry as many as 250 bright pink eggs on their tails for six to eight weeks while waiting for the eggs to hatch.

If your hair grew as fast as giant kelp blades do, it would reach the floor in three to four days.

Mudskippers are fish that use their fins to run across mudflats, climb rocks, and even crawl into mangrove trees.

polar bear

DESCRIBING DIVERSITY

BIODIVERSITY

The variety of life on Earth, from genes to species to ecosystems.
Which parts of the poster represent biodiversity?

MARINE BIODIVERSITY

The diversity of life—genes, species, and ecosystems—found in the saltwater
parts of the Earth. Marine diversity also includes species that depend on
the ocean, such as shorebirds, beach grasses, crabs, and other animals on
shore. **Which parts of the poster represent marine biodiversity?**

sea star

GENETIC DIVERSITY

The genes that create visible and invisible differences among individuals of a
species. Genes are units of DNA and are found within most cells of the body. They
direct how cells develop and what they do, controlling things such as hair, skin, or
eye color, and things you can't see, such as resistance to certain diseases. **Write
the letter G next to a "Sea Bits" fact that describes genetic diversity.**

SPECIES DIVERSITY

The variety of species that inhabit the Earth. Individuals of the same
species can interbreed and produce viable offspring. **Write the letter
S next to a "Sea Bits" fact that describes species diversity.**

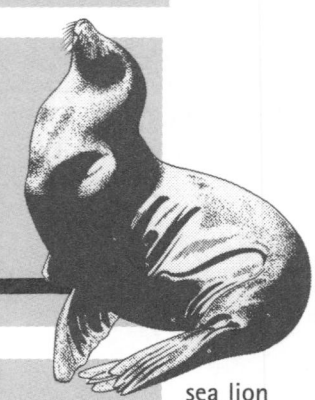

sea lion

ECOSYSTEM DIVERSITY

The variety of ecosystems that make up the Earth. Ecosystems can range in size from
entire forests and coral reefs to tidal pools and rotting logs. **Write the letter E next
to a "Sea Bits" fact that describes ecosystem diversity. (Many different
ecosystems can be found in an ecoregion.)**

POLAR
196 Antarctic Peninsula and Weddell Sea
197 Bering Sea (Canada, Russia, United States)

TEMPERATE SHELF AND SEAS
199 Mediterranean Sea (Albania, Algeria, Bosnia and Herzegovina, Croatia, Cyprus, Egypt, France, Gibraltar, Greece, Israel, Italy, Lebanon, Libya, Malta, Monaco, Morocco, Slovenia, Spain, Syria, Tunisia, Turkey, Yugoslavia)
201 Grand Banks (Canada, St. Pierre and Miquelon, United States)
202 Chesapeake Bay (United States)
203 Yellow Sea (China, North Korea, South Korea)

TEMPERATE UPWELLING
208 California Current (Canada, Mexico, United States)
210 Humboldt Current (Chile, Ecuador, Peru)

TROPICAL UPWELLING
214 Gulf of California (Mexico)
215 Galápagos Marine (Ecuador)

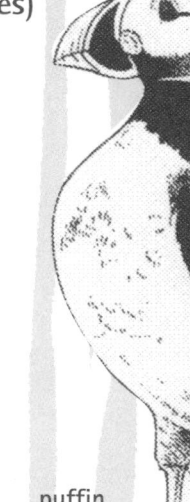

puffin

TROPICAL CORAL
222 Great Barrier Reef (Australia)
226 Tahitian Marine (Cook Islands of New Zealand, French Polynesia)
227 Hawaiian Marine (Hawaii, United States)
231 Red Sea (Djibouti, Egypt, Eritrea, Israel, Jordan, Saudi Arabia, Sudan, Yemen)
233 East African Marine (Kenya, Mozambique, Somalia, Tanzania)
234 West Madagascar Marine (Comoros, Madagascar, Mayotte and Iles Glorieuses, Seychelles)
235 Mesoamerican Reef (Belize, Guatemala, Honduras, Mexico)
237 Southern Caribbean Sea (Aruba, Colombia, Grenada, Netherlands Antilles, Panama, Trinidad and Tobago, Venezuela)

2 | Sea for Yourself

AT A GLANCE
Take a trip to the seashore; visit a natural history museum, science center, zoo, or aquarium; or go on a "virtual tour" of oceans to get more closely acquainted with marine life and the three levels of marine biodiversity.

OBJECTIVES
Learn about the variety of marine life. Find examples of genetic, species, and ecosystem diversity in the marine environment either in your local area, at a scientific institution, or through a virtual field trip.

Marine biodiversity might be a new idea to your students, even if they live near an ocean or an estuary. Before diving into one of the case studies in this module, you can use this activity to inform your students and inspire them to want to learn more about the topic by taking a closer look at the diversity of life that can be found in the ocean. Three options for investigating marine life are suggested in this activity.

If you live near an ocean, Option 1—a trip to the seashore—is a great way to introduce your students to the wonders of marine biodiversity. An experience like this is exciting for students and will give you a chance to explore and reinforce some key concepts about life in the sea. If you don't live near an ocean, you can arrange for Option 2—a field trip to an institution that features examples of marine life; for example, a natural history museum, zoo, aquarium, or science center. Although students won't be seeing marine life in its native habitat, they'll still be exposed to some fascinating life forms and the environments they inhabit. Option 3 is for your students to take a virtual tour of oceans on some lively Web sites. Consider doing more than one option. For example, you could do Option 2 or Option 3 as preparation for Option 1.

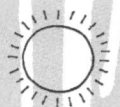

hermit crab

Before You Begin

Option One: Seashore Field Trip

If you're planning to take your students to the seashore, you might want to invite an expert along to help. To find an appropriate person, consult with a local aquarium, nature center, or college or university natural resources department. You could also try a local conservation group or a department of environmental protection or other government agency that deals with the marine environment. If you find an expert, he or she might have ideas (or materials) to help in your exploration of the shore. Be sure to bring field guides, if available.

When choosing a site to visit, you'll want to take into account a variety of factors, including travel time, accessibility, and facilities. Once you've identified a few possibilities, you might consult with local experts about which sites are the least disturbed and offer the greatest opportunities to find a diversity of life. Be aware that usually you are not allowed to collect or move live animals or plants, or even shells. Observe them and then leave them in their natural setting.

Once you've selected a site, check the tide tables (you need to find out at what time of day low tide occurs), and make any necessary travel arrangements (such as transportation, chaperones, permissions slips, and so on). See "Taking Your Class Outside" (pages 78-79) for some practical tips on preparing for outdoor trips with your students.

Option Two: Visit an Educational Institution

If you're taking a trip to a natural history museum, zoo, nature center, or aquarium, call ahead to schedule your visit and make any necessary arrangements (such as transportation, chaperones, permissions slips, and so on).

Be sure to make an advance visit to the institution. (Some places give educators a free pre-visit pass when a group or class reservation is made.) Take along a copy of the "Scavenger Hunt Clues" on page 80 to get a feel for the variety of answers students may find. If you want to alter the clues to make them more appropriate to your group or to the institution you're visiting, retype them on a separate sheet of paper and make copies. Otherwise, make one copy of "Scavenger Hunt Clues" (page 80) for each team of two students. Collect enough clipboards so that each team has one clipboard. (You can also make clipboards out of heavy cardboard and have students attach their papers with clothespins or large binder clips.)

Option Three: Virtual Tour

For each student, make one copy of "Virtual Ocean Explorations" (page 81). Arrange for students to have one or two sessions of computer time with Internet access.

What to Do

Option One: Seashore Field Trip

1. Review the term marine biodiversity.

If you've already discussed the term marine biodiversity with your students, ask them if they remember the three levels of diversity: genetic diversity (the differences among individuals of the same species), species diversity, and ecosystem diversity. Note: If your students are unfamiliar with the idea of biodiversity, you can use the activity "Biodiversity Break-Down" (pages 64-71) to introduce this concept before taking your trip.

Ask the students if they can think of any examples of marine species that live nearby. Do they know of different marine ecosystems in your area? Tell the students that they'll be taking a trip to the seashore and learning firsthand about local marine biodiversity.

2. Give the students pointers on locating samples.

Once you're at your site, tell students that their job is to find as many examples as they can of marine biodiversity in the ecosystem or ecosystems they're exploring. Give them a set amount of time to explore (at least 30 minutes). They should search the shore for examples of marine biodiversity and record their findings in the field notebooks. Encourage them to record plants that have washed ashore, such as seaweed and kelp, and animals (as well as any signs of animals) including seashells, egg cases, crab shells, animal bones, tracks, nests, and droppings.

Students should include the following in their field notebooks: a brief sketch of each item, the name of the organism if they know it (provide field guides, if possible), the number of each species they find, and any differences they see among individuals of a species.

Please Don't Collect Specimens

Many people who work in the marine field believe that students should not be allowed to collect specimens at any seashore. If the students are taught to observe and record their findings in field books instead of collecting specimens, the students' families may learn to do the same.

It is true that collecting shells and other specimens is allowed in some places. Collecting of this kind may be OK, but we recommend that you don't encourage it. If you decide to allow it, learn the local or federal laws that limit what you may collect. Then be sure students are familiar with species that are off limits before you go. And before you leave the seashore, have your students return all live specimens to the areas where your students found them.

3. Find a central area to analyze students' findings.

After the students complete the notes in their field books, have them gather at a central (preferably shady) area. In pairs, have them devise ways to organize the specimens they noted. They may organize them by species and then by higher levels of classification, such as animals with shells, plants that grow on shore, plants that grow in the water, and so on. The students can even try creating a variety of different schemes, reorganizing the specimens as necessary. Ask each pair of students to share with the rest of the class their organizational schemes, as well as anything unusual that they found.

4. Discuss the levels of biodiversity found in the specimens.

- Can the students find examples of genetic diversity in the specimens they've noted in their books?
- Were there visible differences among individuals, or did all the members of the same species look identical?
- Was it sometimes difficult to tell if two individuals were from different species or the same species?
- How many examples of species diversity did your students find? Did they find more or fewer species than they thought they'd find? Were there some categories of plants and animals that had more species in them than others? (For instance, were there more animals with shells than animals without shells?) Why might this be the case?

seaweed

- What kinds of species did the students describe in their notebooks? Are there species that are commonly found in your neighborhood that the students didn't find here at the seashore?
- Were the students able to find any examples of ecosystem diversity in the area you explored? What is the major ecosystem type you're in? Are there any smaller-scale ecosystems found within the larger one (such as tidal pools along a rocky shore)?

If you have time to visit more than one site, it would be interesting to compare the major features of the different ecosystems.

Note: This activity is designed to offer students a relatively unstructured introduction to marine biodiversity—giving them time to explore the seashore on their own. The emphasis is on observation, organization of the species observed, and creative thinking. If you'd like to take a different approach, or if you'd like to use this opportunity to teach your students scientific data collection techniques, there are a variety ways to modify this activity. The following are just two ideas to get you started:

A. Hunt It: Scavenger hunts can be a great tool to help sharpen students' observation skills, and they are also helpful if there are particular things you want students to look for on the seashore. To make use of this technique, you can put together a scavenger hunt that is specific to your region. It can be similar to the one in Option Two, but you may wish to have your students sketch the species they observe, especially if they can't name them. To learn more about local species in advance, you might consult with your local aquarium or nature center.

B. Tie It In: If you're planning to go through one or more of the case studies in this module with your students, you might want to tie in some of the key concepts from the case study while on your trip. Drawing on important concepts from these cases will influence the kinds of things you may want students to do while they're on the shore. The activities, skills, and concepts you focus on during the field trip will most likely depend on how you're planning to teach the case study.

Option Two: Visit an Educational Institution

1. Review the term marine biodiversity.

Introduce or review the concept of marine biodiversity. (See "Biodiversity Break-Down" on pages 64-71 for some ideas.) Can your students remember the three levels of diversity? *(genetic diversity, species diversity, and ecosystem diversity)* Tell the students that the place they will be visiting includes examples of marine biodiversity. They'll be looking for examples of all three levels, as well as the answers to specific questions given to them as part of a marine biodiversity scavenger hunt. You might also have the students check the institution's Web site to find out more about it and to help with the planning.

2. Organize your group for the field trip.

Pair your students with partners for the field trip. Each student should stay with the partner at all times and will work with this person to find examples of the clues listed on the scavenger hunt worksheet. Assign a clear role for the chaperones as they accompany your students throughout the exhibits. For example, you might provide them with some questions they can use to help guide the students in their discoveries.

3. Hand out copies of scavenger hunt.

Once you reach the museum, zoo, aquarium, or nature center, give each team a clipboard and a copy of the "Scavenger Hunt Clues." Make sure every student has a pen or pencil.

Explain to the students that they may not be able to find the answers to all the clues. Their goal should simply be to find as many of the items as possible. When they find an item, they should write its name beside the clue or sketch a picture of it.

You should decide in advance which parts of the institution the students are allowed to explore and whether the teams will move through the area as a group or on their own. If appropriate, select a meeting spot and set a time for regrouping there. Then let the teams pursue their hunts.

4. Review answers.

When students have completed their scavenger hunts, review their results. Which clues were easy to track down? Which ones were more difficult? Which clues led to examples of genetic diversity? Species diversity? Ecosystem diversity? Was it difficult to see genetic differences among individuals of the same species? (You might point out that many important differences among individuals, such as eyesight, hearing, disease resistance, and other traits that are genetically controlled, often can't be seen.)

5. Wrap-up exercise.

As a wrap-up to this activity, have the students select one species they saw and discuss how it's adapted to its ecosystem.

Option Three: Virtual Tour

1. Discuss ocean conditions.

Ask your students if they have any firsthand experience of coastal waters. Coral reefs? The open ocean? The deep sea? Tell the students that, since it isn't easy to visit all these environments, one way for them to get a better sense of life in both deep sea and shallower environments is to go on a virtual tour.

Hand out the "Virtual Ocean Explorations" sheet on page 81, and tell your students that they can select any of these Web sites to visit and explore. All of the sites provide images and descriptions of life in the sea, and many of them discuss what scientists know as well as what they do not yet understand about life in underwater environments. Students should also feel free to search for other Web sites featuring interesting images and information about ocean habitats.

2. Define biodiversity and assign task.

If you haven't already talked about marine biodiversity, do so here, at least in general terms. (You could describe the three levels of biodiversity before starting the activity, if appropriate, or wait until later when discussing the results.) Tell the students that each of them should create a page or two of a field guide or journal that describes his or her exploration of the marine  firefish
Web site. They should include notes telling what is special about the marine ecosystems they explored, and give examples of marine diversity encountered on this site. Have the students make sketches for their pages too.

3. Discuss results.

Have the students share the results of their explorations. Did anyone explore coastal or other shallow regions? Sandy beaches? Deep-sea regions? Have the students compare and contrast the results from each of these regions. What types of ecosystems did they explore? What kinds of species did they find? Did the species differ a lot depending on where they were found? What special adaptations did species have to their particular habitats? Did anyone find examples of genetic diversity?

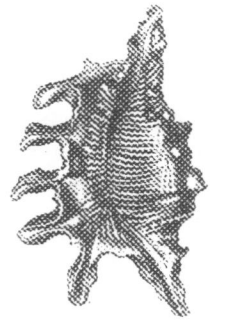
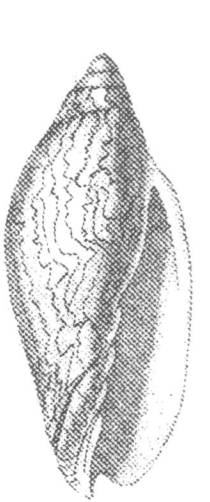
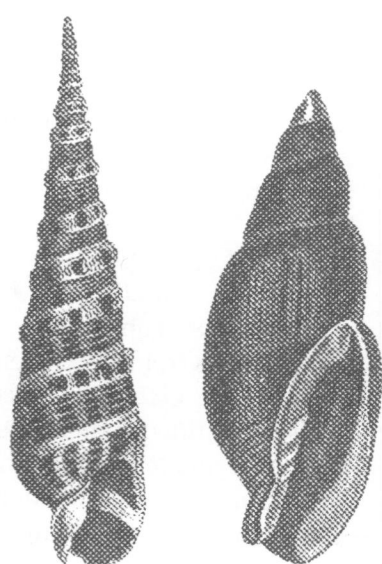

WRAPPING IT UP

Assessment

Have each student write a paragraph explaining how her or his experience revealed genetic, species, and ecosystem diversity in the marine environment. The students should include definitions of these terms along with examples from the experience.

Unsatisfactory—Fewer than three terms are correctly or completely defined or include examples.

Satisfactory—All three terms are correctly defined and include examples.

Excellent—All three terms are correctly defined, include examples, and are connected directly to the experience.

Portfolio

If your class took the seashore field trip or completed the virtual exploration, have students include a field guide or notebook page in their portfolios. Students can include the scavenger hunt sheets if you took a trip to a scientific or educational institution.

Writing Idea

Students can write a journal entry describing their experiences on one of the field trips or a virtual field trip. They should be sure to include the who, what, when, where, why, and how of the trip, describing the activities in which they participated, recording what they learned, and posing several questions for further investigation.

Extension

Give students an outline map of the world and have them mark the locations of a variety of marine ecosystems.

man of war

Taking Your Class Outside

The most powerful multi-media tool available to teachers doesn't even have a plug! The outdoors can provide a dynamic change of pace and place that enriches learning opportunities and also promotes a sense of stewardship for the natural world. To make the best use of your time outdoors, here are some practical "tips and tricks" for working with students in the outdoors.

Before you go outside . . .

- Survey the site before you go out. This is not a big concern when using the school grounds. It becomes very important, however, when using a location that is off site. Trails can be closed or overgrown, beach erosion may have changed an area significantly, or litter and pollution may have rendered a location unsuitable. It's always best to check the site just before you take your group on its trip. Be alert to any potential safety hazards and, if possible, get additional safety information from whoever oversees the property you are visiting. Be sure to brief your students on any potential risks before your group visits the site. For example, if you are visiting the seashore, warn the students about the dangers of walking on slippery rocks or becoming stranded by an incoming tide.

- While still in the classroom, describe in detail the tasks that students will be expected to do outdoors.

- Review the rules and behaviors that will be expected outside. Avoiding a "recess" mindset is critical. It may sound obvious, but behavior expectations need to be emphasized, especially with students who are not used to outdoor instruction. Also, discuss ways to protect the wildlife they will be observing, such as avoiding walking on tidepool creatures at the seashore.

- Pre-arrange groups or pairs before going out. Don't spend valuable outdoor time doing things you could do inside.

- Give background information and define vocabulary in the classroom. Use your outdoor time for doing, rather than telling.

- If students will need to write or record data, consider making simple clipboards. Cut white poster board into pieces measuring about 10 by 12 inches. Laminate the cardboard rectangles twice to improve moisture resistance. Place sheets of paper on top of the boards and use clothespins, binder clips, or bulldog clips to hold them in place. You now have an inexpensive set of lightweight, flat, and very portable writing surfaces.

- Double check to make sure that you have all necessary materials and equipment. Huge amounts of time are wasted when runners have to dash back to the classroom to bring out forgotten items when you're on the school grounds. And forgotten items on a trip can make the experience less effective.

- Begin with very short treks outside (15 minutes or so). Gradually increase your time outside, always being sure to very clearly explain indoors what specific tasks will be accomplished outdoors.

- If you are going to a seashore, park, or nature reserve, you probably will not be allowed to collect or even move live animals or plants from where you find them. Instruct your students to observe these plants and animals and leave them in their natural setting.

fiddler crab

- Use a backpack for your materials rather than a box. And bring a camera to capture the experience.

- Take one large plastic trash bag along for each student. They make great sit-upons when the ground is damp, and raincoats when an unexpected shower comes up. (Cut holes for arms and head.)

- For your materials, buy quart- or gallon-size plastic freezer bags that zip tightly closed. They save space, display the contents, and keep your materials dry.

When you are outside . . .

- Clearly indicate boundaries. Telling students, "Stay on this side of the sand dune" is much clearer than saying, "Don't go too far." It can be helpful to tie ribbons or colorful strings on trees or shrubs that you want to use as boundaries. Without clear boundaries, groups will radiate in all directions, creating time-consuming (and potentially dangerous) situations.

- Always have students stand or sit in a circle while you give directions or debrief activities. Trying to talk to a mass of students standing three or four deep is not very effective.

- As the facilitator, circulate constantly. Because the outdoors is loaded with natural distractions, it is important that you frequently move around to all groups to monitor and refocus attention to the task.

- View the unexpected as a bonus, not as an annoyance. That fantastic shell a student noticed may not have been in your lesson plan, but it can be a great springboard for discussing protective coloration.

- Be patient. Teaching outdoors is a little more time-consuming, but it's well worth the effort.

Back inside . . .

- Use what you discovered outside—tabulate data, make charts, do creative writing, and encourage artistic expressions.

- When possible, post or display something related to the outdoor experience. Pictures of species found on the beach, maps of the area, or student work related to the outing make a statement that the outdoor experience is directly related to the indoor classroom.

- Have field guides and other interpretive resources readily available. Outdoor teaching often encourages students to ask "What was that?"

- If possible, have live plants and animals in the classroom. An aquarium, terrarium, or a classroom pet can go a long way in fostering a sense of wonder and the beginnings of stewardship.

–Herbert W. Broda, Ph.D., Associate Professor of Education, Ashland University

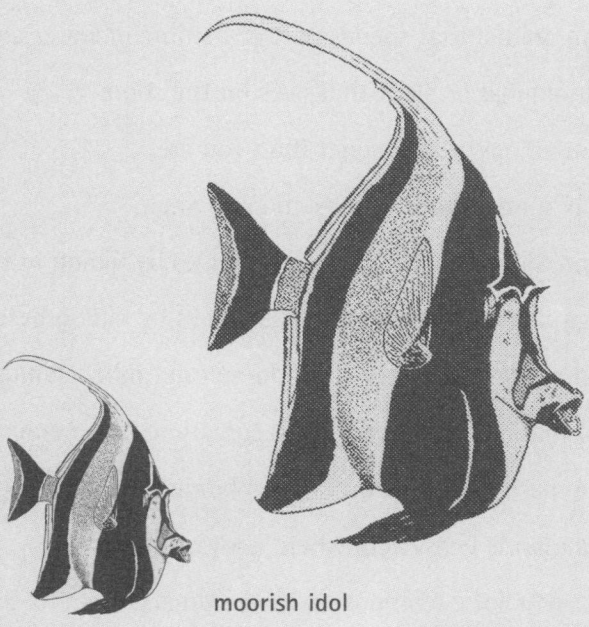

moorish idol

As you walk around, try to find:

1. an animal that blends into its surroundings

2. a species of fish in which the male and female look different from each other

3. a species in which different individuals are different colors

4. a very flat fish

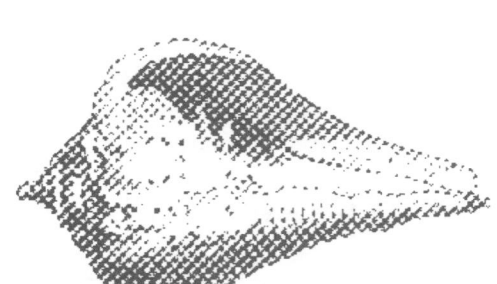

5. an animal that produces light

6. an animal that lives in a shell

7. an animal that's attached to something else

8. an animal that eats other animals

9. an animal that looks like a plant

10. an animal that spends most of its time on the bottom of the ocean

11. a type of seaweed

12. an animal that escapes its enemies by swimming very quickly

13. an animal or plant that drifts near the surface of the water

14. an animal that must go to the surface to breathe air

15. an animal with tentacles

16. an animal that spends part of its time in water and part of its time on land

17. an animal or plant that lives on the shore

18. an animal that's bigger than you are

19. an animal that's smaller than an apple

20. an animal that escapes its enemies by hiding in the sand

21. an animal that escapes its enemies by hiding between rocks

22. a marine ecosystem that doesn't get much sunlight

23. a marine ecosystem that is sometimes submerged by water and sometimes not

24. a marine ecosystem that is a home for sharks

25. a marine ecosystem where a lot of plants grow

26. a symbiotic relationship (two animals or plants that depend on each other)

Activity adapted from "Ocean Odyssey" in *Ranger Rick's NatureScope: Diving into Oceans.* ©2002 National Wildlife Federation.

VIRTUAL OCEAN EXPLORATIONS

1. www.ocean.washington.edu/people/grads/scottv/exploraquarium/vent/intro.htm
This Web site is part of the University of Washington's School of Oceanography Exploraquarium. "Choose your own adventure" as you explore deep-sea vents—geysers on the ocean floor.

2. seawifs.gsfc.nasa.gov/squid.html
Throughout history, giant squids have inspired countless myths and legends. What do we now know about these real-life giants? What are we still trying to find out? This Web site lets you know!

3. www.mbayaq.org/efc/efc_se/se_mod.asp
Just off the coast of Monterey, California, lies a deep canyon inhabited by a fantastic variety of sea creatures. This Web site takes you down into Monterey Canyon to find out what lives there and gives you a behind-the-scenes look at the Monterey Bay Aquarium's exhibit on this fascinating habitat.

4. www.divediscover.whoi.edu
Travel with marine biologists as they explore underwater habitats throughout the Pacific and Indian Oceans. The Web site features daily field notes, slide shows, and lots of other information about these research expeditions.

5. www.pbs.org/wgbh/nova/abyss
Discover the odd landscape and strange life forms of the abyss—an ecosystem with black smokers lying one and one-half miles below the surface of the water off the Pacific Northwest coast.

6. www.cyberlearn.com/zones.htm
This site offers you a brief tour of the different zones of life in the coral reefs of Micronesia.

7. www.worldwildlife.org/expeditions/reef
Dive to the depths of the largest Atlantic Ocean coral reef system and come face-to-face with 30-foot whale sharks and more than 500 other species of fish.

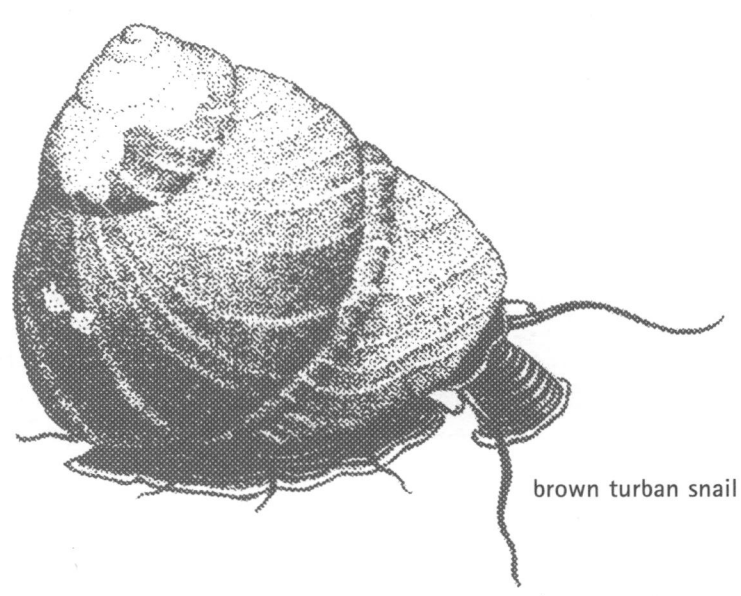

brown turban snail

SUBJECTS
science

SKILLS
gathering (simulating), organizing (plotting data, graphing), interpreting (generalizing, relating, inferring, reasoning)

FRAMEWORK LINKS
3, 23.1, 53.1, 53.2

VOCABULARY
biosphere, ocean trench, seafloor plain, water column

TIME
one session

MATERIALS
two hats with brims, construction paper, 12 flags or cones, about 90 wrapped candies

CONNECTIONS
After this activity, use "Going Under" (pages 94-98) as a way to help students better understand scientists' (and nonscientists') perspectives of what it's like to be under the sea.

AT A GLANCE
Conduct an outdoor simulation to demonstrate why it's difficult to discover new species in the ocean.

OBJECTIVES
Explain some of the major differences between ocean conditions near the shore and in the deep sea. Explain how those differences can affect ocean research.

Nearly 99 percent of the habitable part of the planet, called the *biosphere*, is in the ocean. With its sunny coastal waters, open seafloor plains, deep trenches, and vast water column, the sea contains a wide variety of places for life to thrive—and provides much more livable space than the land does. And although hundreds of thousands of species in the ocean have been discovered, the potential for undiscovered life in the sea is extremely high. That's because up to 95 percent of the ocean remains to be explored with the same level of detail that we've explored the landmasses of the Earth.

In this activity, your students will get the chance to simulate the conditions of underwater exploration and understand why it's tough to find new species in the sea.

BIOFACT
Tidal waves are caused by earthquakes and other abrupt movements of the ocean bottom, not by tides.

Before You Begin

Using hats and construction paper, fashion two hoods that can be placed over students' heads to restrict their vision to the area directly around their feet. A hat with a brim that extends all the way around is best. Attach the construction paper lengthwise around the brim, using as many pieces as necessary to encircle the hat with no open spaces.

In your schoolyard, a nearby park, or other open area, set up three "Expedition Areas" for the simulation. Use flags, cones, or other markers to delineate the boundaries of the three areas, which should measure 20 feet by 40 feet. In each area, slightly conceal 30 wrapped candies for students to find. Try to conceal each set of candies with an equal level of difficulty, but there is no need to make them very hard to find.

What to Do

1. Discuss ocean conditions.

Ask the students to imagine underwater life in coastal waters. You might suggest that they imagine a coral reef. What does it look like, and what are the water conditions?

Many of us have seen pictures or movies of coral reefs and know that they are bright, colorful places filled with life. The waters are warm and clear and are home to a variety of species such as corals, fish, sponges, clams, algae, and dolphins. Most corals are found in shallow waters relatively near the shore.

Now ask your students to imagine what it's like at the bottom of the deep sea, at 10,000, 20,000, or even 30,000 feet below the surface. What's the temperature? How much light is there? Do they think it would feel just like being under water near the surface of the ocean, or is there more pressure from the heavy water column? What kinds of animals might live at those depths?

2. Discuss ocean exploration.

Of all the species that have been identified on Earth, only about 15 percent are marine species. Many scientists think that the number of species in the ocean is much higher than we now know. Do your students have ideas about why scientists haven't identified more marine species? What might the conditions in the ocean (and especially in the deep ocean) have to do with the low number of identified species? *(The conditions in the ocean are very different from those on land, where humans are adapted to life. For one thing, there's no air, so scientists must carry it with them when they dive or travel in submarines. They can only dive safely to a depth of 200 feet. If the scientists want to study organisms at a greater depth, they have to use a vehicle. Also, the high water pressure creates problems for humans. When scientists dive under water and move up and down in the water column, they must take time to adjust to the pressure. To go very deep, strong submarines must be built to withstand the high pressure. In the deep sea, scientists can't see the organisms or habitats without the use of powerful lights, and the temperature is very cold. All this makes for tough conditions for discovering new marine species, particularly in the deepest parts of the ocean.)* The following simulation should help students appreciate how scientists are limited in their exploration of ocean environments.

BIOFACT

Some species of sponge can live to be more than 100 years old.

3. Assign students to exploration groups.

Take the students outside, but not so close to the "expedition areas" that they can see the candies. Tell the students that they're going to be recreating what it's like to be a scientist searching for species. They'll be put into one of three groups of scientists searching for new species. One group of about 18 students will search for species on land. Explain that this group will have five minutes to search for species (represented by wrapped candies, with each candy indicating a different new species) in an area that you've designated for them. All students in this group should try to find as many new species as they can in five minutes.

Another group will be searching for new species in the coastal ocean. Since it's harder to do this because of the need for diving equipment and the limited time available to divers, a smaller number of students will be looking in this realm. Assign about eight students to this group. This group will also have less time to search because of restrictions in the amount of air they can take with them in their diving equipment. This group has only two minutes to search for species.

A third group of students will represent the deep ocean explorers. This is an even smaller group because of the advanced equipment and training needed. Assign about four students to this group. Because the deep ocean is dark and explorers can see only the short distance lit by their lights, these explorers will be limited in what they can see. From the deep-sea group, select two students to be scientists and two students to be navigators. The scientists will wear hoods over their heads to limit what they can see, simulating the situation of scientists who travel to the ocean bottom in submarines and who are very limited in what they can see. The navigators' job is to make sure the scientists don't run into anything or get hurt (but navigators can't help the scientists find the species—their job is just to lead the scientists away from hazards). These deep-sea students have only one minute to complete their exploration and find species, because time is very limited when explorers travel to the ocean bottom in submarines.

The simulations should be run one after the other, rather than simultaneously. This will give all of the students a chance to observe the conditions for each of the different groups. Have each group keep track of the number of species they find. Before you return to the classroom, try to collect any undiscovered wrapped candies so that they're not found by local wildlife.

4. Graph the simulation results.

Once the simulation is complete, return to the classroom to analyze the results. Tell the students the actual number of species that were in each area. Based on that information, have them calculate the percentage of species they found. Then have them create a bar graph representing the results of their finds. Have the students consolidate their bar graphs and compare the results. How many species did each group find? Did the students searching on land find the most species? How many did the coastal ocean explorers find? The deep-sea explorers? Do the students think the simulation conditions were like those in real life? How might they be the same or different? Do the students think that the percent of undiscovered species could be the same in reality? Why or why not? Given the results of this simulation, do the students think there could be more marine biodiversity that we don't yet know about?

Tell your students that the oceans of the world remain relatively unexplored. Some scientists believe that there may be millions more unnamed and unstudied marine species!

> *"Probably the greatest enticement for those who today are devoting their lives to the study of the sea is the lure of the unknown, the challenge of the undiscovered, the thrill of discovery on what is truly the last frontier on Earth."*
>
> **—Harris B. Stewart, oceanographer**

WRAPPING IT UP

Assessment

Have the students imagine that they are magazine reporters. Ask them to write a short article on the differences between research done with scuba equipment near shore versus research conducted using a vehicle designed for diving into the deep ocean. The students' stories should creatively and convincingly describe what would be seen in the shallow parts of the ocean as well as in the depths, the types of life likely to be found at each level, and some of the challenges to conducting research at each level.

Unsatisfactory—Story does not correctly distinguish between what is seen near shore and what is seen in the deep sea. The story does not include at least two types of life or explanations of research challenges; story is informational but lacks creative and convincing descriptions.

Satisfactory—Story describes the differences between what is seen at both levels and correctly names at least two types of life likely to be found at each level. Story explains the research challenges at each level but lacks creative and convincing descriptions.

Excellent—Story creatively and convincingly portrays an explorer's journeys to both ocean areas and discusses types of life and issues related to research challenges at each level.

Portfolio

Have students include their creative stories (see Assessment), their newspaper stories (see Writing Idea), or their "Ocean Explorations" articles (see Extension) in their portfolios.

Writing Idea

Have students research and write a short story for the school newspaper focusing on a recent marine discovery. Topics of interest may include sea vent species of the deep ocean, shipwrecks, or underwater volcanic or tectonic activity.

Extension

Have your students research ocean explorers and write a short article on one person or one particular career related to ocean exploration. What work do these people do? What training do they have? What conditions do they work under? What discoveries have they helped make?

4 | Services on Stage

SUBJECTS
social studies, science, language arts, art (drama)

SKILLS
applying (composing, creating), presenting (writing, illustrating, acting, persuading), citizenship skills (working in a group, defending a position)

FRAMEWORK LINKS
30, 30.1, 33.1, 33.2, 33.3, 34.1

VOCABULARY
phytoplankton

TIME
two to three sessions, depending on amount of time allotted for research, skits, and ad campaign projects.

MATERIALS
copies of "Sea Services Cards" (pages 90-91) and "Sample Biodiversity Ads" (pages 92-93), four slips of paper

CONNECTIONS
For more activities on ecological services, try "Biodiversity Performs!" and "Secret Services," both in Biodiversity Basics.

AT A GLANCE
Act out four short skits that demonstrate some of the many services marine biodiversity provides, then design a print or video ad that educates people about the importance of marine biodiversity.

OBJECTIVES
Explain how marine biodiversity affects people's everyday lives. Describe the role of marine biodiversity in balancing the gases in the air, providing a source of new medicines, making up a large portion of the world's food supply, and providing enjoyment for millions of people.

The wild array of life in the ocean makes it a fascinating topic of study. But how important is marine biodiversity? How would our lives be different if the oceans weren't home to such a wide variety of life?

Among other things, we wouldn't have the kind of atmosphere that we need. (Tiny ocean-dwelling plants, called *phytoplankton*, produce the majority of Earth's oxygen.) Also, we'd be without a huge source of animal protein that comes from seafood, and we'd have to give up medicines from chemicals found in marine life.

In this activity, your students will use their creative abilities to put on four short skits that highlight some of the many reasons that marine biodiversity is important.

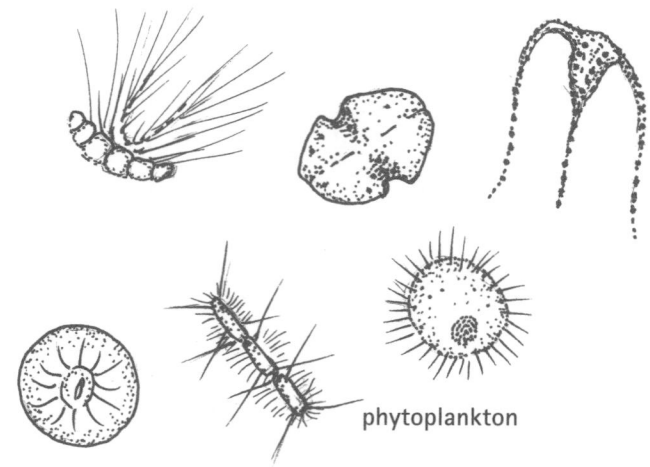

phytoplankton

Before You Begin

Copy the four "Sea Services Cards" (pages 90–91). On four strips of paper, write "comedy," "mystery," "drama," and "drama." Make enough copies of "Sample Biodiversity Ads" (pages 92-93) for each team of three to four students.

What to Do

1. Ask students how they think marine biodiversity affects their everyday lives.

Do your students think that marine biodiversity affects them? How? How do they think life would be different at home and at school without marine biodiversity? Have the students share some of their ideas. They may be surprised to know that marine biodiversity affects everyone, even those who live thousands of miles from the coast. Tell them they'll be putting on four short skits that will help demonstrate some of the reasons that marine biodiversity is important.

2. Assign students to four groups and give each group its subject and style.

Divide the class into fourths. Explain that each group is going to get an explanation of a different service that marine biodiversity provides to people. Then they'll have to create a short skit that demonstrates that service for the rest of the class in one of three styles—comedy, drama, or mystery. Have one representative from each group choose a subject card, which you've folded so they can't see the subject they're choosing. Once each group has its subject, have another representative choose the group's style. Again, fold these so students can't see which style they're picking.

3. Create and practice skits.

Give the students a set amount of time to create and practice their skits. The amount of time you give them will depend on the amount of time you have for the activity. If time is short, the activity can be completed in one class session. Students should

prepare skits of about 5 minutes in length and should have at least 20 minutes to create them. If you have more time, students can do more research on their subjects and can create longer skits.

4. Present the four short skits.

Once students are ready, have them present their skits. After each skit, you may want to give the group some time to make any additional comments that they think will help explain the service they demonstrated in their skit. Also, allow students in the class to ask questions. You may need to fill in some missing details to help clarify the marine service the group has portrayed in the skit.

5. Discuss the importance of marine biodiversity.

Once all of the groups have presented their skits, hold a discussion about the services marine biodiversity provides. (Refer to Chapter Two of the background [pages 30-37] to help lead this discussion.) Were students surprised by the many services that marine biodiversity provides? Can they think of other ways marine biodiversity is important? (When discussing the topic of seafood, you might want to mention that there are problems associated with seafood harvesting that they will be learning about in future lessons.)

6. Assign ad campaign project.

Do your students think that members of the general public are aware of how important marine biodiversity is to them? Tell the students that, to wrap up this activity, they will be designing an ad campaign to help spread the word about marine diversity.

Organize your class into teams of three to four students. If you want to keep the assignment simple, have the teams design a poster ad that could raise awareness about marine biodiversity. If your students are more ambitious, you can have them make short video ads instead of print ads. You should allow the teams several days to plan and script those ads, and a week or more to produce them.

Before designing their ads, tell your students that they should decide who their audience is. For example, they could target a particular age range (peers, parents, or senior citizens) or a particular interest group (consumers, sports fans, or music lovers). They should also decide what their message is and whether they need examples to illustrate the message. Explain that, by doing these things first, they'll find it easier to come up with an interesting design.

Have the students brainstorm characteristics that make ads effective. Among other things, a good advertisement

- presents a clear message;
- relates the message to people's lives;
- is attractive, clever, and interesting enough to grab people's attention; and
- connects to people's values and what they care about most.

For examples of print and other media ads, give your students copies of "Sample Biodiversity Ads."

7. Share results.

When the teams have completed their ads, have them share the ads or videos with the entire group. Encourage the students to make concise, clear, and interesting presentations. After the presentations, display the posters where other people can see them and show the videos to a larger audience.

WRAPPING IT UP

Assessment

On the chalkboard, draw the trunk of a tree and label it "Marine Biodiversity." Then add four main limbs labeled "Food," "Medicines," "Air," and "Enjoyment." Have students expand their trees with branches growing from each of the main limbs. Each of the branches should provide examples that help explain the service that each main limb provides, such as antibiotics, sunscreens, painkillers, and cancer treatments on the "Medicine" limb, and shrimp, cod, tuna, and salmon on the "Food" limb. Only one branch (oxygen) will be needed for the "Air" limb.

Unsatisfactory—Student has provided only one correct extension per limb for food, medicine, and enjoyment.

Satisfactory—Student has provided two or three correct extensions per limb for food, medicine, and enjoyment

Excellent—Student has provided four or more correct extensions per limb for food, medicine, and enjoyment.

Portfolio

In the portfolios, include students' biodiversity ads.

Writing Idea

Students can write radio public service announcements (PSAs) based on their ad campaign posters. The PSAs should be less than one minute long, include pertinent information on the services provided by their particular marine species, and use snappy, attention-getting language. If possible, tape record or videotape the students making their announcements, then listen to the taped messages and have the class note any that sound "ready for prime time."

Extension

Ask students to do some research on how marine biodiversity provides employment to millions of people—a marine service that was not highlighted in the sea services cards. Once students have gathered information, have them write up short paragraphs on those services, modeled on the format of the sea services cards. When they are finished, make a list on the board of all of the marine-related occupations that they thought of. Then encourage the group to expand on the list as other ideas occur to them. Be sure to include occupations of people who live far from the sea, such as a professor of marine biology, a server in a seafood restaurant, and a seafood seller in a supermarket.

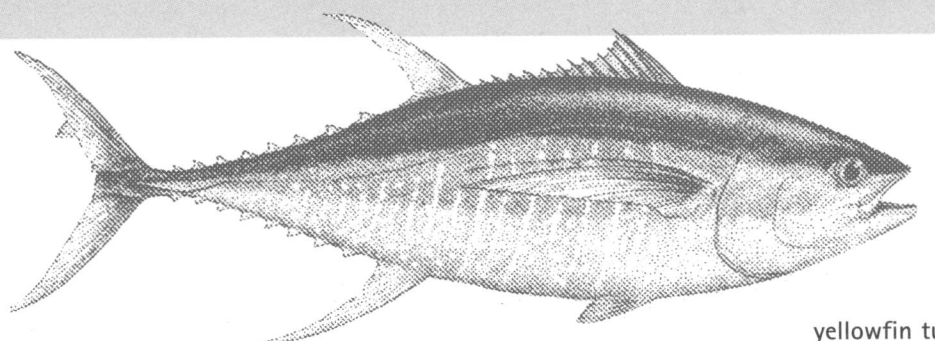

yellowfin tuna

BIOFACT

The yellowfin tuna is one of the fastest fish in the sea. It has the ability to become "super-streamlined" by laying its fins flat against its body, creating a sleek, torpedo-like effect.

Marine Biodiversity Is a Source of New Medicines

Have you ever been in pain? Had an infection? Gotten a sunburn? Then marine biodiversity may have come to the rescue. Drug companies are testing, and already using, drugs that have come from different kinds of sea life. Drugs derived from sea life include antibiotics, cancer treatments, pain relievers, skincare products, and sunscreens.

The ocean is a great source of new medicines because the diversity of marine life contains a wide variety of chemicals. In places like coral reefs, many species have powerful chemicals that they use to discourage predators from eating them. And many of those same chemicals can be used in medicines.

Not only are the chemicals important to medicine, but other parts of organisms can be important too. For example, the blood of the horseshoe crab can be used to test for bacterial infections in people, and some doctors are using corals to help repair damaged bones. So, the next time you're swimming in the ocean, just think of all the new cures that could be swimming around you!

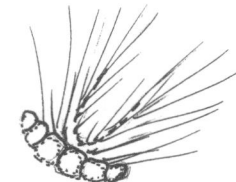

Marine Biodiversity Helps Balance the Gases in the Air

People and other animals take in oxygen and give off carbon dioxide. So, what helps keep the overall amount of oxygen from decreasing in the atmosphere and the amount of carbon dioxide from increasing? Plants perform this vital service—they take in carbon dioxide and give off oxygen, helping to balance the mix of gases in the air. And most of the plants responsible for doing this are found in the ocean. Tiny marine plants and plant-like organisms called phytoplankton (FIE-toe-plank-ton) float near the surface where sunlight can reach them, and they are responsible for making the majority of the oxygen in the air. Most kinds of phytoplankton are so small they can be seen only with a microscope.

How can such tiny organisms do so much work? Even though they're much smaller than most plants on land, there are many more phyto-plankton. In fact, billions can be found in just a gallon of sea water. So, the next time you take a breath of fresh air, thank those little ocean plants!

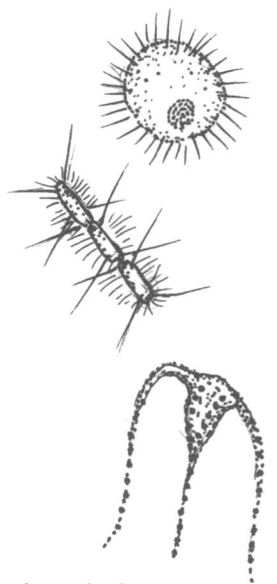

phytoplankton

Marine Biodiversity Is a Major Food Source

Have you ever eaten a fish stick? Tuna sandwich? Shrimp? Clams? Seaweed? There are lots of different kinds of seafood, and billions of people who eat it.

Seafood provides an average of almost 20 percent of the animal protein that people eat worldwide. (Animal protein comes mostly from red meat, chicken, and seafood.) In the United States, the average person eats 14 pounds of seafood every year.

Many Americans eat seafood because it's nutritious. It's high in protein, vitamins, and minerals, and it contains oils that help reduce the risk of heart attack. It's also low in fat. In some countries, seafood is the people's main source of animal protein, so it's a very important part of their diet. Even if seafood isn't your favorite thing to eat, remember that billions of people around the world depend on it for a healthy diet.

Marine Biodiversity Influences Our Culture

Whale watching. Deep-sea fishing. Snorkeling on coral reefs. Birding in marshes. Exploring the wonders of tidepools. When you stop and think about it, people have found many ways to enjoy marine biodiversity—but this is nothing new: The Ancient Greeks decorated their pottery with images of marine creatures. And many tales have been told through the ages about the amazing and mysterious life of the sea.

People everywhere gravitate toward the seacoast, drawn by its beauty, resources, and biodiversity. Today, 3.8 billion people worldwide live within 100 miles of the coast. In the United States, more than half the population lives in coastal areas. Many of these people, and even many inland inhabitants, visit the seashore to observe or otherwise enjoy the animals and plants living there. Thousands of others appreciate biodiversity by visiting public aquariums.

With so much richness to offer—colorful creatures, tasty seafood, opportunities for adventure, and untold mysteries—it's no wonder the ocean has come to shape our lifestyles in such meaningful ways.

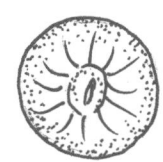
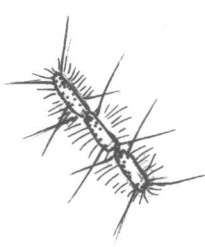

Amazing Grace...

Don't let it vanish without a trace.

1.800.CALL.WWF
www.worldwildlife.org/act

Get your free World Wildlife Fund Action Kit and help leave our children a living planet.

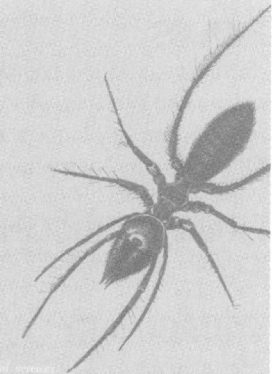

Everybody wants

to save creatures that are

warm, fuzzy and cute.

Would you settle for fuzzy?

SUBJECTS
language arts, science, social studies

SKILLS
gathering (reading comprehension), presenting (writing, describing, articulating, explaining, making analogies and metaphors)

FRAMEWORK LINKS
1.1, 34.1, 53.2, 56

VOCABULARY
perspectives

TIME
one session

MATERIALS
copies of "Blue Views" (pages 97-98), video of ocean life (optional, see Resources, pages 367-368)

CONNECTIONS
Encourage students to further explore their own perspectives on marine creatures by trying "What Do You Think About Sharks?" and "Rethinking Sharks." To look at various perspectives on other biodiversity-related topics, check out "A Wild Pharmacy" in Wildlife for Sale and "Spice of Life" in Biodiversity Basics.

AT A GLANCE
Explore different perspectives of what it's like to be under the sea.

OBJECTIVES
Articulate personal feelings about the ocean and ocean exploration.

Many of your students may have been to the beach but may not have noticed the life all around them. And some may have had the chance to snorkel or dive to discover what's under the surface of the sea. Those that have may be eager to share their experiences.

In this activity, students will gain a new perspective on what it's like to be under the sea. They'll read two different accounts about exploring life in the ocean—one from someone new to the undersea world and another from a seasoned professional. Then they will get the opportunity to describe life under water.

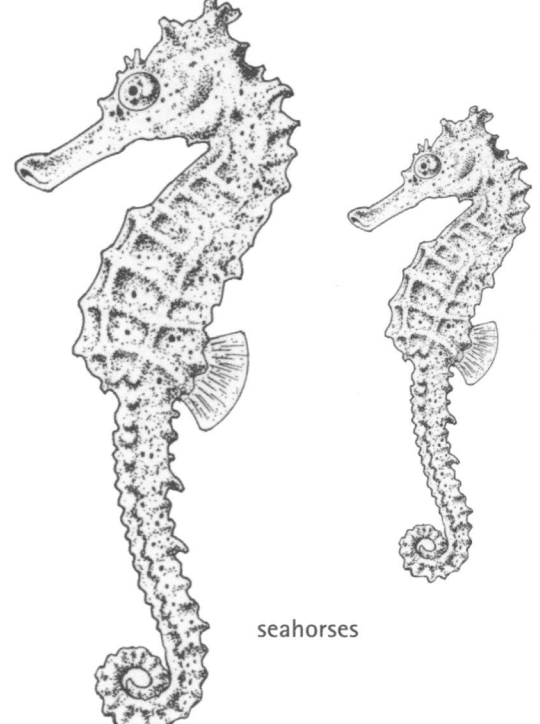

seahorses

What to Do

1. Share personal underwater experiences.

If your students have spent any time under water, they can probably appreciate how different the land is from the sea. Have any of your students ever been snorkeling or diving? Where? What was it like? How does that experience compare with experiences other students may have had on the shore or on boats?

2. Read "Blue Views."

Tell the students that they'll be reading two essays that talk about different kinds of underwater experiences in the ocean. The first is writer and humorist Dave Barry's account of his first scuba trip to the reefs off the Florida Keys. The second is about the experiences of marine scientist Sylvia Earle, who has studied life in the ocean for many years and even lived in an underwater research station for several weeks. You may want to have the students take turns reading Dave Barry's essay out loud for dramatic effect.

3. Discuss the "Blue Views."

What did the students think about the different perspectives of ocean exploration? Could they identify with writer Dave Barry's fears of the ocean? What did they think of marine biologist Sylvia Earle's experiences living under the sea? Did either of the readings change their views about life under the sea?

4. View a video (optional).

Show a video with scenes of ocean life to give your students a better sense of what it's like to explore the ocean.

5. Write blue views of your own.

Students who have spent time under water can write blue views that describe the kinds of life they've seen. What were they afraid of and what was especially intriguing? Students who haven't spent time under water might write about what they saw in the video, places they'd like to go, things they'd like to see, and things they would rather not see. They might also reflect on what they think it would be like to live under water the way Sylvia Earle did. How long would they like to stay under the sea, and what kinds of things would they like to investigate while there? This is an opportunity for students to be creative and use their imaginations about what it's like to interact with marine biodiversity in the ocean rather than on TV or behind glass in an aquarium.

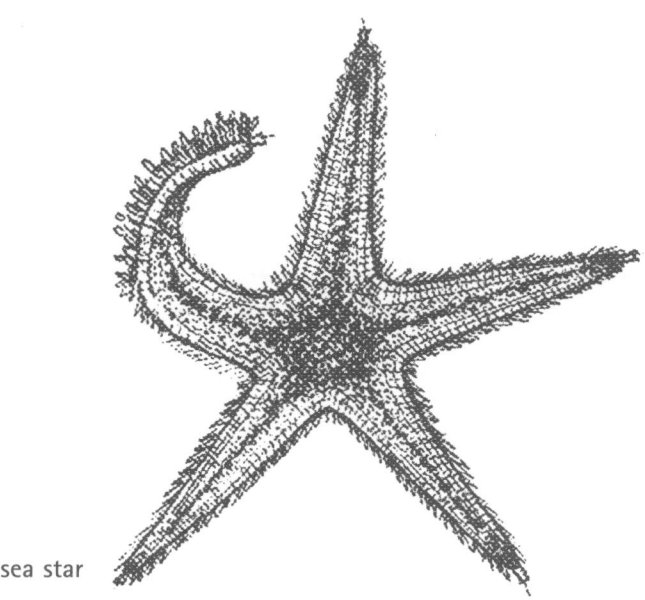

sea star

BIOFACT

After losing an arm, a sea star can regrow it through a year-long process of regeneration.

WRAPPING IT UP

Assessment

With your students, build a rubric for assessing their blue views stories. For example, columns might include such things as (1) personal feelings, (2) reasons for their positions, and (3) multiple perspectives offered.

Unsatisfactory—Identifies the feelings but fails to offer reasons for any of them.

Satisfactory—Identifies the feelings but fails to offer reasons for all of them.

Excellent—Identifies the feelings and offers reasons for all of them.

Portfolio

Students should include their own blue views in their portfolios.

Writing Idea

Ask students to imagine that they have the opportunity to interview one of the two authors whose pieces they read for this activity (Sylvia Earle or Dave Barry) or other authors who have written about ocean explorations. They should conduct research on the individual they have chosen, specifically focusing on their marine experiences. Based on the students' research, they should develop a list of 10 questions to draw out more information from the interviewees on why they are interested in oceans, what other marine-related experiences they have had (other than those described in this activity's readings), and what concerns they have about threats to marine diversity.

Extension

Have your students read a book that deals with life in or around the ocean. Possible titles include:

- *Moby Dick* by Herman Melville
- *The Old Man and the Sea* by Ernest Hemingway
- *Island of the Blue Dolphins* by Scott O'Dell
- *Caught by the Sea: My Life on Boats* by Gary Paulsen
- *A Whale for the Killing* by Farley Mowat

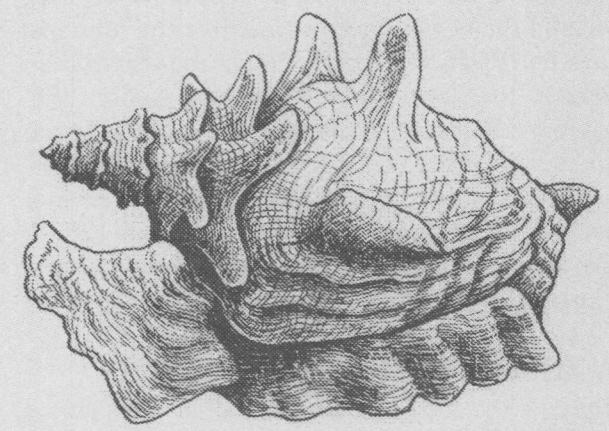

giant conch

Note: For general resources on marine biodiversity, please see the list on pages 360-369.

I'm swimming about 20 feet below the surface of the Atlantic, a major ocean. I'm a little nervous about this. For many years my philosophy has been that if God had wanted us to be beneath the surface of the ocean, He would never have put eels down there.

But I'm not panicking. That's the first thing you learn in scuba class: Don't panic! Just DON'T DO IT! Even if a giant eel comes right up and wraps around your neck and presses its face against your mask and opens its mouth and shows you its 874,000,000,000,000 needle-sharp teeth, you must remain COMPLETELY CALM so you'll remember your training and take appropriate action, which in this case I suppose would be to poop in your wet suit.

Also there is the whole issue of shar . . . of sha . . . of sh . . .

There could be s--ks down here, somewhere.

But so far, all the marine life has appeared to be harmless. Mostly it has consisted of what I would describe, using precise biological terminology, as "medium fish," many of which are swimming right up and giving me dopey fish looks.

So anyway, I'm swimming along the reef, with my nervousness gradually being replaced by a sort of high—a combination of fascination and amusement—when suddenly I hear my scuba instructor, Ray Lang, make the following statement: "Brnoogle." Everything anybody says through an air regulator underwater sounds like "brnoogle."

When I look at Lang, he's pointing excitedly off to my right, so I turn and see a large ray, which looks sort of like a giant underwater bat. This is a major test of my ability not to panic.

But the ray pays no attention to me. It just cruises by. And as it passes by, I find myself, without really thinking about it, trying to follow it—me, a major weenie when it comes to dealing with the Animal Kingdom—here I am, flippering through the blue-green Semi-Deep in pursuit of this nightmare-inducing thing.

Swimming next to me, Lang points toward the surface, up above the ray. I look, and there, silhouetted against the surface, is a large school of: barracuda. Yes! The ones with the teeth! In person! They're looking long and lean, looking very alert, all pointing in the same direction.

But for some reason, the barracuda don't seem scary, any more than the ray does. For some reason, none of this seems scary. Even the idea of maybe encountering a smallish s--k doesn't seem altogether bad. It's beginning to dawn on me that all the fish and eels and crabs and shrimps and planktons who live and work down here are just too busy to be thinking about me. I'm a traveler from another dimension, not really a part of their already event-filled world, not programmed into their instinct circuits. They have important matters to attend to, and they don't care whether I watch or not. And so I watch.

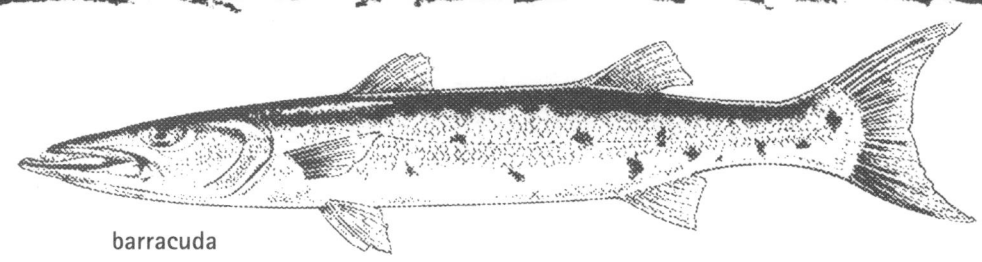

barracuda

I know how important diving can be as a way to observe life in the sea directly. I use scuba in the same way I use microscopes and other equipment—as a tool that makes it possible to explore. But I've found that the biggest problems with scuba diving are that I can't go as deep or stay as long as I want to. I've often dreamed of being able to live under the sea. My dream of staying under water for days at a time came true in a spectacular way.

I was selected to become an aquanaut to head the first team of women who would live in an underwater laboratory called Tektite for two weeks. In addition to scuba, we'd use special equipment called rebreathers to explore and study the reefs. Like the life-support system of astronauts, the rebreathers aquanauts use recycle air over and over, removing carbon dioxide chemically and automatically adding oxygen from a special tank as needed.

Located just offshore in Great Lameshur Bay, St. John Island in the U.S. Virgin Islands, the Tektite laboratory seemed to defy the laws of nature.

Imagine sitting warm and dry at a table 50 feet under water, munching on a sandwich, and talking with your friends while fish peer in the window! Imagine washing dishes, then taking off your T-shirt and shorts, putting on a bathing suit, flippers, mask, and air tank, then stepping through a round hole in the floor and swimming off into the sea. That's what my four companions and I did several times a day for two weeks in 1970.

We often got up before dawn to be on the reef when the sun came up. We took note of which fish were early risers and which ones liked to sleep in. Some, such as parrotfish, surgeonfish, and wrasses, are active all day and tuck into crevices and crannies at night. Other fish rest during the day—sometimes in the same crevices vacated by night sleepers. Dawn and dusk are great times to watch the changeover between creatures active at different times and to see which fish are solitary, which ones join up with others, and whether or not the same fish get together repeatedly as their day—or night—begins.

Living under water has many advantages, but most important is the gift of time. We could and often did make long excursions to depths of more than 100 feet. We spent as much as 12 hours a day in the water, taking time out now and then to eat and sleep or examine samples using microscopes inside the laboratory. Of course, at the end of our mission under water, we had to decompress. It took 19 hours in a special chamber to return safely and slowly to surface pressure, but it was worth it to have been completely submerged for two weeks.

Reprinted by arrangement with the National Geographic Society from the book Dive! My Adventures in the Deep Frontier *by Sylvia A. Earle. Copyright 1999 by Sylvia A. Earle.*

surgeonfish

WWF-Canon/Catherine Holloway

"Knowledge of the oceans is more than a matter of curiosity.
Our very survival may hinge upon it."

—John F. Kennedy, Thirty-Fifth President of the United States

Case Study
Coral Reefs

Most people are aware of the beauty and variety of life found in a tropical coral reef. But fewer people realize how many services reefs provide, from buffering coastal communities against storms to sheltering young fish. In this case study, your students will take a global tour of coral reefs, discovering their beauty and benefits as well as the threats faced by these amazing underwater ecosystems.

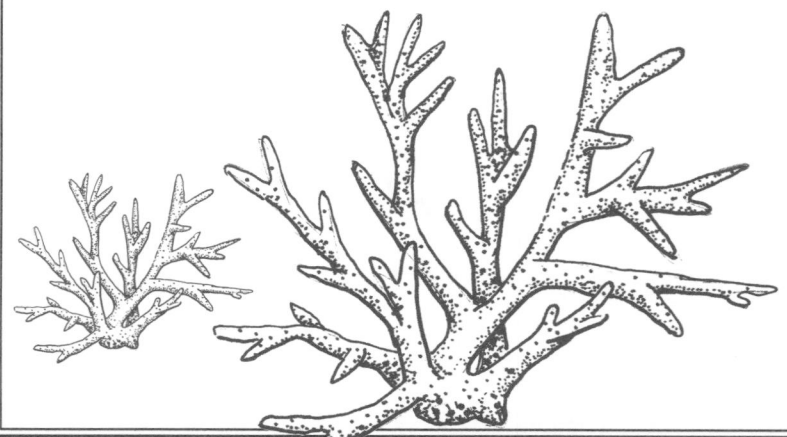

WWF-Canon/Catherine Holloway

"The sea that gives us life is the coral reef sea. The life proceeding from the tiny coral polyps flows into the life of the island people— in this is found the sacred universe."

—Uozumi Kei, Representative, Women and the Sea Association

Background Information

Snorkel along a tropical coral reef and you'll discover an underwater paradise. Swaying pink anemones, bright yellow sponges, and delicate feather stars cling to the surface of the reef. Tangs, parrotfish, and other tropical fish dart through the warm, clear water. A spiny lobster scrambles after a snail, and a green sea turtle slowly glides past. But the wonders of a coral reef go well beyond its beauty and brilliance. Few habitats support as many wild species and offer such far-ranging benefits to humankind as do coral reefs.

Rain Forests of the Sea

Coral reefs provide shelter and food to thousands of species worldwide, from shellfish to sea horses to sharks. In fact, reefs are so rich and ecologically important that some people call them the "rain forests of the sea." And like rain forests, coral reefs are threatened habitats. Human activities around the globe are harming coral and, in turn, the diversity of life that depends upon it. What's more, coral reefs offer so many benefits to humans that threats to coral could have serious consequences for people living in coastal as well as inland communities throughout the world.

BIOFACT

More than half of Earth's 6.3 billion people live in coastal areas, where many coral reefs are easily accessible.

All About Coral

What Is Coral?

To really understand coral, it's important to separate the living organisms from the reefs that so many of those organisms build. Each individual coral animal, called a *polyp*, consists simply of a tube-shaped body with tentacles surrounding the mouth. Sea anemones, relatives of coral, are also polyps and have a similar anatomy. All polyps are cnidarians, named from the Greek word "*cnides*," which means "nettle." And all polyps share a common trait—they have stingers on their tentacles that they use to kill prey.

Coral polyps are divided into two basic groups, soft coral and reef-building coral, and their skeletons distinguish the two groups from each other. Soft coral polyps don't build reefs. They have little bits of limestone (calcium carbonate) within their soft bodies and a flexible skeleton that lets them bend and sway in the current. (These soft corals are an interesting group of animals that your students may wish to study, but this case study focuses on the reef builders.)

Reef-building coral polyps make stony skeletons of limestone under themselves. As the polyps of a specific species grow and reproduce, they form a colony. The colony develops a shape that's specific to that species, and many species have been named according to the shape of the colony—brain, elkhorn, staghorn, flower, and so on. Together these colonies form coral reefs.

By day, a colony's polyps lie low on the skeleton, but at night, the polyps extend their tentacles and snag tiny plankton floating by. Since the tentacles can reach only a small distance, corals rely on constant water motion to bring a steady flow of food close enough for them to snatch.

The types of coral polyps that are the focus of this case study live in warm, shallow water and cultivate their own internal "farms" of algae, called

coral skeleton

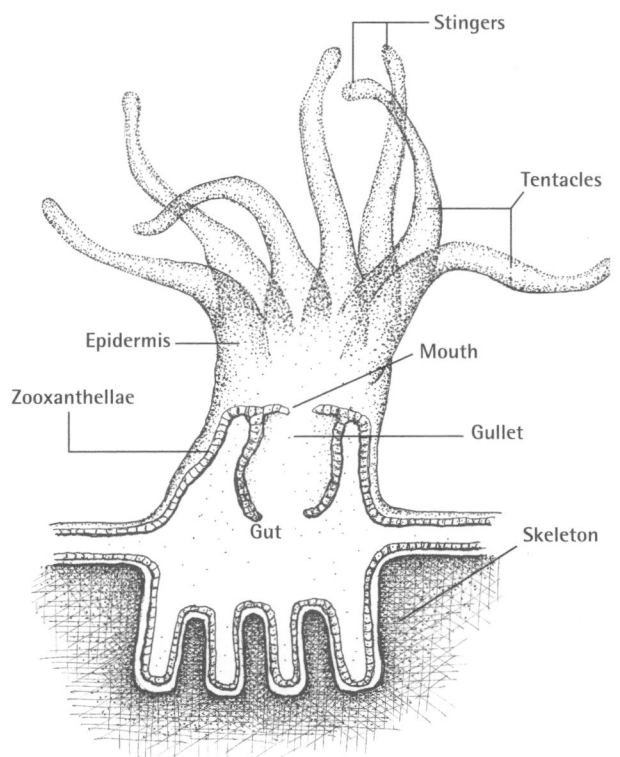

The zooxanthellae, shown above, lie just beneath the thin epidermis. In some coral species, zooxanthellae are also found beneath the surface of the tentacles. The epidermis, zooxanthellae, and gut connect from one polyp to the next.

zooxanthellae (zo-zan-THELL-ee), to make additional food. The algae are packed into the polyps' tissues and give the coral its color. Like green plants on land, zooxanthellae use the sun's energy to make sugars through the process of photosynthesis. But the algae don't retain all the sugars they make: Some of the sugars provide the polyps with energy for basic survival. In turn, the coral polyps provide the algae with a safe home and a regular source of nitrogen and carbon dioxide, which are created as waste products of coral polyps. To sustain this symbiotic partnership with algae, corals must live in warm water that is clear enough for sunlight to penetrate and support the photosynthetic process.

Coral polyps reproduce asexually when new polyps branch off from old ones, somewhat like a bud sprouting from a tree. The difference is that the new polyp is a separate individual and will create its own skeleton. Each new polyp is genetically identical to the polyp from which it sprouted.

Shape Names

Reef-building corals create colonies in all sorts of shapes, earning them such descriptive names as **brain**, **elkhorn**, **staghorn**, *fire, leaf, broccoli, pillar, starlet, finger, rose, cactus, and flower coral. In fact, hundreds of species of reef-building corals (and many more that don't build reefs) inhabit the Earth's oceans.*

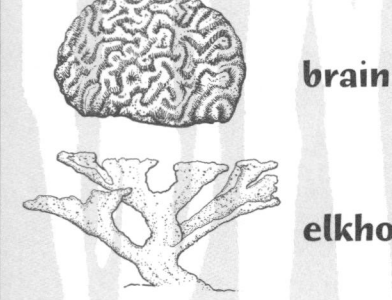

brain

elkhorn

staghorn

Coral Reef Construction

Branching coral colonies grow about three or four inches a year, and other coral colonies grow even more slowly. But after hundreds or even thousands of years of continuous limestone deposits, a huge coral colony can form. Colonies of different kinds of coral species growing in the same area for centuries eventually create a coral reef. Australia's Great Barrier Reef, the world's single largest reef complex, is hundreds of feet thick and almost as large as California. Scientists believe that the coral skeletons that form its initial base started growing more than 600,000 years ago. To determine the age of a coral reef, scientists can bore through coral rock and evaluate the skeleton growth lines in much the same way that they can examine the annual rings of trees.

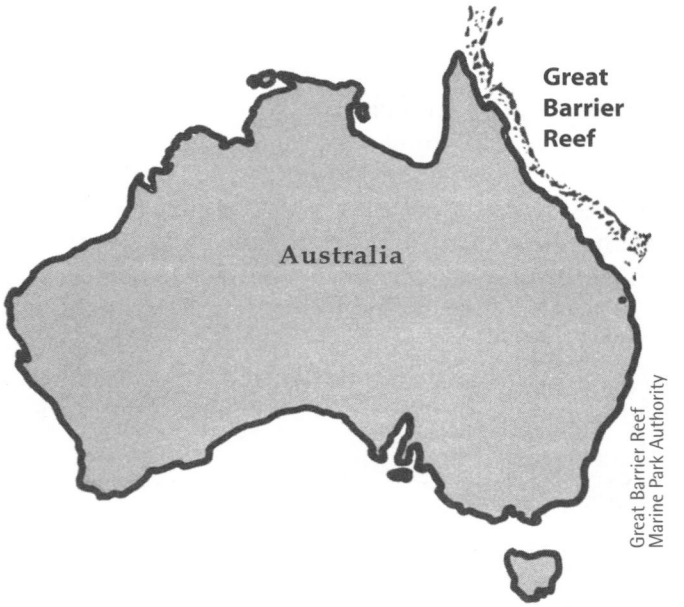

Great Barrier Reef

Australia

Great Barrier Reef
Marine Park Authority

But corals also reproduce sexually—they release eggs and sperm into the water. The resulting baby corals (called *planulae*) leave the adults and float on currents. At this stage they are considered to be plankton. If they reach a hard surface, they may become attached to it and start to grow. Many planulae never make it, becoming a meal for hungry reef dwellers. But by producing hundreds of thousands of eggs and sperm at a time, each coral polyp increases the chances that at least some of its offspring might grow and begin a new colony.

Where's the Reef?

About 110,000 square miles of coral reefs exist in the world, totaling an area approximately half the size of France. Although that probably sounds like a lot of coral, it actually covers less than one percent of our planet's sea floor. (See page 120 for a world map highlighting coral reefs.)

One reason coral reefs aren't more common is that coral polyps are very selective about where they live. If you look at a map of coral reefs, you'll discover that all of the reefs seem to be located in the middle of the world map. That's because most reef-building corals favor shallow, continuously warm water. These conditions are most often found on the eastern shores of continents, around small islands, and on the tops of underwater mountains that are just beneath the surface of the sea. Corals grow best in water with an average temperature of between 72° F and 78° F, and they are usually no deeper than 325 feet, with the majority of the reefs being located in water that is less than 120 feet deep. All of these conditions are right for coral, in part, because they're right for the zooxanthellae that live inside them. If the water weren't shallow, the zooxanthellae would not receive enough sunlight to carry out photosynthesis.

Three Kinds of Coral Reefs

- **Fringing reefs** grow in shallow ocean water attached or directly adjacent to a coast. They are found mainly in the Indian Ocean. There is a large, almost-continuous fringing reef along the coasts of Kenya, Tanzania, and northern Mozambique.
- **Barrier reefs** grow parallel to the coast. A barrier reef is separated from land by a lagoon. The reef shelters the mainland from wave action, except in the worst of storms. The calm water in a lagoon provides ideal growing conditions for seagrass beds and mangrove forests—two rich marine habitats. The Great Barrier Reef of Australia is a well-known example.

- **Atolls** are reefs that develop at or near the surface of the sea when volcanic islands that were once surrounded by reefs sink below the surface. The ring of coral that remains near the sea surface is known as a coral atoll. Many coral atolls are found in the South Pacific Ocean. The Marshall Islands, for example, have a series of famous ones, including the Bikini Atoll.

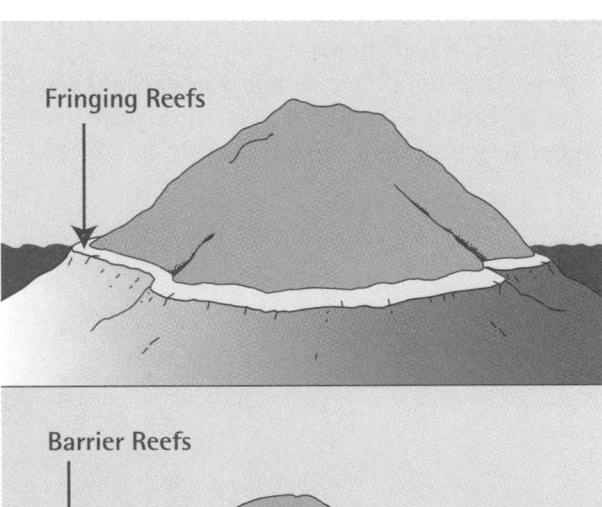

Fringing Reefs

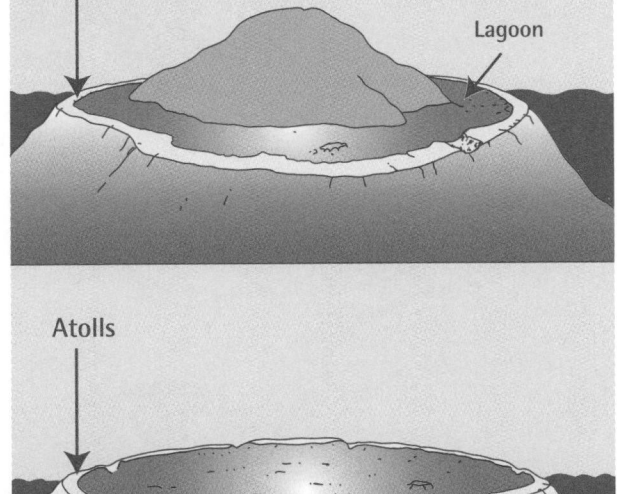

Barrier Reefs

Lagoon

Atolls

Creature Feature

Anyone who has spent time in the vicinity of a coral reef knows that they are crawling, swaying, and swimming with life. In fact, corals support about 25 percent of all oceanic species, making them the most diverse of all marine ecosystems. Some reef habitats are more diverse than others. The ones in the central Indo-Pacific, such as the reefs in the Sulu-Sulawesi Seas (surrounded by the Philippines, Malaysia, and Indonesia), have more than 10 times as many coral and fish species as reefs in the eastern Pacific.

Because they have so many nooks and crannies, coral reefs provide living spaces for a large variety of animals. Octopuses and squirrel fish hide in and under the reef to escape the hungry mouths of groupers and eels. Nurse sharks sleep in coral reef caves. Brittle stars slip into cracks in the reef to escape the light of day. Many predators—including groupers, eels, sea turtles, and sharks—patrol the reef in search of their next meal. On a reef, just as on land, one set of creatures is active by day and another set at night.

Some reef-dwellers feed on coral. Fire worms crawl along the reef, slurping up the polyps. And parrotfish bite off and grind up chunks of coral to get at the algae living there. The crunched coral forms the sand around the reef.

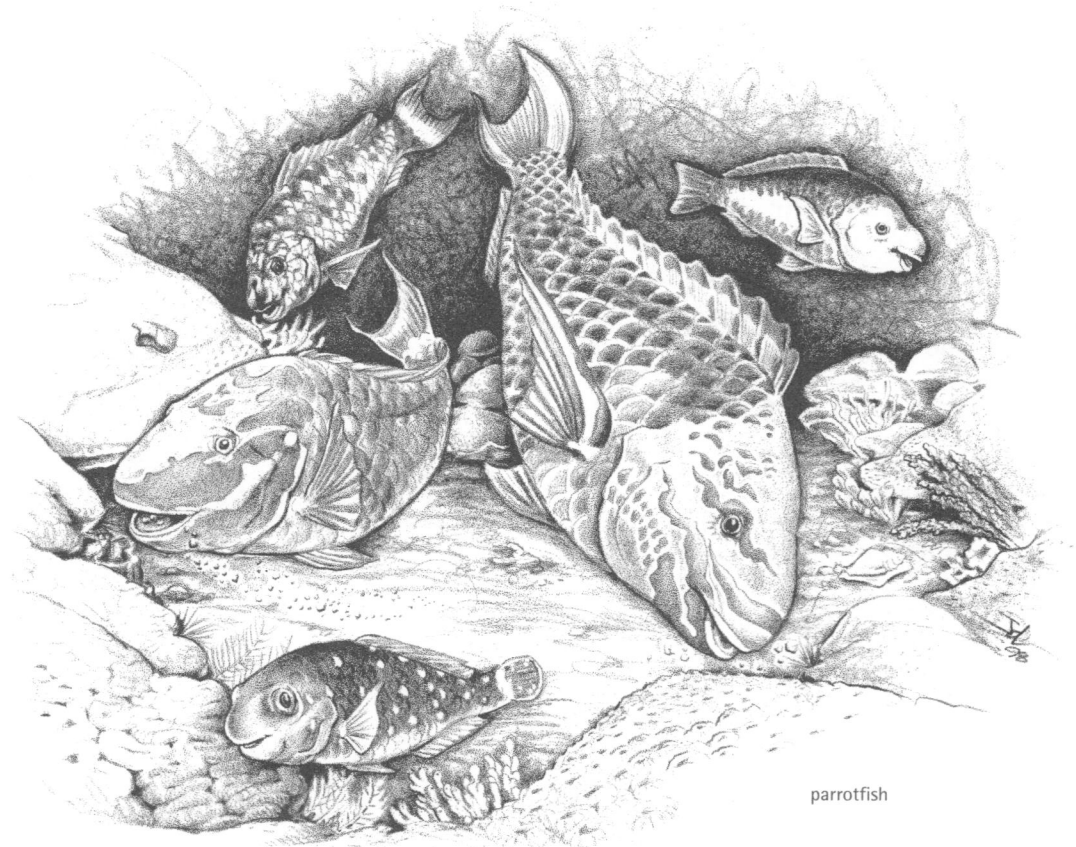

parrotfish

BIOFACT

Parrotfish eat algae that they scrape off of coral reefs using their beak-like mouths. Inadvertently, the parrotfish bite off small bits of coral, which they then grind up using their powerful jaws. The ground-up coral passes right through and is deposited on the ocean floor as white sand, which eventually ends up on tropical beaches.

Our Coral Connection

Coral reefs are essential to most of the living things that feed or seek shelter in and around them, and they're important to many others. Reef habitats also benefit people in a multitude of ways. For one thing, many species that inhabit coral reefs for at least part of their lives are important sources of seafood, including crabs, lobsters, fish, and other organisms. These species provide an income for the fishers who capture them. Reefs also generate income for individuals and businesses involved in local tourism. After all, reefs are captivating destinations. Many people will spend a great deal of money for a chance to snorkel and scuba dive in these tropical ecosystems.

In addition, a surprising number of products come to us from coral reefs. For example, people have used coral to make an additive for concrete and mortar. Coral is also used to make compounds for violin varnishes. But far more importantly, a remarkable number of medicinal products are derived from coral, including a substance used to replace broken human bones and treatments for such ailments as viral diseases, arthritis, asthma, and cancer.

Coral reefs as a whole are an important component of the coastal landscape. They protect coastal communities from strong waves generated by storms, helping to prevent soil and sand erosion on shore. And broken coral provides a steady source of the white sand that attracts so many people to tropical beaches.

Cold-Water Coral

In this case study we describe coral reefs that live in tropical waters. But about a half-mile down in the Atlantic Ocean, where fish glide through dark, chilly waters, thriving communities of cold-water corals also exist. These corals can form reefs similar to their tropical coral cousins, or they can grow skyward like trees to heights of up to 10 feet. Unlike tropical corals, cold-water corals can exist without sunlight. Instead of depending on microscopic algae to survive, cold-water coral capture food particles floating in the water around them. Unfortunately, when humans use bottom-trawling nets to catch fish and other sea creatures, these fragile habitats are easily destroyed.

angler fish

Threat to Coral Reefs

Coral reefs grow so slowly that, in some ways, they're like old-growth forests. If the reefs are destroyed, it can take centuries for them to return to their previous condition. Some natural phenomena have always destroyed, and will always destroy, sections of coral reefs. Hurricanes and other tropical storms can be so violent that they tear up reefs or cover them with so much silt and mud that the zooxanthellae are unable to photosynthesize. Also, powerful rainstorms can dilute the salt water around corals so much that the corals are not able to survive. Storms are not all bad, though. When an old reef is torn apart, unoccupied places are opened up for the colonization of new arrivals, and this may promote species diversity in coral reefs.

But coral reefs have come under new and far greater threats from people. Hardly anyone harms coral on purpose, yet people cause the most serious problems faced by coral reefs and the species that

Bob Fitzpatrick, USFWS

live there. For example, coral polyps die when tropical fish collectors use cyanide around the reef to stun the fish they hope to capture. And overfishing is causing the decline of many reef species, from spiny lobsters and groupers to sea turtles. Even the reef structure is threatened because coral is in demand for the aquarium trade, and enormous amounts of coral rock are used for construction in some parts of the world. Divers also cause damage to coral reefs, often inadvertently. Simply touching or standing on coral can damage it. And boat operators sometimes allow their anchors to drop on coral reefs, striking and often breaking the coral.

These local effects on coral are not as destructive, though, as the damage caused by pollution. Half the world's population currently lives within 40 miles of the coast, causing intense stress to coral reefs and other coastal habitats. Coastal development increases the runoff of sediments and nutrients from the land into offshore waters, harming or killing nearby coral reefs. Pesticides, sewage, and other forms of pollution from coastal communities can also harm and kill coral reefs.

Currently, an even greater threat to reefs is coral bleaching due to global warming. As global temperatures rise, so do ocean temperatures. The zooxanthellae that inhabit coral (and give the coral its color) cannot survive if the water is much warmer than normal. When temperatures increase, the zooxanthellae leave the coral, depriving it of a key source of food. Then, the coral polyps slowly whiten, or "bleach," and die. If ocean temperatures return to normal, the zooxanthellae may return to the coral and it can recover. However, if the temperatures stay high, the coral will experience a massive die-off. Cases of widespread coral bleaching have been recorded across the globe, but the most devastating effects to date have been recorded in the Indian Ocean where scientists estimate that 50 to 95 percent of all corals have died.

> *"The wonder is not that coral reefs are in danger—and they are—but rather, that they have tolerated so much for so long."*
>
> **–Sylvia Earle, marine biologist**

Reef Relief

All around the world, people are realizing that coral reefs are in trouble and need our help. Many of those people are conservationists. Some are scientists who conduct ongoing research about life in coral reefs. But others are divers who recognize that they have been "loving coral to death." In addition, on many tropical islands, villagers who fish in local waters have observed reef-related changes that are the result of pollution, the growing tourist trade, and irresponsible fishing practices. Many fishers worry that their way of life is at risk. And they are very much a part of efforts to protect coral reefs for their children, as well as for future generations around the world.

Because coral reefs are dying at unprecedented rates, many people believe we need to take immediate action to help protect them. Some scientists are mapping the exact location of coral reef systems and documenting the diversity of life found there, as well as the threats to the systems. These efforts are helping to set priorities for conservation. As part of this effort, conservationists are calling for a significant reduction in the production of greenhouse gases around the world, in an effort to help slow global warming.

Scientists also believe we need to expand the system of marine protected areas—which are the equivalent of underwater wildlife refuges. For example, the Florida Keys National Marine Sanctuary protects some of the coral reefs in the Caribbean. And the Great Barrier Reef Marine Park Authority has been a model of marine protection management for years. Many people have proposed a global network of coral-reef marine protected areas, but a major challenge will be developing clear guidelines for establishing and managing these areas.

Finally, many scientists are looking into the potential for coral reef restoration and rehabilitation. By finding ways to create new reefs or restoring existing ones, we may be able to counteract a little of the damage caused by pollution, global warming, and other threats. (See Mini Case Study #10, Shipwrecks: Refuse to Reefs, pages 316-317, for one way scientists are constructing new reefs.) Of course, it is still far better to do all we can to prevent the damage in the first place and to save the reef habitats for brittle stars, shrimp, parrotfish, sea turtles, sharks, and millions of other reef-loving creatures.

Coral reefs form natural barriers that protect shorelines from the sea. Without coral reefs, parts of Florida would be under water.

1 | Build-a-Reef

 **AT A GLANCE**
Draw or build models of a coral colony and display them in a classroom "coral reef."

 **OBJECTIVES**
Define "reef-building corals" and name several factors that limit their growth. Describe adaptations of coral. Translate information about corals into a drawing or a three-dimensional model.

Take a look at close-up pictures of corals and you'll be dazzled by the intricacy and diversity of their form and structure. It's no wonder people have long collected corals for display and that many enjoy snorkeling and scuba diving around coral reefs.

Coral reefs are made up of colonies of coral formed when tiny coral polyps secrete a limestone exoskeleton beneath their living tissue. This skeleton grows throughout the life of the colony. By building new skeletons on top of old ones, coral polyps gradually create huge reefs that offer homes and sources of food to hundreds of other species. (See pages 103-104 for more on how coral reefs are formed.)

By making their own coral art project, your students will learn more about the anatomy of coral polyps, the variety of coral colonies, and the formation of coral reefs—and they'll have a chance to enjoy the aesthetic appeal of coral.

WWF-Canon/Sylvia Earle

Before You Begin

For each student, make one copy of the unlabeled "A Coral Polyp" (page 115). (For your reference, see labeled polyp diagram on page 103.) Gather art materials for students to use in making their coral reef.

What to Do

1. Discuss corals.

Ask your students to share some of their current knowledge about corals and coral reefs. Do they know what coral is? Where do they think coral reefs are found? Do they have any idea why it might be important to study coral? (Don't worry if students cannot provide thorough answers to these questions. Reassure them that they will be able to answer all of these questions and more by the end of the unit or case study.)

2. Assign art activity.

Explain to your students that there are two main kinds of corals—soft corals and reef-building corals. In this activity, they'll be exploring some of the 795 species of reef-building corals that scientists have so far identified. Can the students define the word *species*? (*A species is a group of organisms with unique characteristics that distinguish them from other organisms. If they reproduce, individuals within the same species can produce fertile offspring. For example, mako sharks, elkhorn coral, and loggerhead sea turtles are all species.*) The students will explore this incredible diversity of coral by searching on the Internet and reviewing the books and magazines that you have gathered. Then they will make a model (preferred) or drawing of one species of coral based on the information they find.

Write on the board the common names of several different kinds of reef-building corals including elkhorn, leaf, brain, broccoli, staghorn, pillar, whip, starlet, rose, cactus, and flower. Tell the students to choose one of these species, or a different one altogether, as a subject for their model. Ask the students to search for information about their coral species in the books or magazines you've provided or on the Internet (see Coral Reef Resources on pages 148-149).

Provide the students with a variety of art materials, such as modeling clay, wire, and paint, and have them create their models. Encourage them to represent their coral colony with as much detail as possible. What color is the colony? What texture? If any students made drawings instead of 3-D models, have them make little cardboard stands for their drawings to lean on to help create a 3-D effect.

3. Discuss coral polyp anatomy.

When the students have completed their models, ask them what they have learned about coral colonies and the polyps that form them. Distribute copies of the unlabelled "A Coral Polyp" and write the following words on the board:

- Epidermis
- Gullet
- Gut
- Mouth
- Skeleton
- Stingers
- Tentacles
- Zooxanthellae

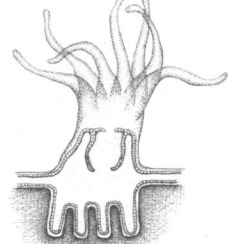

Using the information they have already gathered, ask your students to work alone or in small teams to label the diagram. While they are working, sketch the diagram on the board or put up an overhead projection of it. Together, fill in the appropriate labels, having students share their responses. (Answers are on page 103.) If no one comes up with the right answers, help your students fill in the correct labels.

Ask the students if, based on this information, they think brain coral, for example, is a rock, a plant, or an animal. (*It's a structure created by a colony of animals.*) What clues do they have to substantiate that fact? (*The polyps that form the colony have a gut, mouth, and tentacles. Each coral polyp creates a rocklike "skeleton" under itself, which combines with the skeletons of all the other polyps in the colony to create a rocklike structure, which is called a reef.*) What functions do the different parts of the coral polyps and colony perform? (*Tentacles secure food, zooxanthellae make food, the skeleton provides shelter for the polyp, and so on.*) Ask if anyone discovered whether coral polyps look different at different times of day. (*The coral polyps expand— usually at night—stretching out their tentacles to catch plankton. Since coral polyps are so tiny, the reefs have a fuzzy look at night, when the tentacles are extended and a smoother look during the day, when the polyps retract their tentacles.*)

4. Discuss reefs.

Now that your students have a better understanding of individual coral polyps and colonies, encourage them to think about coral reefs. Explain that a coral reef is formed as the numerous coral polyps secrete their underlying hard skeletons. Over hundreds of thousands of years, these skeletons form large reefs. Under the right conditions, coral reefs grow at an average rate of about one-half inch a year. Based on that rate, scientists estimate that Australia's Great Barrier Reef started growing more than 600,000 years ago! Ask the students to think about why this may make coral reefs very vulnerable. (*They recover very slowly from any damage or destruction.*)

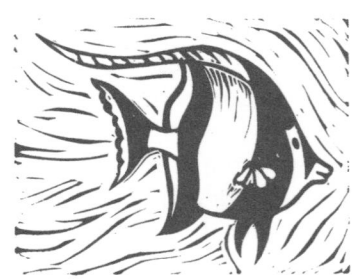

"[Corals] are to the giants of the reef what the more slender parts are to the lords of the forest, adding the elegance and delicacy of slighter forms to the strength, power, and durability of their loftier companions."

–Louis Agassiz, paleontologist, 1852

5. Research reef habitats.

Tell the students that they are now going to combine their individual coral sculptures into a classroom coral reef. You might point out that a coral reef is an important habitat for many species. Can the students define habitat? (*A habitat is a place where an animal, plant, microorganism, or other life form lives and finds the nutrients, water, sunlight, shelter, living space, and other essentials it needs to survive.*) To create their reef, the students need to go back to their sources of information to see if they can find out in which part of the reef their model belongs. Explain that each species of coral has special adaptations to different reef conditions. The students should investigate how the shape of their coral colony makes it suited to a particular location on a reef and how this and other unique traits help the different species survive. (For example, branching elkhorn coral is found in heavier wave zones as it is stronger than more delicate forms and is able to withstand the buffeting.) This may not be easy information to find for all types of coral. If a student can't find much information on a coral's locations on the

The porous limestone skeletons of corals have been used for human bone grafts.

reef, at least he or she should try to find out in which parts of the world the coral grows. Have the students create a small label describing their coral and list the type of habitat, any special adaptations or traits (if possible), and the place or places where it can be found. The labels should be similar to what visitors might see in a museum or aquarium.

6. Create a reef with the students' artwork.

Draw a large reef on poster-sized paper (for a guideline, use the "Coral Reef Outline" below) and place it on the floor. Then have the students take turns placing their models or drawings on the reef. The students who found out something about their corals' habitats should place their models first. As they do so, they should provide an explanation for why their coral is suited to that part of the reef. Then the other students can place their models on the reef.

When the students have finished, have them look at their completed reef, and encourage them to compare a coral reef to a forest. In what ways are the two habitats similar? (*Both of the habitats have many layers of life; both coral species and plant species are adapted to particular niches; both contain a wide variety of species, colors, and shapes.*) Be sure to remind students that not all of the coral colonies that they have depicted in class would be found growing on the same reef (unless you had asked the students to choose corals common to one region). But the completed reef is still a helpful representation of coral diversity and corals' adaptations to specific niches.

Coral Reef Outline

WRAPPING IT UP

Assessment

Assess the quality of work that went into completing the assignments for this activity.

Unsatisfactory—Diagram of polyp is incorrectly labeled, model is poorly constructed, and label for model is inadequate.

Satisfactory—Diagram is properly labeled; model and label are satisfactory.

Excellent—Diagram is properly labeled; model is well constructed, and label is exceptionally informative.

Portfolio

In the portfolios, have students include their labeled diagram of a coral polyp and field guide entries (from the "Writing Idea" section).

Writing Idea

Have the students write a field guide entry that describes their coral species. This description should include information on where the coral is found, species that live near it, color variations it may have, its scientific name, and other natural history details. This field guide entry should also include a sketch that reflects the coral's unique physical characteristics.

Extension

Have the students research the reproductive strategies of coral. Do corals produce sexually or asexually? As related to coral, what do these terms mean? What are the advantages and disadvantages of each form of reproduction?

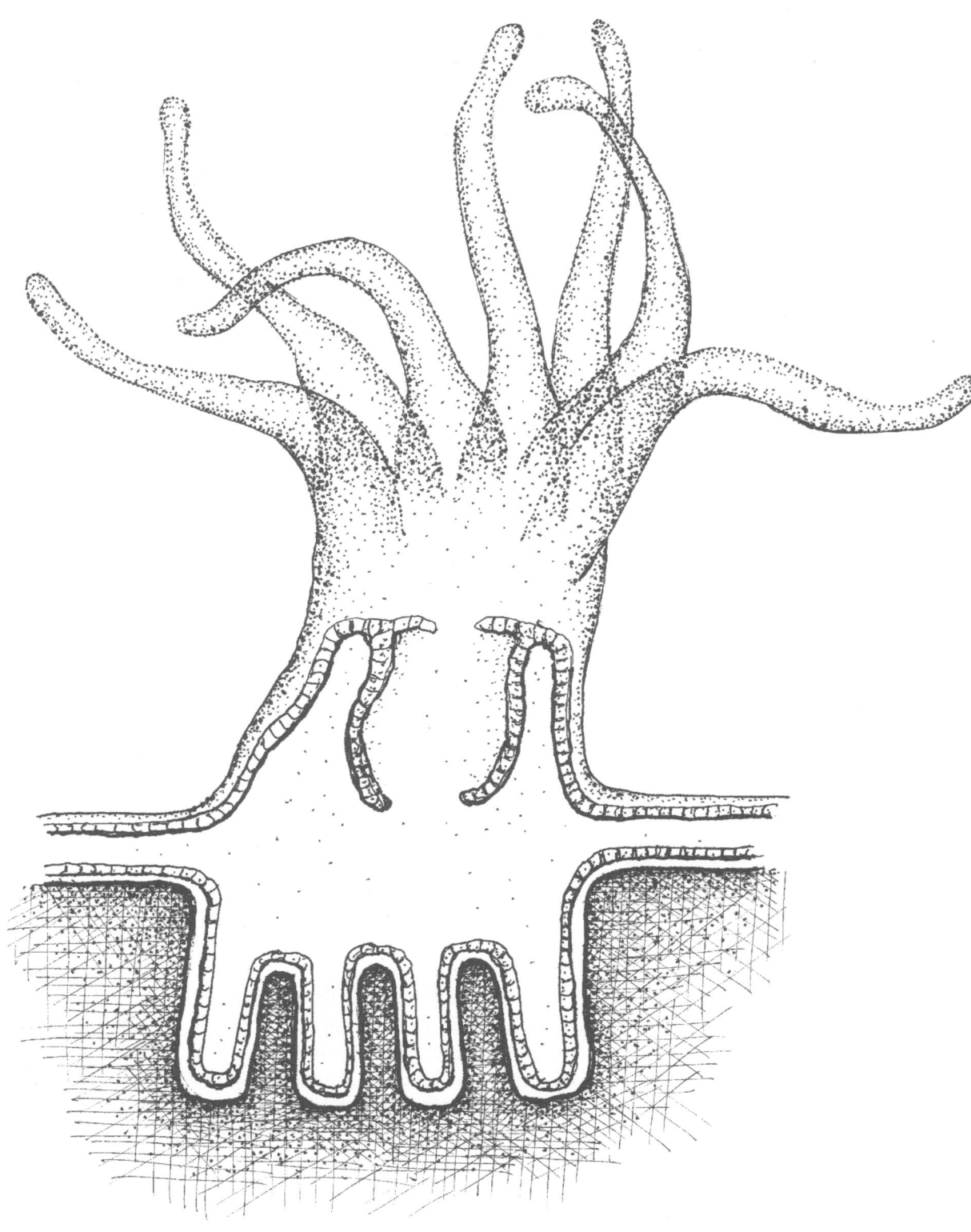

SUBJECTS
science, social studies (geography)

SKILLS
gathering (reading comprehension), organizing (mapping), presenting (describing, explaining)

FRAMEWORK LINKS
10, 13, 17, 17.1, 30.1, 33.2, 34.1, 35, 47.1, 52.1

VOCABULARY
equator, latitude, longitude, silt, siltation, Tropic of Cancer, Tropic of Capricorn, zooxanthellae

TIME
three sessions

MATERIALS
copies of "Postcards from the Reef" (pages 123-127) and "Coral Reefs of the World" map (page 120), atlases

CONNECTIONS
Use "Coral Bleaching: A Drama in Four Acts" (pages 140-143) to further explore climate change, one of the greatest threats to coral reefs and marine diversity. To learn about hot beds of terrestrial biodiversity, try "Mapping Biodiversity" in Biodiversity Basics.

 **AT A GLANCE**

Gather clues provided in a series of fictional postcard messages to determine the location of coral reefs around the world. Learn more about the benefits of healthy coral reefs to people and wildlife, as well as some of the threats that coral reefs face.

 **OBJECTIVES**

Describe several locations around the world where coral reefs are found. List at least three necessary conditions for coral reef formation. Name several benefits of coral reefs, and articulate activities that threaten corals.

I f you look at a map showing the worldwide distribution of coral reefs, you'll quickly come to see that they aren't scattered evenly all over the globe. Instead, they are concentrated near the equator, normally between the Tropic of Cancer and the Tropic of Capricorn. Reefs frequently grow along eastern shores—the ocean currents bring warmer water there than western shores. Corals prefer the warm, clear, and shallow waters of these tropical areas. They grow best in 72° F to 78° F temperatures and where the sunlight can reach them. Coral reefs are also more abundant where there are waves that carry food, oxygen, and nutrients to the reef. (See page 105 for more about coral reef distribution.)

Coral reef habitats are extremely valuable to people and other living things, so their limited range makes them all the more precious. Unfortunately, a number of natural events and human activities threaten coral reefs worldwide.

In this activity, your students will gather clues from postcards sent by fictional friends to determine where the world's major coral reefs are located. Then they'll analyze the results to determine what these areas have in common. At the end of the activity, they'll go back to their friends' postcards to see if they can find clues to some of the benefits of coral as well as information on some of the threats coral reefs face.

red beard sponge

Before You Begin

For each team of two students, make one copy of the "Postcards from the Reef" and one copy of the "Coral Reefs of the World" map. If you don't think each team will have time to solve all the postcard mysteries, make fewer copies of the "Postcards" and cut them apart so that each team receives two to four postcards. Gather enough atlases so that each team of students has one. If you don't have enough for each team, encourage teams to share the same atlas. (It may be confusing for teams to go from one atlas to another because the atlases can vary quite a bit in their portrayal of geographic formations.)

What to Do

1. Divide the class into teams of two students each.

Distribute the atlases and take time to explain the various parts and sections of an atlas, if needed. Then give each team of students a copy of "Postcards from the Reef." Tell the students that the postcard messages are from fictional travelers who have sent them from areas of the world that have coral reefs. Each postcard message has clues to the author's whereabouts, which is also the location of a major coral reef. The underlined words on the cards are clues to the types of places (city, country, or island chain) from which the postcard was sent. Using the atlases provided, the students should fill in the locations in the blank spaces provided at the top of each message.

2. Distribute "Coral Reefs of the World" maps.

Explain to the students that the dark areas on their maps represent the coral reefs of the world. Using their atlases and the information from their postcards, have the students label each of the dark areas with the correct corresponding number from the "Postcards from the Reef" page. Also, have the students label the equator, the Tropic of Cancer, and the Tropic of Capricorn on their maps.

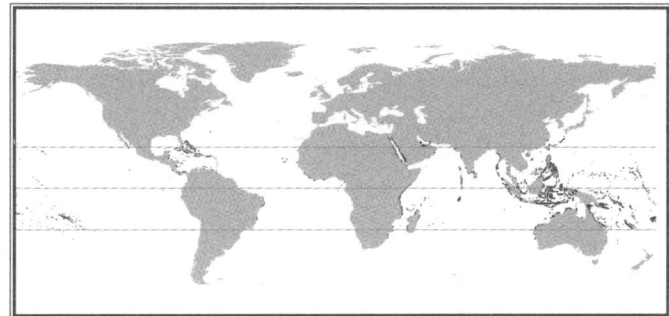

3. Analyze maps.

After the students have completed their labeling, have them take a closer look at the "Coral Reefs of the World" maps. Are there any generalizations they can make about the location of coral reefs in the world? (*Coral reefs are concentrated near the equator, between the Tropic of Cancer and the Tropic of Capricorn—and often along eastern shores, since the ocean currents bring warmer water there rather than on western shores.*) What conditions exist in the tropics? (*Temperatures don't vary widely over the course of the year and are usually mild. Ocean waters in this region are generally warm.*) What does this say about the conditions most corals need to survive? (*They require a relatively stable environment with warm conditions.*)

BIOFACT

Many reef fishes are clever at changing their appearance. The hogfish is usually orange during the day, blending in with the colorful reef. At night, sleeping on white sand, it turns white, becoming less visible to nighttime predators.

4. Discuss other conditions necessary for coral survival.

Explain that corals need to live in clear, shallow water so that sunlight can reach them—they grow best in 72° F to 78° F temperatures. Coral reefs are also more abundant where there are waves that carry food, oxygen, and nutrients to the reef. Coral polyps host tiny algae inside them, called *zooxanthellae*, which use sunlight to synthesize sugars. Can the students name the term that describes this process the zooxanthellae carry out? *(The students should recognize this as photosynthesis.)* The zooxanthellae share their sugars with the coral, providing a critical food source.

The other way that corals get food is by snatching passing plankton with their tentacles. The corals depend on water currents to bring the plankton within reach. Therefore, corals also need moderate wave action to survive—strong enough to bring them plankton, but not so strong that it breaks the coral.

5. Review postcards for clues to the benefits of corals and threats corals face.

Tell the students that the postcard messages contain more information than simply the location of the traveler; they also include hints to ways that coral benefits many of Earth's ecosystems. A few messages also refer to threats corals face. Have the students review the postcard messages and see if they can figure out the benefits and threats that the travelers have alluded to. If the students cannot find a benefit or threat on each postcard, that's OK: Several of the issues are quite difficult. If time permits, however, you may want to have students use the Internet to see if they can figure out the benefits and threats described. For example, by plugging in the key words "coral" and "asthma" and "arthritis," they may discover Web sites discussing the potential medicinal uses of coral.

Activity adapted from "A Chance of Success" in *Corals and Coral Reefs 4-8 Teachers Guide.* © SeaWorld, Inc.

> *"All of us who care about the future of the planet's coral reefs must find new and innovative ways to work together to reverse the tide of destruction. Together, we can make a difference."*
>
> **–Vaughan R. Pratt, President, International Marinelife Alliance**

6. Review answers.

As a group, review the answers to "Postcards from the Reef" (see page 122). Were the students surprised to learn that coral reef habitats provide people with so many benefits? Review some of the threats mentioned. Tell the students they will learn more about these threats and their effects on coral communities in later activities.

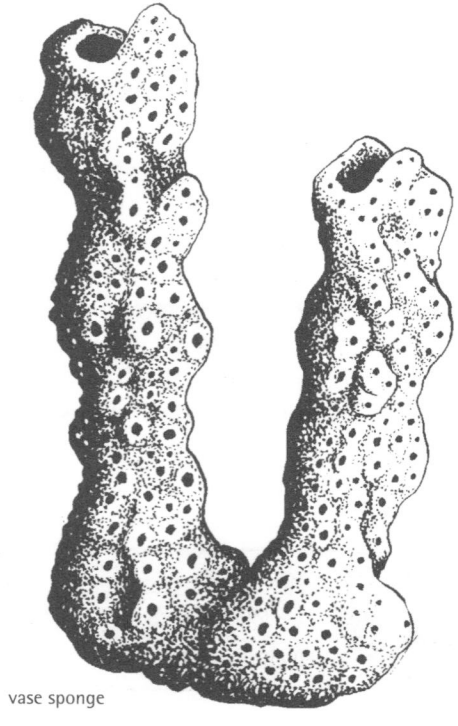

vase sponge

WRAPPING IT UP

Assessment

Do a "quick quiz" with your class. Have them write their responses to the following four questions on a note card or scrap of paper:

1. What conditions allow coral reefs to develop?
2. Where are most coral reefs located?
3. What are some of the benefits of coral reefs?
4. What are some of the threats coral reefs face?

Unsatisfactory—Named one or two conditions, areas where reefs exist, benefits of coral reefs, and threats to coral reefs.

Satisfactory—Named three conditions, three or four areas where reefs exist, three benefits, and two or three threats.

Excellent—Named more than three conditions, five or more areas where reefs exist, four or more benefits, and four or more threats.

Portfolio

Students should include their "Coral Reefs of the World" maps in their portfolios.

Writing Idea

Students can choose one of the postcard locations and do more research on the species that live in that area. Based on this research, they should expand the postcard into a longer email that they would write to friends and family, describing the reef, its location, its species, and what activities they would participate in if visiting there.

Extensions

- Have students label the rest of the reefs on their map using information from atlases, the Internet, and other resources.

- Have students research marine protected areas (MPAs) that include coral reefs. Ask the students to draw in these MPAs on their maps, using colored markers. Which reefs of the world are protected by MPAs? Which are not?

grouper

BIOFACT

Similar to other species of hard corals, leaf corals secrete a mucus that removes sediment from their surfaces and prevents their polyps from being smothered.

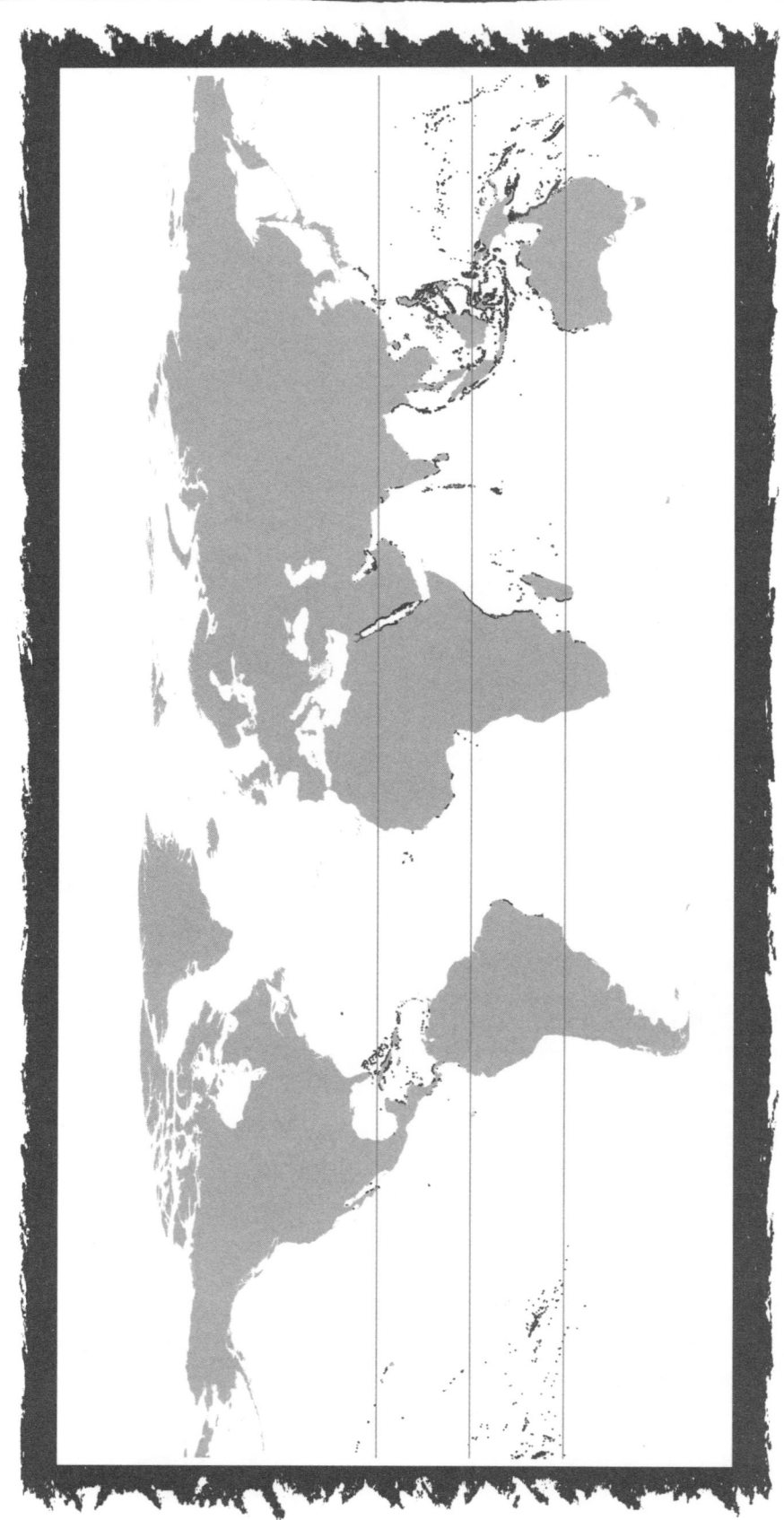

Key

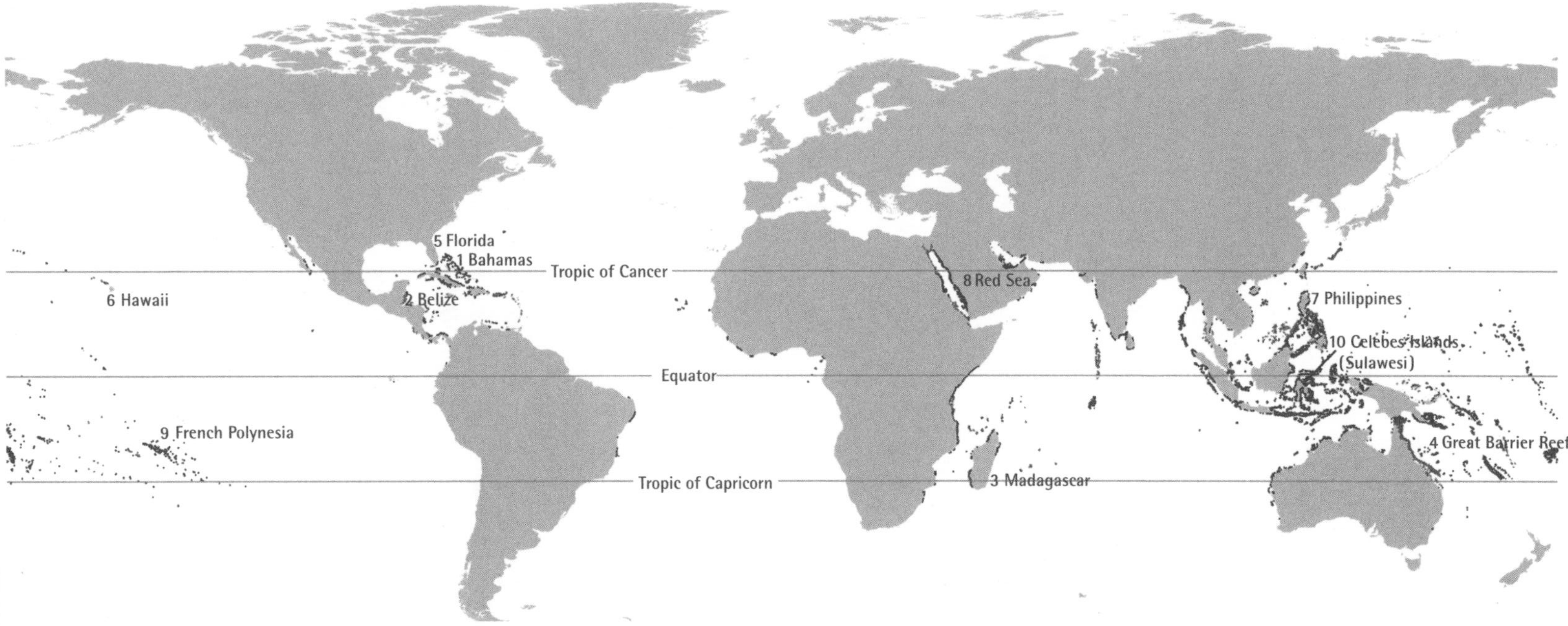

5 Florida
1 Bahamas
2 Belize
6 Hawaii
9 French Polynesia
8 Red Sea
3 Madagascar
7 Philippines
10 Celebes Islands (Sulawesi)
4 Great Barrier Reef

Tropic of Cancer
Equator
Tropic of Capricorn

Map provided by ReefBase: A Global Information System on Coral Reefs. ReefBase is a project by ICLARM—The World Fish Center, with support from the International Coral Reef Action Network (ICRAN). www.reefbase.org

REEFBASE

ANSWERS TO "POSTCARDS FROM THE REEF"

1.

Bahamas. *Benefits:* Coral reefs provide a home to many marine species that people consume, including lobsters, fish, crabs, and oysters. In this way, reefs not only provide millions of people with important nourishment, they also help create a source of income for local fishers. *Threats:* Some reef species, such as lobsters, groupers, and snappers, are overfished.

2.

Belize. *Benefits:* Recreation around coral reefs provides a livelihood to thousands of people living in coastal communities. Boat operators, tour guides, restaurant owners, hotel owners, and all the people they employ collectively earn millions of dollars through the tourist industry. *Threat:* This message alludes to one recreationally related threat that corals face: Boat anchors dropped onto coral reefs can damage them.

3.

Madagascar. *Benefit:* The breakdown of coral skeletons provides tropical beaches with their white, soft sand.

4.

Great Barrier Reef. *Benefits:* Coral reefs draw tourists and therefore provide both recreational and economic benefits. *Threats:* In some parts of the world, people are simply "loving coral reefs to death." Touching and standing on reefs can damage them permanently.

5.

Florida. *Benefits:* Corals have many existing and potential medicinal benefits. Scientists have used coral reefs to make a substance to replace broken bones. And they're investigating the potential use of coral compounds to treat arthritis, viruses, asthma, and cancer.

6.

Hawaii. *Benefits:* Many communities celebrate new hotels and other forms of development in coastal areas near coral reefs because these developments attract tourists and tourist dollars. *Threats:* There is a down side to this development: Dredging for new construction increases siltation of coastal areas, smothering coral reefs and making it difficult for the zooxanthellae to photosynthesize. And the pollution created by the people who use these hotels and other buildings can also harm corals.

7.

Philippines. *Benefits:* Corals provide a natural buffer between the open ocean and coastal areas, reducing the impact of heavy wave action associated with tropical storms. In this way, coral reefs reduce beach erosion on the mainland. *Threat:* Sometimes, heavy waves can destroy corals.

8.

Red Sea. *Benefit:* Coral reefs are a hotspot for marine biological diversity, providing food and shelter to about 25 percent of all marine species.

9.

French Polynesia. *Benefits:* Coral reefs are beautiful and have long impressed people who see them in the wild. *Threat:* Tourists and collectors sometimes break off chunks of coral as souvenirs or for display in aquariums, causing permanent damage to the reefs.

10.

Celebes Islands (also known as Sulawesi). *Benefit:* Coral reefs are home to many of the most colorful, interesting tropical fish that people collect in their home aquariums. *Threats:* Some people who collect tropical fish from coral reefs use cyanide to stun the fish, which can cause serious damage to coral reefs. That's why it's better to purchase fish that are certified as having been caught in an environmentally friendly way or have been raised for the aquarium trade.

1.

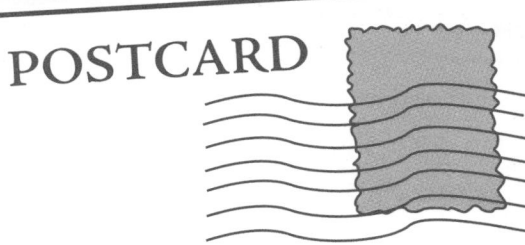

POSTCARD

We've been on Great Abaco Island for the last three days, and tomorrow we're off to Andros. Every day we snorkel for hours, and every night we go to one of the local restaurants and eat the freshest, most delicious seafood I've ever tasted. These <u>islands</u> are truly paradise.

2.

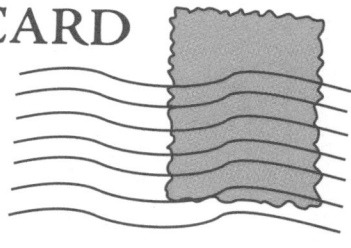

POSTCARD

After we left Guatemala, we cruised up the coast toward the tip of the Yucatan Peninsula. But before we got to Mexico, we had to stop in this small <u>country</u>. It's awesome. A tour boat operator took us out into the bay, dropped his anchor overboard, and let us dive around the reefs for hours.

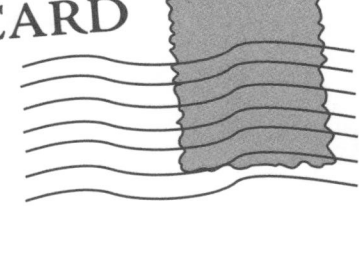

3.

POSTCARD

From the capital city of Atananarivo, we took a bus across the <u>country</u> to the coast. Tomorrow we'll go snorkeling around the reefs, but today we just sat out on the white sandy beaches for hours.

4.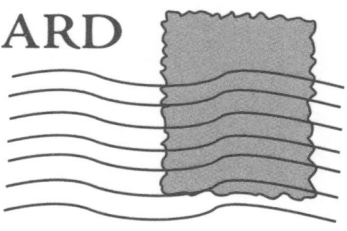

POSTCARD

We flew to Brisbane, then drove up the coast. From just about any coastal town, you can head out to explore this <u>reef</u>—it's the largest one in the world! Even though it's so big, some parts are crawling with tons of divers. We saw people hanging onto the reefs, peering into every nook and cranny, and sometimes even standing on them.

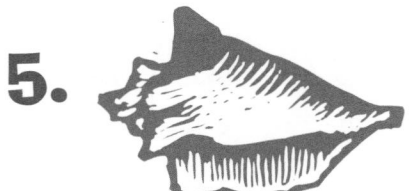

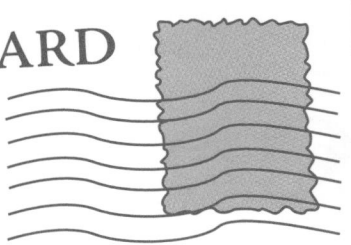

5.

POSTCARD

_ _ _ _ _ _ _ _ _ _

After spending a few days in the swampy Everglades, we drove over to the Keys. A bunch of doctors were having a conference—something about potential new treatments for asthma, arthritis, and cancer. As for me, I was outside—appreciating the wonders of this <u>state</u>'s beautiful reefs!

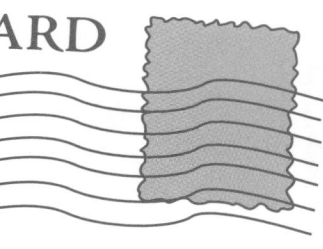

6.

POSTCARD

_ _ _ _ _ _ _ _ _ _

This U.S. <u>state</u> made up of islands is even more beautiful than it looks in pictures! We have a room in a hotel right on the beach where I go snorkeling every day. Maybe next year we'll stay in the new hotel they're building next door.

POSTCARD

7.

I'm writing to you from a café in Manila. I wish I were snorkeling right now in the reefs off the coast, but instead a huge tropical storm has kept us in this <u>country's capital city</u> for a few more days. The wind is raging, and they say there are high waves out at sea.

POSTCARD

8.

Cairo is a beautiful, fascinating city, but I'm still glad I headed southeast to the coast and spent time snorkeling in this incredible <u>sea</u>. We saw a fantastic variety of tropical fish swimming around the reefs, including manta rays and gray reef sharks. They seem drawn to the reef just as we are!

9.

POSTCARD

We sailed from Fiji to Tonga to the Cook Islands before arriving at this French island chain. The reefs are gorgeous! I'm going to bring you home a piece of coral rock so you can have a memento from these beautiful <u>islands</u>.

10.

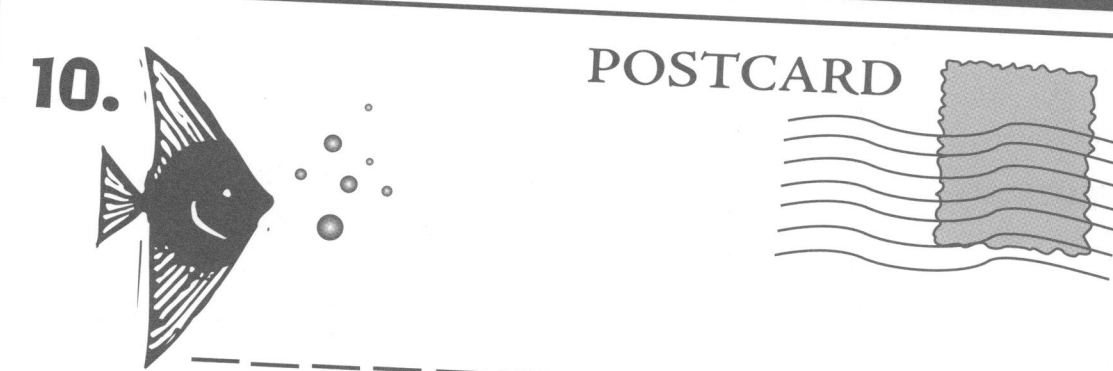

POSTCARD

Today we snorkeled around the coral reefs found off these <u>islands</u> at a latitude of 2° S and a longitude of 121°10' E. The tropical fish that dart around the reefs are spectacular! No wonder collectors sell them to aquarium owners back home.

3 Coral's Web

SUBJECTS
science

SKILLS
organizing (arranging), analyzing (identifying components and relationships among components), interpreting (identifying cause and effect, relating, inferring, reasoning)

FRAMEWORK LINKS
1, 1.1, 3, 10, 10.1, 12, 13, 16, 23, 25, 29, 30.1, 33.1, 35, 47.1, 52.1

VOCABULARY
food web, habitat

TIME
two sessions

MATERIALS
one copy of the "Coral Reef Cards" (pages 132-138), one copy of "Outside Forces" page 139), large roll of string or spool of thick thread (300 to 600 feet in length), pushpins or tape, markers, bulletin board, large piece of poster-sized paper with "Coral Reef Outline" (page113) drawn on it, chalkboard and chalk

CONNECTIONS
"Build-a-Reef" (pages 110-115) is a good activity to get students exploring the ocean's diversity of coral reefs and their inhabitants. For more on food webs and the cascading effects of stresses on them, see "All the World's a Web" in Biodiversity Basics.

 **AT A GLANCE**

Use information cards to create a coral reef food web, and then explore some of the ways that natural and human forces affect this web of life.

 **OBJECTIVES**

Describe some of the ways that species in a coral reef interact with one another. Explain three specific food web relationships in a coral reef. List some of the natural and human events that can affect coral reef species and discuss why food webs magnify the effect of any such event.

Coral reefs provide food and shelter for about 25 percent of all oceanic species—which is more than any other marine habitat. As in all ecological communities, the species that inhabit a coral reef do not simply coexist in the same geographic area. Instead, species interact in many complex ways. For example, the zooxanthellae that live in coral polyps gain protection and nutrients while providing the coral with a major source of food. Coral reefs create hiding places for a variety of species, from lobsters to sharks. And all the species in a coral reef are connected to one another through food webs. For example, sharks and moray eels feed on octopuses, which feed on lobsters, which feed on mussels and crabs, and so on.

In this activity, your students will learn about the diversity of life on a coral reef by creating a food web. Then they will explore the effects of outside forces on this interconnected reef community.

THE FAR SIDE® BY GARY LARSON

© 1993 FarWorks, Inc. All Rights Reserved/Dist. by Creators Syndicate

Oo! Igor!... Come here and look at this brain coral!

The Far Side® by Gary Larson © 1993 FarWorks, Inc. All Rights Reserved. Used with permission.

Dr. Frankenstein vacations in Hawaii.

Before You Begin

Make one copy each of the "Coral Reef Cards" (pages 132-138) and one copy of "Outside Forces" (page 139). On a sheet of poster-sized paper, draw the outline of a coral reef system as shown on page 113. (Be sure to make the drawing large enough so that there will be room to tape the cards in place on the diagram.) Post the reef drawing on a wall or bulletin board. Have tape or pushpins on hand.

What to Do

1. Discuss a coral reef community.

Introduce the activity by asking your students to think about the words "coral" and "coral reef." Can someone explain the difference between them? (*Corals are a kind of animal; coral reefs are the habitats created by many coral colonies.*) What do your students know about a coral-reef community? Can anyone name some specific species that live there? (*The "Coral Reef Cards" name some species.*)

Tell your students that they're going to begin an activity that looks at coral-reef species. But first, ask them to think about species that live as a community

NOAA Coral Kingdom

in any ecosystem, such as a forest, desert, river, or pond. Do those species all live totally unconnected to one another? (*No, most interact with each other in many ways.*) What are some of the ways that species interact? (*They eat each other for food, some animals pollinate plant species, some organisms help decompose dead organic matter and thus make the nutrients available to other living things, some*

Activity adapted from "The Coral Reef Community." *Coral Reef Teacher's Guide.* © World Wildlife Fund, 1986.

species provide shelter for others, and some living things coexist in symbiotic relationships.)

Just as in any other ecological community, the species in a coral reef interact with and depend upon one another in many ways. This activity focuses on food webs within a Caribbean coral reef community. Be sure that your students recognize that food webs are just one subset of the many kinds of interactions in a coral reef. For example, a crab that uses a snail shell for its home has a critical relationship with snails, even though one does not eat the other.

2. Hand out the "Coral Reef Cards."

Have students form a circle in the middle of the room and give each student one "Coral Reef Card." If you have more students than cards, some can stand together and share a card. If you have fewer students than cards, you may want to select in advance the cards you will be using to ensure that every student will end up having at least one connection to another student. **Note:** If you have to exclude some of the cards, be sure to include the cards numbered 9, 18, and 26 (coral polyp, zooxanthellae, and human being).

Ask the students to read the cards to themselves, noticing two things that are described on each of the cards:

- What other living things help the species named on their cards survive?
- What other living things need their species to survive?

3. Form a coral reef web.

Begin with one student and ask that student to read his or her card aloud. When the other students hear information related to the species on the cards they hold, they should raise their hands. Then have the students use string to connect any interacting species. For example, when the person holding the parrotfish card reads the card aloud, the people with algae and barracuda cards should raise their hands. Then use string to connect the parrotfish to the algae (parrotfish prey) and to the barracuda (parrotfish predator). Walk around the outside of the circle as you complete the web. Repeat the process with each student in the circle until all of the related cards/students have been connected using the string.

4. Discuss the coral reef food web.

A big "web" of relationships has formed within the circle of students. While they are still holding the string, discuss the following questions. (*Answers may vary, but some possible responses follow the questions below.*)

- Which species has the most relationship strings? (*Humans have the most.*) What does that mean? (*Humans are highly diversified in what we capture and eat. It also means that we can have the greatest effect on the health and diversity of the reef community.*)
- Which species has the least? (*Name any that are connected by the string to only one other species.*) Why? (*Because those species eat or are eaten by only one species.*)
- Do any of the species seem especially important for the health of the reef community? (*The coral polyps are crucial because, without them, the reef community wouldn't exist. Humans are also important because of our potential effect—not only in what we take to eat but because of all the other effects our activities can have on the reef community.*)

moray eel

5. Examine the effects of outside forces on the food web.

After you have discussed all the relationships on the cards, tell the students that you are going to bring some outside forces to their coral reef food web. These forces may directly affect one species and indirectly affect others.

One at a time, read aloud the "Outside Forces" on page 139. Based on the descriptions of the forces, any students holding cards of species that are directly affected by a force should lower their heads and gently tug their strings. The tugging indicates that a species has been affected in a negative way. Any other students connected to a student who tugs on a string should also lower their heads and tug their strings. If the action has a positive effect, the students affected should each raise the hand that isn't holding the string. Continue reading aloud from the list of "Outside Forces" until you have finished the set.

6. Discuss the simulation.

When you have finished reading through the "Outside Forces" list, have the students return to their seats. (If you like, you can have them color in their species as you talk.) What kinds of things did they learn about coral reefs in this exercise? (*Answers will vary, but students may mention the diversity of life found in reefs, the interdependence of this life, and threats to coral reefs.*) Has anyone in the group heard of the word *biodiversity*? If necessary, explain that the term biodiversity refers to the variety of life on Earth (see page 26 for more about marine biodiversity). What is the relationship between coral reefs and biodiversity? (*Coral reefs help sustain an incredible diversity of marine life—also known as marine biodiversity. You might mention to the students that about 25 percent of all marine species are believed to rely on coral reefs for food or shelter.*) In what ways do threats to coral reefs and individual species within them threaten marine biodiversity? (*Depletion of one species can affect a myriad of other species because of the interconnections of organisms throughout the food web. And damage to the entire coral reef degrades the habitat that many marine species depend upon.*)

Ask the students to think about some of the ways that a diverse ecosystem, such as a healthy coral reef, might be more resilient in the face of outside forces than an ecosystem that is less diverse. (*The more species in the ecosystem, the greater the possibility that some will survive changes in environmental conditions.*)

7. Have students place their picture cards on the coral reef diagram.

To conclude this activity, place the large coral reef diagram on a bulletin board. Then ask the students to attach their species to an appropriate location on the coral reef. Once all species have been attached, have the students come back up to the diagram and draw lines to connect species that eat one another in the food web. (**Note:** You may want to consider using the Extension activities below before doing this last step.)

WRAPPING IT UP

Assessment

Have students create concept maps starting with the central stem of "coral." Encourage them to branch from the stem (with connectors and labels) with the concepts of food, natural events, and human-caused events. Wherever possible, show how the concepts overlap and interrelate.

Unsatisfactory—The concept map includes one or two stems of one or more levels; at least one idea is related to two or more stems.

Satisfactory—The concept map includes at least three stems of two or more levels each; at least one idea is related to two or more stems.

Excellent—The concept map includes at least three stems of two or more levels each; at least two ideas are related to two or more stems.

Portfolio

Have each student write several paragraphs describing the coral species on the card he or she received and what role that species plays in its marine ecosystem. Include those descriptions in the portfolios.

Writing Idea

Many threats to coral reefs may seem far away for students living in North America (particularly areas that are not located on the coast). To explore how our actions can have far-reaching effects, have students research ways in which the coral reef threats discussed in the activity are linked to everyday actions taken by people living in countries such as the United States. (Students may want to look at topics including climate change and trade in corals and other marine species.)

Extension

Have your students research their coral reef species and write a short report on its life history. What do scientists know about this species? What don't they know? Does the species rely on other coral reef species in ways that have nothing to do with food? Explain.

2

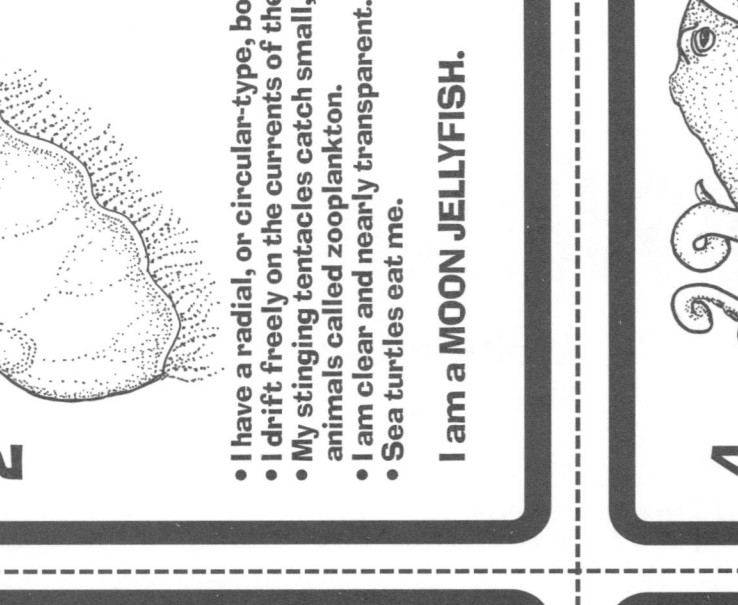

- I have a radial, or circular-type, body outline.
- I drift freely on the currents of the ocean.
- My stinging tentacles catch small, drifting animals called zooplankton.
- I am clear and nearly transparent.
- Sea turtles eat me.

I am a MOON JELLYFISH.

4

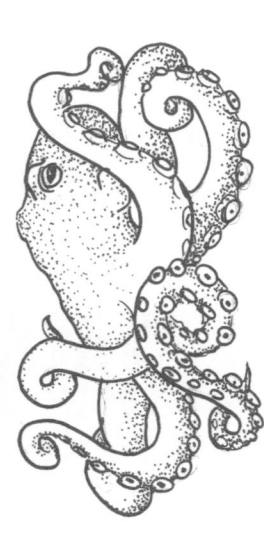

- I do not have a shell or backbone.
- I crawl along the ocean floor and hide in cracks and holes in the reef.
- I can change color quickly and hide in a cloud of inky water.
- I catch clams and snails.
- Eels and groupers eat me.
- I have eight arms.

I am an OCTOPUS.

1

- I live in a hard tube that I build for myself.
- I draw myself quickly into my tube if I need to hide from an animal that is trying to eat me.
- With my gills, I catch tiny drifting animals called zooplankton.
- On my head, I have fine, thin gills that filter my food.
- I am a type of worm with bristles.

I am a FEATHER DUSTER WORM.

3

- I am made up of an entire colony of animals that work together as if they were one animal.
- I grow into a fan-shaped creature that waves back and forth in the water.
- With my tentacles I catch zooplankton, which I eat.
- Fireworms eat me.
- I am a type of "soft" coral.

I am a SEA FAN.

6

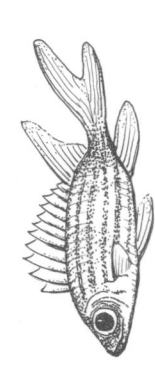

- I have a backbone, four flipper-like legs, and a hard shell.
- I breathe air.
- I am more closely related to lizards and snakes than I am to fish.
- I visit coral reefs and seagrass beds.
- There, I eat sponges and sea grasses, especially turtle grass.
- People kill many individuals of my species for our meat and shells.
- People often dig up and eat our eggs, which we lay on beaches.
- We are often caught accidentally in nets and are in danger of extinction.

I am a GREEN SEA TURTLE.

8

- I have a backbone, scales, and fins.
- I am bright red with big, round eyes.
- I have sharp spines on my top fin.
- I hide in crevices in the reef during the day and come out at night.
- I swim through the water and eat shrimp and small fish.
- Groupers and eels eat me.

I am a SQUIRREL FISH.

5

- My body is divided into many segments.
- Each segment has legs below and bristles above.
- My bristles sting.
- I crawl around the reef and eat coral polyps.
- I am a type of worm.

I am a FIREWORM.

7

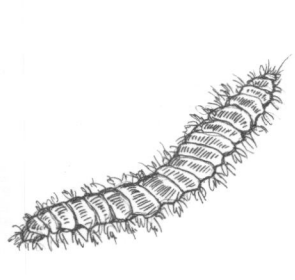

- I am a "jointed-leg" animal, with a hard outer shell for a skeleton.
- I have 10 limbs.
- Two of my limbs are much larger than the others, and they have claws that I use to catch and crush my food.
- I eat small fish, pieces of sea animals, and other things I find on the sea floor.
- I prefer to eat sea urchins and snails.

I am a CORAL CRAB.

CORAL REEF CARDS (Cont'd.)

10

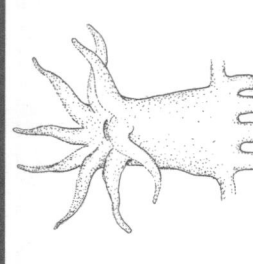

- I live in a beautiful spiral shell.
- I move along the sea floor and eat algae.
- I lay my eggs in the sand.
- Spiny lobsters eat me when I am small.
- When I am bigger, people catch me for food.
- I am a type of snail.
- In the past, there were many like me in the Caribbean, but now we have become harder to find.

I am an QUEEN CONCH.

12

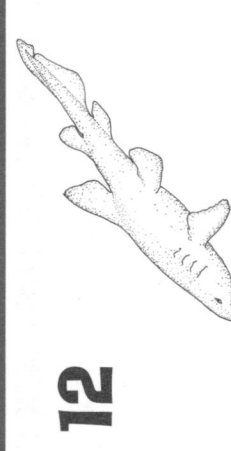

- My backbone and the rest of my skeleton are made of cartilage.
- I am a fish, but I'm not a bony fish.
- I have a good sense of smell and two whisker-like "barbels" near my mouth.
- The barbels help me find food.
- I eat clams, crabs, and lobsters.
- I rest in coral reef caves.
- Many people are afraid of me, but I am seldom dangerous to them.

I am a NURSE SHARK.

9

- I am one individual in a colony of animals that look just like me.
- I have tentacles with stingers.
- I deposit a stony skeleton beneath my body.
- I catch and eat small, drifting animals called zooplankton.
- Colonies of animals like me make up a coral reef.
- Parrotfish and butterflyfish eat me.

I am a CORAL POLYP.

11

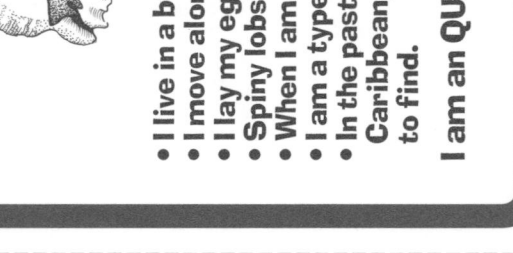

- I have a radial, or circular-type, body form and I don't have any stingers.
- I drift freely through the water, though you may find me washed up on the beach.
- I feed on small animals called zooplankton.
- I am colorless and nearly transparent.
- Jellyfish eat me.

I am a SEA WALNUT.

134 *World Wildlife Fund*

14

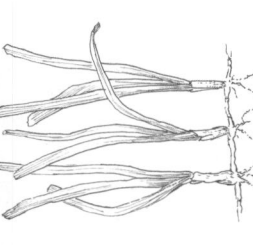

- I don't need to catch or hunt food because I make my own, using energy from the sun.
- I grow on the sandy ocean floor between the reef and land.
- I am a plant.
- I have long, thin leaves.
- Many young fish, shellfish, and other animals find shelter among my leaves.
- Turtles eat me.

I am TURTLE GRASS.

16

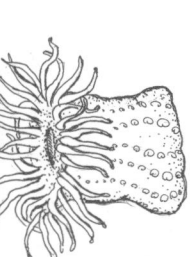

- I have a tube-shaped body with stinging tentacles.
- I usually grow attached to a solid surface, such as a rock or seashell.
- My tentacles sometimes catch small fish, but I eat mostly zooplankton.
- Sometimes I grow on seashells that crabs are living in.
- I steal bits of food from the crab and protect the crab from octopuses and other crabs.
- I am eaten by starfish and sea slugs.

I am a SEA ANEMONE.

13

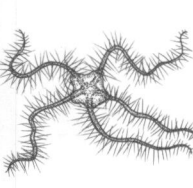

- I am a spiny-skinned animal with a star-shaped body.
- I have five long, thin arms. I move on many tiny feet, which are located on the bottom of my arms.
- If I lose an arm, I can grow another one in its place.
- I eat algae as well as bits of dead plants and animals on the reef.
- I feed at night. During the day I hide in dark crevices on the reef.

I am a BRITTLE STAR.

15

- I have a backbone, fins, and scales.
- I have a long, smooth body, and very sharp teeth.
- I swim very quickly.
- I eat many small fish such as four-eyed butterflyfish and parrotfish.
- Few other animals bother me, but humans sometimes catch me.

I am a BARRACUDA.

18

- I belong to a "soup" of creatures that drift through the reefs' waters.
- To see me, you usually need to use a magnifying glass or a microscope.
- Some members of my group grow up to be larger animals, but some stay tiny.
- Sea anemones, sea fans, and various other creatures eat me.
- Some members of my group eat tiny algae, while others eat members of our own group.

I am ZOOPLANKTON.

20

- I am a jointed-leg animal with a hard outside skeleton.
- I have 10 legs.
- After I lay my eggs, I carry them under my curled tail.
- I have two large antennae, which I use to defend myself.
- I eat snails, worms, and crabs.
- Groupers eat me.
- People catch and eat so many individuals of my species that not many of us are left.

I am a SPINY LOBSTER.

17

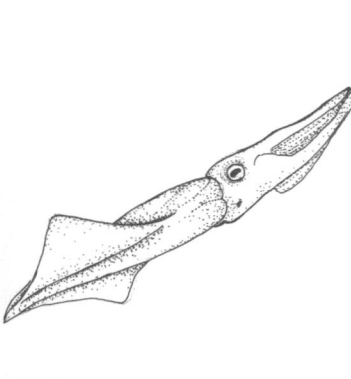

- I have a soft body with ten arms.
- Two of my arms are long tentacles that I use to catch small fish, which I eat.
- I can change color quickly.
- Sharks and people eat me.
- I can swim very fast.

I am either a SQUID or a CUTTLEFISH.

19

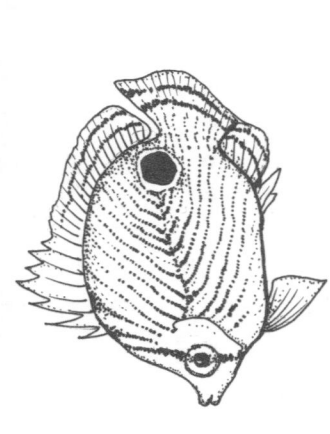

- I have a backbone, fins, and scales.
- I am round-shaped almost like a wheel.
- I eat zooplankton, the soft polyps of corals, and various worms.
- I have two large spots near my tail. The spots fool bigger fish—such as barracudas—that try to eat me. (They think the spots are big eyes!)

I am a FOUR-EYED BUTTERFLYFISH.

22

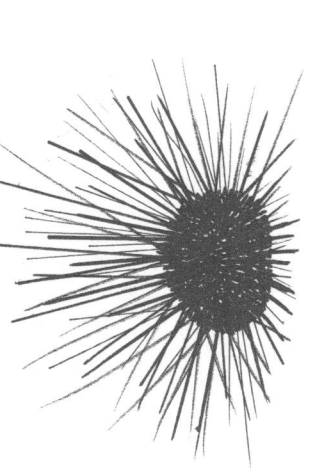

- I have a backbone, fins, and scales.
- My mouth looks like the beak of a bird.
- I am brightly colored.
- I am one of the largest reef fish.
- I eat algae that grow on dead coral and inside coral polyps.
- Barracudas eat me.

I am a PARROTFISH.

24

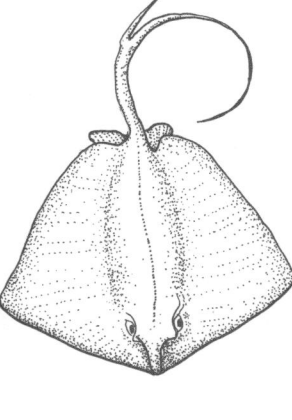

- I have a backbone, fins, and scales.
- I have a soft skeleton, like my shark relatives.
- I have a barb on my tail. It has a nasty sting.
- My body is very flat, and I spend most of my time lying partially buried on the sandy ocean floor.
- I eat snails, crabs, and clams.

I am a STINGRAY.

21

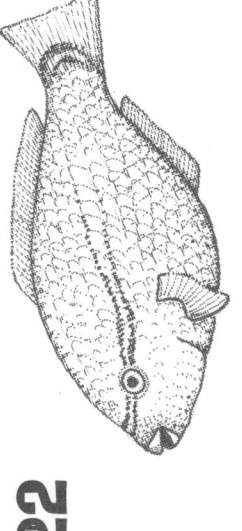

- I am a spiny-skinned animal with a circular-shaped body.
- I eat algae that grows along the reef and ocean floor.
- I have long spines that help protect me.
- Turbot (queen triggerfish) eat me.

I am a LONG-SPINED SEA URCHIN.

23

- I have a backbone and fins. I am quite big.
- I am not a shark or a fish.
- My body is warm, like yours.
- I breathe air.
- I come in from the open sea to visit the edge of the reef.
- I often travel in schools, or groups.
- I eat sea stars and various other fish that swim in schools, such as sardines and squirrelfish.
- I am a type of whale.

I am a DOLPHIN.

26

- I have a backbone, am an air-breather, and live on land.
- I eat groupers, turtles, squid, parrotfish, conch, lobsters, and many other sea animals.
- I often catch so many animals on the reef that some species have a hard time surviving.
- Sometimes, things that I do in the water and on land harm coral reefs and the animals and plants that live there.
- I have many ways to help coral reefs and to appreciate their beauty.
- I sometimes use coral to decorate my home, my aquarium, and my body.

I am a HUMAN BEING.

28

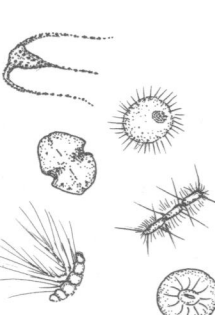

- Some species in my group are so small that they drift in the water without being seen.
- Other species in my group grow large and are leafy or grasslike.
- Some species in my group grow on stones or dead coral.
- I need only sunlight, water, and substances dissolved in the water to live.
- Parrotfish, queen conch, snails, and other sea creatures eat me.

I am ALGAE. When I drift in the water, I am called **PHYTOPLANKTON.**

25

- I have a backbone, fins, and scales.
- I have a big mouth, and I'm marked with spots and stripes.
- I can swim, but usually I keep still and try not to be seen.
- I am eaten by sharks and caught by humans.
- I eat small fish, such as squirrel fish.

I am a GROUPER.

27

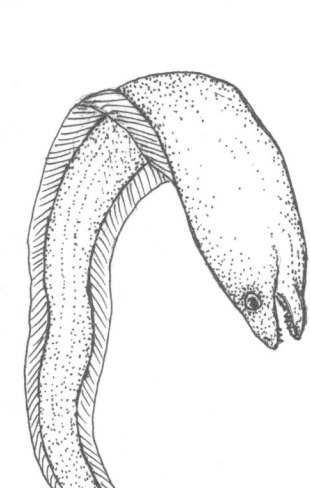

- I have a backbone, gills, fins, and tiny scales.
- I am well known for my large, fierce-looking jaws.
- I am long and snakelike.
- I eat octopuses and squirrel fish.

I am a MORAY EEL.

1. I own a factory that manufactures electrical goods. My factory is located on the shore near the coral reef. I dump chemicals into the river that flows near the reef. The chemicals kill the baby fish on the turtle grass flats of the reef.

2. I am marine algae, and when fertilizer and sewage flow from the island into the water, I grow very quickly. I can block sunlight so it doesn't reach the coral polyps below, which causes oxygen levels to decline in the water and harms the coral. Without sunlight, the zooxanthellae living in the polyps are not able to make food for the living coral.

3. I am a starfish that eats coral polyps. I eat large areas of coral polyps until I'm eaten by a sea urchin.

4. I am a fisheries agent who puts a limit on the number of spiny lobsters that can be caught on the reef.

5. I am a builder who develops hotels along the beach. I dredge sand from the flats around the reef to create the hotels' beautiful, sandy beaches. But this dredging can cause harm to the coral.

6. I collect fish and coral for U.S. home aquarium owners. Sometimes I use cyanide to stun and capture the fish, which can also end up killing other fish and corals. But people are willing to pay me a lot of money for the saltwater fish and corals that I collect.

7. I run an ecotourism business that brings divers and snorkelers to the reef. I tether my boats to a buoy so my anchor won't hurt the reef. And I teach all my passengers to help conserve the reef by not touching anything and watching where they step when wearing fins.

8. I am a scuba diver with a spear gun, and I take many groupers from the reef.

9. I am a hurricane that brings torrents of fresh water onto the reef, and my winds stir up big waves that break off corals and dump silt on the reef.

10. I am a teacher in an island school that educates people about coral reef conservation and the importance of marine biodiversity.

Coral Bleaching: A Drama in Four Acts

SUBJECTS
language arts, science, art (drama)

SKILLS
analyzing (identifying components and relationships among components, discussing), applying (restructuring, synthesizing), presenting (demonstrating, acting), citizenship skills (working in a group)

FRAMEWORK LINKS
10, 12, 13, 26, 35, 52, 52.1

VOCABULARY
algae, coral bleaching, global climate change, photosynthesis, symbiosis, zooxanthellae

TIME
one to two sessions

MATERIALS
copies of "Coral Bleaching: A Drama in Four Acts" (page 143); pieces of cardboard, colored paper, markers, and other materials students can use to make props in their performances

CONNECTIONS
To get students thinking about the many ecological services that healthy marine habitats provide, start with "Services on Stage" (pages 86-93), which will also make use of their theatrical skills. To help empower students and encourage them to think about how they can make a difference in slowing the effects of global warming, have them create educational Web sites using the "Reef.Net" (pages 144–147).

AT A GLANCE
Perform skits that show the interdependence of corals and zooxanthellae, as well as the devastating effects of rising ocean temperatures and coral bleaching.

OBJECTIVES
Describe the relationship between corals and zooxanthellae, and define symbiosis. Explain some of the causes and effects of global climate change, including coral bleaching.

As you probably know, global warming describes the worldwide phenomenon in which temperatures are on the rise. Almost all scientists attribute global warming, at least in part, to the burning of fossil fuels and the release of carbon dioxide and other greenhouse gases into the atmosphere. You've probably heard that global warming could change weather patterns, alter growing seasons, raise sea levels, and allow tropical diseases to spread into previously unaffected areas. But did you know global warming also poses a major threat to coral reefs?

Tiny algae called *zooxanthellae* inhabit coral polyps, and these algae give coral polyps their color at the same time as supplying them with a steady source of food. But zooxanthellae are extremely sensitive to rising temperatures: If ocean waters become too warm, the zooxanthellae will leave the polyps. And if the zooxanthellae don't return, the coral will bleach and eventually die. (See page 103 for more on the relationship between coral and algae.)

In the last two decades, just as average recorded global temperatures have hit all-time highs, corals have been bleaching at unprecedented rates. The Eastern Pacific Ocean, Caribbean, Indian Ocean, and many other marine ecosystems have experienced major bleaching events. This bleaching is occurring at a time when corals are already faced with numerous threats, including water pollution and the mining of reefs for construction rock. Coral bleaching is an especially frustrating problem because it's just as likely to affect coral reefs that are being protected through conservation efforts as those that are not, and the problems extend beyond political boundaries, so it may take the cooperation of many nations to do something about them. In other words, global warming puts *all* coral reefs at risk. Because coral bleaching is such a serious, worldwide concern, people need to understand the issues and be willing to explore possible solutions. And because it's a complex topic, we've divided the issues of global warming and coral bleaching into manageable pieces that your students can dramatize for one another in creative skits. Through these dramatizations, your students will learn about a challenging scientific topic and have fun while they're at it.

Before You Begin

The class will be divided into four groups and each group will perform one act of "Coral Bleaching: A Drama in Four Acts." Make copies of page 143 and cut into separate strips (each member of the group needs only the strip describing his/her act). Gather the materials listed (page 140) for students to use as props.

What to Do

1. Introduce activity.

Tell the students that this activity will introduce them to coral bleaching, which represents a major threat to coral reefs. Understanding this concept is critical for acknowledging key reasons that coral reefs are in trouble around the globe. Because the concept is complicated, your students will take turns acting out scientific information about coral bleaching for the other members of the class. By the end of the performances, the students should be able to explain to others the basics of coral bleaching.

2. Divide class into four equal-sized teams.

Give all members of a team the same "act" to read. In other words, one team's members will read "Act I," another team's members will read "Act II," and so on. Tell the students that their job is to figure out a way to perform the concepts described in their "act" for the other members of the class. One student can narrate the performance, but the rest of the group should play the part of the species and processes described. Students may use any of the materials you have gathered to make props. For example, if their act contains a reference to carbon dioxide, they can decide whether to have a person play the part of carbon dioxide, whether to write the words "carbon dioxide" on a piece of cardboard, or whether to use another prop to represent carbon dioxide in their performance. The point of the exercise is to demonstrate the concept clearly, while showing creativity in their efforts.

3. Take turns presenting concepts.

After the students have had 10 or 15 minutes to prepare their performances, have them take turns presenting them to the rest of the class. Because the concepts build on one other, you should start with "Act I," then have student groups present the remaining acts in order. Ask the audience to listen and watch carefully so they will be able to tell the whole story when the play is finished.

4. Discuss issues presented in the performances.

Ask the students to reflect on what they learned through the performances. Why are zooxanthellae so critical for coral reef survival? *(They manufacture a significant portion of the coral's food.)* Do the zooxanthellae benefit from coral? *(Yes, they gain shelter, nutrients, and a source of carbon dioxide.)* What word do biologists use to describe an ecological relationship between two species? *(Symbiosis, or a symbiotic relationship.)* Is zooxanthellae the coral's only source of food? *(No.)* Where else do corals get their food? *(They also capture some of their own food with their tentacles, but not enough to survive on.)*

Now ask your students to share what they know about ocean temperatures. Are ocean temperatures stable? *(All ocean waters experience seasonal temperature shifts, although tropical waters typically experience less temperature variation than other areas.)* What is significant about ocean temperatures and coral survival? *(Coral polyps are adapted to a particular temperature range. If temperatures rise above that range for an extended period of time, the zooxanthellae leave the corals. If the zooxanthellae don't return, the corals will bleach and soon die.)* What causes ocean temperatures to rise? *(Students may know that temperatures fluctuate naturally with the seasons, rise periodically with El Niño, and that they have been rising overall because of global warming. El Niño is a periodic disruption in normal ocean-weather patterns across the tropics.)*

You might explain to your group that El Niño alone is capable of causing some coral bleaching. But the last two El Niños have been particularly devastating for corals—causing massive die-offs across large areas. Many scientists believe that global warming is now exacerbating normal El Niño events and pushing corals past their limits. What's more, as mean temperatures rise because of global warming, even seasonal high temperatures may be enough to cause coral bleaching.

Ask your students to define global warming. *(Global warming refers to a rise in global temperatures that scientists have been recording over the last 100 years. Global warming has been accelerating in recent decades.)* What causes this rise in temperature? *(Most scientists believe that the root cause of global warming is fossil fuel combustion. As we burn fossil fuels to drive our cars and power our homes, we release carbon dioxide and other "greenhouse gases" into the atmosphere. As these gases accumulate, they form a kind of blanket that traps heat, much as the glass of a greenhouse lets in sunlight and keeps the heat in.)* You might ask your class if a group of students would like to perform a dramatization of global warming for the rest of the group.

Ask your students to reflect on what they know about threats to coral reefs. Can they think of any ways that coral bleaching is different from other problems related to coral reefs? *(The scale of the problem is greater and the severity of the effects is greater. What's more, bleaching can harm corals in protected areas as much as those in areas that aren't protected.)* Tell your students that scientists have recorded major bleaching throughout the world. The effects have been especially devastating in the Indian Ocean, where 50 to 95 percent of all corals have died in some areas.

5. Discuss ways to combat coral bleaching.

Ask your students if they can think of ways that people might be able to combat coral bleaching. Reassure them that there definitely are ways people can help protect coral reefs. Tell them that the last activity in this unit will give them a chance to think through these ideas in greater detail.

WRAPPING IT UP

Assessment

Have each student draw a picture that explains the symbiosis of coral polyps and zooxanthellae. Ask them to show what happens to the relationship when the ocean warms up. Have each student label his or her picture, including the concepts learned in the skits.

Unsatisfactory—The drawing includes only illustrations without demonstrating the interdependence between corals and zooxanthellae. Labels are either incomplete or do not address food sources, climate change, or coral bleaching.

Satisfactory—The drawing includes corals and zooxanthellae and incorporates labels that explain at least one of the concepts learned relating to climate change, food sources, or coral bleaching.

Excellent—The drawing includes corals and zooxanthellae with labels that reveal an understanding of the concepts learned relating to food sources, climate change, and coral bleaching.

Portfolio

Have students include the Assessment activity in their portfolios. They can also include their newspaper article (see "Writing Idea").

Writing Idea

Have students select a reef that has been affected by coral bleaching and write a short newspaper piece about it. They should try to find information on the history and location of the reef, what species live there, what it was like before bleaching, how it has been impacted by bleaching, and what scientists are predicting will happen in the future to that particular reef.

Extension

Have your students research recent news articles about coral bleaching. Where has it been recorded? What have scientists said about the situation?

Act I

Zooxanthellae (zo-zan-THELL-ee) are tiny algae that live inside the tissues of coral polyps. The coral polyps give shelter to the zooxanthellae, protecting them from getting eaten. They also provide the zooxanthellae with carbon dioxide, which the zooxanthellae use to manufacture their food. One thing that zooxanthellae provide for corals is their color. The tissue of corals is actually clear.

Act II

Zooxanthellae (zo-zan-THELL-ee) are tiny algae that live inside the tissues of coral polyps and give the coral its color. These algae use sunlight to help manufacture sugars through the process known as photosynthesis. Zooxanthellae use some of the sugars for their own growth. But an enzyme in the coral causes the algae to share some of their sugars with the coral. These sugars are a major food source for corals. By night the corals catch their own food—plankton—with their tentacles.

Act III

Coral polyps are inhabited by tiny, colorful algae called zooxanthellae (zo-zan-THELL-ee). These algae use sunlight to make food for themselves and the coral. But if the water around the coral reef becomes cloudy with silt after a storm or because of coastal development or pollution, the zooxanthellae can't get enough sunlight to make sugars. Then they begin to suffer and start to leave the coral. When the zooxanthellae leave, the coral loses its color (which is a process called bleaching) and may even die. In past years, coral bleaching has been caused by severe weather events, sedimentation, or pollution.

Act IV

Corals are inhabited by tiny, colorful algae called zooxanthellae (zo-zan-THELL-ee). These algae gain protection from the coral and in turn provide the coral with a key source of food. But this relationship is threatened by rising sea temperatures associated with global warming. Corals grow best in temperatures between 72° F and 78° F, and they can normally tolerate temperatures up to 84° F. When sea temperatures stay above about 89° F, the zooxanthellae are expelled by the coral or leave it on their own. Without the zooxanthellae, the corals weaken and lose their color, or bleach (whiten). If the temperatures return to normal, the zooxanthellae may return to the coral. Then it is possible that the coral will recover. But if the temperatures stay elevated for a long period of time, the zooxanthellae will not return to the coral and the coral will die.

5 Reef.Net

SUBJECTS
science, social studies, language arts, art

SKILLS
gathering (researching, collecting), applying (synthesizing, restructuring, composing), presenting (writing, reporting, explaining, clarifying), citizenship skills (working in a group)

FRAMEWORK LINKS
40, 62, 63, 67, 67.1, 68, 71, 72

VOCABULARY
coral bleaching, cyanide fishing, global warming, marine protected areas (MPAs), public service announcements (PSAs)

TIME
one or two nights for research, two or three in-class sessions (stretched over the span of a week)

MATERIALS
copies of the "Coral Reef Web Sites" handout (page 147), computers, materials for creating brochures, sample public service announcements and other forms of advertising

CONNECTIONS
Use "Coral's Web" (pages 128-139) and "Coral Bleaching: A Drama in Four Acts" (pages 140-143) to give students background on threats facing coral reefs. For suggestions on ways to engage audiences in biodiversity-related issues, see "The Biodiversity Campaign" in Biodiversity Basics. *And if your students are interested in further exploring how wildlife is portrayed by the media, try "Animal Magnetism" in* Wildlife for Sale.

AT A GLANCE
Create a Web site or educational materials that focus on coral reef management and conservation.

OBJECTIVES
Explain why reefs are important. Name some of the major threats to coral reefs and marine biodiversity. Explain some of the ways organizations are working to protect reefs. Describe what individuals can do to help protect reefs.

Most people agree that studying coral reefs can be great fun—they are beautiful, biologically rich realms with fascinating natural history. But studying coral reefs can also be depressing because human activities are threatening reefs in so many ways.

Fortunately, much can be done to preserve coral reefs and the wealth of life they harbor. Establishing protected areas around more reefs, controlling pollution and reef fishing, and doing everything we can to reduce global warming are among the many ways we can help sustain them. What's more, we can spread the word about reefs—why they are valuable and why they're at risk—to build wider support for their protection.

In this activity, your students will have a chance to channel their concern for reefs into positive action. They'll be developing a strategy for raising awareness about coral reefs and letting people know how to help keep them thriving for generations to come.

WWF-Canon/Jurgen Freund

Before You Begin

Make one copy of the "Coral Reef Web Sites" handout for each student. Also make copies of samples of public service announcements (see pages 92-93) and other forms of advertising. You might also want to have students bring to class printouts of articles from the Internet that highlight the problems facing coral reefs, which will encourage them to begin thinking about reef-related issues in advance of the activity.

What to Do

1. Describe assignment.

Tell the students to imagine that they have been asked to develop a Web site, television or radio public service announcement, or printed materials about some aspect of coral reef conservation. For example, they might create a Web site concerning the benefits of coral reefs and the many problems they face. Students might also create a brochure for divers describing coral reefs and the strategies divers can use to ensure that the reefs are not harmed. Or they might create a radio public service announcement, teaching listeners about coral reef bleaching.

2. Organize students into teams of four to six.

Distribute a copy of the "Coral Reef Web Sites" on page 147 to each student. Tell the students that the Web sites on the sheet have a wealth of information that will help round out their understanding of coral reefs. They may either spend some time browsing these sites before they choose a topic, or they may choose a topic first and then use the Internet to find additional information. Based on their chosen topic and area of interest, the teams should decide on what kind of educational materials they wish to create. Have the teams outline their project and list its goals and objectives (for example, to communicate the message of why reefs are important or to describe threats facing reefs, ways to protect reefs, ideas for ways people can help, and links to other resources). Then have the teams develop a tentative schedule for research, writing, and graphic design (if printed material) or recording (if radio or video). They should submit their outlines and schedules to you in advance for approval and suggestions.

3. Give students a predetermined length of time to prepare their materials.

Depending on group composition, you may want to have periodic progress reports to find out how the teams' projects are progressing. Remind the students that they should be sure to present comprehensive and accurate information. To help ensure accuracy, they should ask themselves a few questions, including: How reliable are our sources? Is the information backed up by scientific research, or is it based on opinion? Is there more than one good source for each fact? Can we find experts to verify our information?

4. Wrap up and report out.

When the students have created their materials, have them present their results to the rest of the class. Then see if you can find a way to put these materials to good use. Your teams might visit younger students' classrooms and share their knowledge about coral reefs. Or they might put their work on display at a local library, bookstore, or shopping mall.

WRAPPING IT UP

Assessment

See section #2 under "What to Do." When students submit their goals and objectives, encourage them to cover these three topics in their chosen way:

1. Why reefs are important
2. Threats facing reefs
3. Ideas for ways people can help

Unsatisfactory—Met only one of the goals.

Satisfactory—Met two or three of the goals satisfactorily.

Excellent—Met all three goals with a lively, convincing presentation.

Portfolio

Include copies of the students' Web site layouts or educational materials in their portfolios.

Writing Idea

Ask students to research coral-reef-related Web sites and write critiques of how well the sites communicate about the problems facing reefs. Students should also provide suggestions for improvement and expansion of the sites.

Extensions

■ Ask the teams to research marine protected areas (MPAs). They should pay particular attention to the different agencies and conservation organizations that are currently involved in managing and creating these areas. Have the groups make a list of the criteria that are used in establishing an MPA. The students should also research how MPAs affect people living near them and what impacts these areas can have on people who make a living from reef-related professions, such as fishing, scientific research, or tourism.

■ Many tropical fish species that live in coral reefs have been captured by using cyanide and sold to aquarium owners. The cyanide stuns the fish so they can be easily captured, but it also damages or kills corals and other reef species. There is now an international "seal of approval" for fish that have been caught in an environmentally friendly way. Have your students conduct a community investigation to find out where local pet store owners and aquarium suppliers obtain their fish. Are they certified by the Marine Aquarium Council—a practice that ensures the fish or corals were taken sustainably by fishers who manage and protect the reefs? Or were the fish bred in captivity, another practice that helps protect coral habitats?

NOAA Coral Kingdom

A Sampling of Web Sites Related to Coral Reefs

Australian Marine Conservation Society
www.amcs.org.au

Reef Environmental Education Foundation
www.reef.org

Coral Reef Alliance
www.coral.org

Reef Relief
www.reefrelief.org

Coral Reef Fishes
www.geocities.com/RainForest/2298

Waikiki Aquarium: Coral Research
waquarium.otted.hawaii.edu/coral/index.html

Florida Keys National Marine Sanctuary
www.fknms.nos.noaa.gov

International Marinelife Alliance
www.marine.org

NOAA's Coral Reef Online
www.coralreef.noaa.gov

Coral Reef Resources

Here are some more resources to help you design and enhance your Coral Reef Case Study. Keep in mind that this resource list includes materials we have found or used; however, there are many other resources available on coral reefs. For a more in-depth list of resources about marine biodiversity in general, see the Resources section on pages 360-369.

Organizations

The Center for Ecosystem Survival works in partnership with schools, universities, zoos, aquariums, botanical gardens, natural history museums, and science centers worldwide to conserve wildlife and nature and preserve biodiversity. Through its "Adopt a Reef" program, students can become Coral Reef Crusaders and help preserve and protect coral reefs. Adopt a Reef Program, 699 Mississippi, Ste. 106, San Francisco, CA 94107. (415) 648-3392. **www.savenature.org**

The Coral Reef Alliance (CORAL) is a member-supported, nonprofit organization dedicated to conserving coral reefs worldwide. 2014 Shattuck Ave., Berkeley, CA 94704-1117. (510) 848-0110. **www.coral.org**

Coral Reef Information System (CoRIS), a NOAA Web site, provides a comprehensive database for coral reef information. Visitors can access a glossary of coral-related terms, take part in coral reef discussion forums, or learn about NOAA's role in protecting these important natural resources. **www.coris.noaa.gov**

The International Coral Reef Initiative (ICRI) is an informal network of governments and international agencies working with scientific and conservation institutions to improve coral reef management practices, increase political support for coral reef protection, and share information on the health of coral reef ecosystems. **www.environnement.gouv.fr/icri**

Marine Aquarium Council (MAC) is an international nonprofit organization working to conserve coral reefs and other marine ecosystems by educating those engaged in the collection and care of ornamental marine life, from reef to aquarium. 923 Nu'uanu Ave., Honolulu, HI 96817. (808) 550-8217. **www.aquariumcouncil.org**

Reef Relief is a nonprofit membership organization dedicated to preserving and protecting living coral reef ecosystems through local, regional, and global efforts. Reef Relief Environmental Center, Historic Seaport, 201 William St., Key West, FL 33040. (305) 294-3100. **www.reefrelief.org**

Books

The Blue Planet, derived from the BBC/Discovery Channel series of the same name, reveals the diverse wildlife found in six distinct habitats that make up the watery parts of our world. The book includes chapters on tropical seas with coral ecosystems. With a forward by Sir David Attenborough and 400 full-color photographs, this is the first complete and comprehensive portrait of the entire ocean system. (DK Publishing, Inc., 2001).

Coral Seas by Roger Steene is a book of coral reef photographs, which range from extremely close-up, microscopic shots to large-scale cross-sectional shots. (Firefly Books, 1998). $50.00

A Field Guide to Coral Reefs: Caribbean and Florida is a Peterson Field Guide to coral reefs and their residents. (Houghton Mifflin, 1999). $21.00

Great Reefs of the World by Carl Roessler includes photographs and descriptions of coral reefs around the world. (Pisces Books, 1992). $19.95

The Incredible Coral Reef by Toni Albert is an award-winning book about coral, geared toward students in grades 3 through 8. (Trickle Creek Books, 2001). $10.95

The World Atlas of Coral Reefs by Mark D. Spalding, Corinna Ravilous, and Edmund P. Green gives an overview of the planet's coral reefs using color maps and photographs that illustrate reefs and reef organisms. Taken from outer space by NASA astronauts, 85 photographs of coral reefs provide an aerial perspective. (University of California Press, 2001). $55.00

Curriculum Resources, Posters, Videos, and Web Sites

The Bridge–Ocean Sciences Education Teacher Resource Center is a growing collection of marine education resources available on-line. It provides educators with a convenient source of accurate and useful information on global, national, and regional marine science topics, and gives researchers a contact point for educational outreach. Under "Ocean Science Topics," choose "Ecology" then "Coral Reefs." Sea Grant Marine Advisory Services, Virginia Institute of Marine Science, College of William and Mary, Gloucester Point, VA 23062. **www.vims.edu/bridge**

Coral 2000 (Volumes 1 and 2) by Reef Relief are two videos that document changes in Florida coral reefs over time. Available from the Reef Relief's online store. **www.reefrelief.org/main.html**

Coral Reef is a poster by artist Larry Duke illustrating reef ecosystems. Available from the Monterey Bay Aquarium, 886 Cannery Row, Monterey, CA 93940. (831) 648-4800. **www.mbayaq.org**. $15.00

Coral Reefs: An English Compilation of Activities for Middle School Students by Sharon Walker, R. Amanda Newton, and Alida Ortiz is a collection of coral reef activities and information, including activities about coral biology, reproduction and growth, distribution, reef life, and conservation. (1997). This teaching manual is also available in Spanish (EPA 160-B-97-900b). Free. (800) 490-9198.

NOAA's Coral Reef includes information on NOAA's coral reef initiative, coral health and monitoring program, and Year of the Reef (1997), plus coral reef photos, news releases, links, and the Great American Fish Count. **www.coralreef.noaa.gov**

Reef Environmental Education Foundation is an organization devoted to preserving coral reef life through education, research, and involvement. Its Web site includes a fish gallery with images and information on a variety of tropical fish. **www.reef.org**

SeaWorld/Busch Gardens Animal Information Database is a gateway to discovering marine life through animal fact pages, teacher resources, and information on educational programs. Click on "Animal Resources" and then "Corals and Coral Reefs" to learn about coral, from a hands-on experiment about how reefs develop to the eating habits of coral polyps. The site also offers a monthly newsletter for educators featuring a new marine topic in each issue. **www.seaworld.org**

The Shape of Life: Life on the Move—Cnidarians. The second hour of this eight-part television and CD series uses breakthroughs in scientific discovery to reveal the rise of the animal kingdom, focusing on cnidarians, corals, and their cousins. *The Shape of Life Activity Guide* features explorations, activities, and lessons to accompany the series for informal science educators, teachers, students, and families. A companion book to the series is also available. (Monterey Bay Aquarium Press, 2002.) $19.95. **www.pbs.org/kcet/shapeoflife/resources/index.html**

Case Study
Shrimp

Shrimp cocktail, popcorn shrimp, shrimp scampi. Say the word "shrimp," and most people think of a favorite meal or recipe. But there's more to these popular crustaceans than cocktail sauce and hors d'oeuvres. In this case study, your students will learn more about the economic, cultural, and ecological benefits of shrimp. They'll study bycatch, shrimp farms, turtle excluder devices, and other shrimp-related topics. After exploring these topics, your students will be challenged to imagine possible pathways to more sustainable shrimp fisheries.

World Wildlife Fund

WWF-Canon/Anthony B. Rath

"People who enjoy eating shrimp don't know that natural resources are being destroyed to bring it to them."

—Saul Montufar, President, Honduran Committee to Defend the Flora and Fauna of the Gulf of Fonseca

Background Information

Did you know that many shrimp mothers carefully brood their young in a pouch? That some shrimp can make sounds as loud as a popgun to warn other shrimp of danger? Or that cleaner shrimp carefully pick parasites and debris off of fish? These amazing creatures flourish in a variety of habitats—salt water and fresh water, polar seas, and thermal springs—where they provide a critical food source for fish, turtles, birds, and people. Over decades, fishers have hauled tons of shrimp from the sea. Today, shrimp have become a plentiful mainstay on the seafood menu. And that's good news for shrimp lovers.

But, unfortunately, there's bad news too. Shrimp fishing can take a big toll on the environment. Some shrimp trawlers drag their heavy nets across the ocean floor, destroying everything in their paths. In fact, some scientists liken the damage from trawling to that of clear-cutting a forest. And that's not all: For every pound of shrimp that makes it to your table, an average of four pounds of bycatch (accidentally caught sea turtles, fish, and other marine animals) are tossed back into the sea, dead or dying. In addition, aquaculture—raising shrimp on farms, which many people hope someday will ease pressure on wild populations—has environmental impacts too. Shrimp farming has resulted in the loss of mangrove forests and has increased coastal pollution.

All this makes shrimp a perfect example of how difficult it is to eat our seafood and catch it sustainably too.

Shrimp Natural History

The Incredible, Edible Shrimp

With their hard shells and segmented bodies, shrimp are *crustaceans.* Most commercial species of shrimp, some of which are called prawns, are 10-legged crustaceans, or *decapods*—a group that also includes their equally edible relatives, lobsters and crabs. Nearly all of the edible shrimp belong to the suborder Natania, or the "swimmers." For example, white shrimp belong to this group. They are found off the coasts of Florida, South America, and Australia, as well as in other warm parts of the world, including the Mediterranean Sea.

Lots of Variety

There are nearly 2,000 different kinds of shrimp found worldwide. They range in size from less than an inch long to the 13-inch-long black tiger prawns in the tropical waters off Australia. Cleaner shrimp, painted shrimp, mantis shrimp, spot prawns, and brine shrimp all have different ways of making a living. (See the "Shrimp Lifestyles" box on page 154 for more on different shrimp species.)

Because of their great diversity, it's hard to make generalizations about shrimp as a group. But there's one thing they do have in common: They're critical parts of the food web. For example, some shrimp are *scavengers.* They break down dead animals with their strong mandibles and are, in turn, eaten by larger animals. Others are *filter* feeders. They swim along on their backs, collecting and eating tiny organisms such as diatoms, bacteria, and flagellates. Then these shrimp become dinner for cod and other fish.

My Life as a Shrimp

When shrimp are ready to reproduce, the males and females spawn, releasing thousands of eggs and sperm into the open sea. A single female can release as many as 500,000 eggs at a time. Within 24 hours, the eggs that are fertilized hatch into larvae, which soon start feeding on plankton—tiny floating plants and animals.

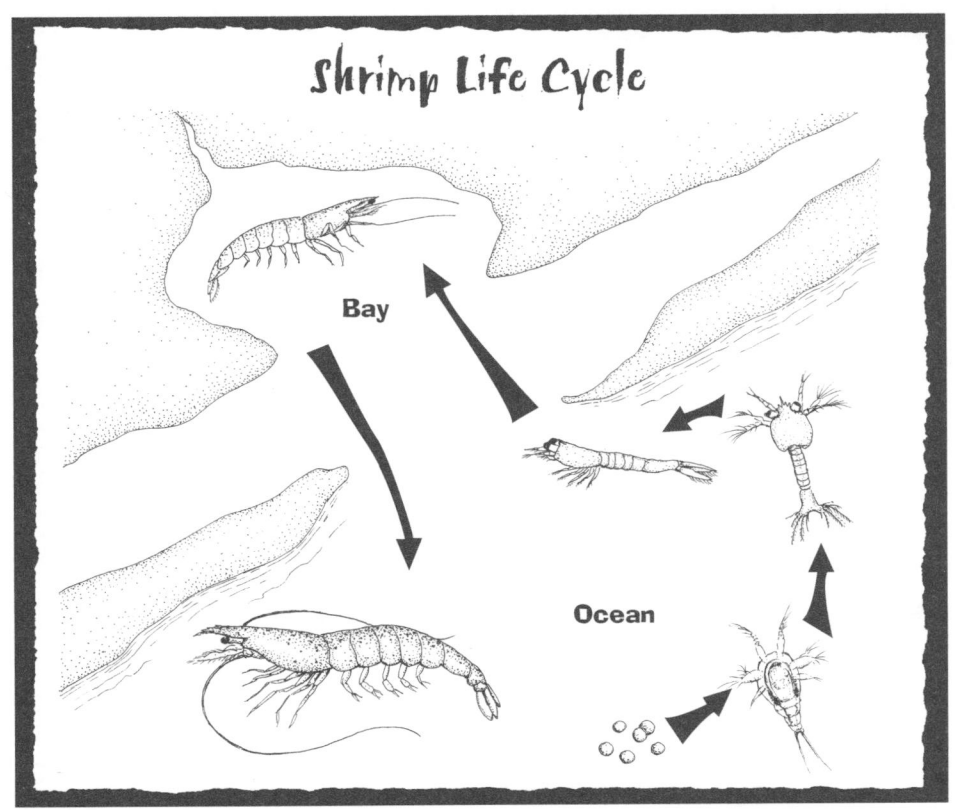

Shrimp Life Cycle

Bay

Ocean

After the larval period ends (about 12 days later), the young shrimp migrate from the open ocean into nutrient-rich bays and other coastal areas. There they find shelter and food and grow into larger juveniles. Weeks later, they return to the sea to mature and mate. Some species live for only 120 days while others can live as long as 6 years. A few species of shrimp give birth to live young rather than lay eggs; a few others can produce young from unfertilized eggs.

As adults, shrimp live in the open ocean in relatively shallow waters—usually less than 200 feet deep. They tend to stay near the bottom, where they hide in sand and mud during the day and come out to feed at night.

Shrimp Lifestyles

There's more to shrimp than the little creatures that end up on our dinner tables. Here is just a sample of the many varieties of crustaceans that we call "shrimp."

You don't want to mess with the **mantis shrimp!** With its strong claws, the mantis will tackle creatures bigger than itself. To feed, it burrows in the sand or mud, where it awaits prey such as worms or fish. Some mantis shrimp spear their victims with a jackknife-like claw that unfolds with lightning speed. Others hide in holes in rocks and club their prey rather than spear it. Mantis shrimp live in the shallow waters of tropical and subtropical seas.

Brine shrimp are one of the few creatures that can survive—and flourish—in super-salty inland waters such as the Great Salt Lake in Utah and other highly saline lakes around the world. Their thick-walled eggs are very hearty and can withstand summer drying and winter freezing. When the eggs are dropped in water, they hatch quickly and develop into adults that are sometimes called "sea monkeys." Many aquarium owners are familiar with this shrimp species, because the adults or their dried eggs are often sold in pet shops as a convenient fish food.

Opossums have a brood pouch, and so do **opossum shrimp**. The female shrimp's brood pouch is formed by special plates on the inner sides of her thoracic feet. The young stay sheltered there until they're old enough to fend for themselves. Most species of opossum shrimp live in the sea, but a few live in fresh water.

Cleaner shrimp do a big favor for certain fish. Some species of cleaner shrimp wait at special cleaning stations near anemones or coral in the Indo-Pacific and Red Sea. When a fish comes near, the cleaner shrimp will climb aboard and begin picking off parasites and cleaning wounds. Once the shrimp have cleaned the outside, the fish opens its mouth. The shrimp then move into its mouth and over its gills to remove internal parasites. The shrimp get a meal and the fish gets a good cleaning.

As their name suggests, **snapping shrimp** use their large front claw to make a loud popping sound. This sound not only scares away potential enemies, but it is also strong enough to stun passing prey. For these creatures of the tropical seas, catching a meal is a snap!

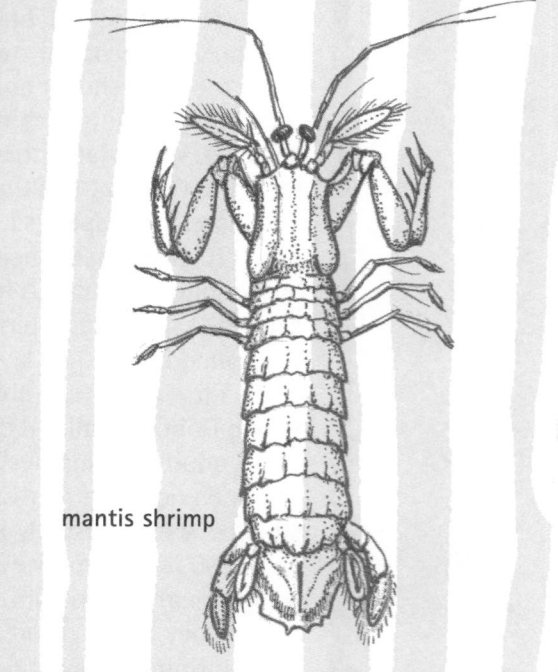

mantis shrimp

Shrimp Issues

People around the world love to eat shrimp. Almost one billion pounds a year are consumed in the United States alone. But catching and raising shrimp have serious consequences for ocean food webs and marine habitats. Here's more about the problems.

Trawling Troubles

One of the major problems associated with shrimp trawling is bycatch. Because shrimp nets have a very small mesh size to catch their tiny quarry, the nets catch and kill enormous amounts of unwanted species. For every pound of shrimp that is sold, approximately four pounds of bycatch—sea turtles, fish, and other marine animals—may be thrown back into the sea, dead or dying. That means that, for the 250 million pounds of shrimp caught by trawlers in the Gulf of Mexico and the Atlantic Ocean near our southeastern states, roughly a *billion* tons of other animals are destroyed. Experts think that about 35 million young red snappers—a fish that's popular as seafood—are killed each year as bycatch.

Bycatch isn't the only problem associated with trawling. Because shrimp rest near the ocean bottom during the day, that's where the trawl can get the most shrimp. Heavy shrimp gear (which often weighs more than a ton) hauled across the seafloor may pull up everything in its path—rocks, animals, and plants. In heavily fished areas, trawls may pass through several times per year. Bottom trawling reduces biological diversity not only by killing animals and plants, but also by destroying the places where these species live and altering entire ecosystems. It's estimated that bottom trawling

shrimp trawling

NOAA William B. Folsom, NMFS

for shrimp and other marine animals affects an area equivalent to nearly half the world's continental shelf each year, leaving behind damage that can take centuries to recover. Bottom trawling also stirs up

> *"If there are no mangrove forests, then the sea will have no meaning. It is like having a tree with no roots, for the mangroves are the roots of the sea."*
>
> **–Andaman Sea fisher, Trang Province, Southern Thailand**

sediments, releasing nitrogen and other nutrients into the water. This increase in nutrients can reduce the amount of free oxygen in the water, making it harder for marine animals to survive.

Wetland Woes

Seafloor destruction isn't the only habitat problem plaguing shrimp: The loss of coastal wetlands also harms vulnerable young shrimp. As more and more development takes place on the coasts, the mangrove forests and other wetlands where young shrimp find shelter and food are disappearing. It's estimated that half of the world's mangroves have been destroyed. Losing these and other fragile coastal wetlands harms more than the shrimp. The loss also affects many other species and greatly reduces the biological diversity of the marine ecosystems. Mangroves filter impurities out of water, prevent erosion, shelter coastal areas, and provide vital fish habitat.

A Dead Zone

In some areas, coastal pollution is so bad that it has created "dead zones" for many marine animals. In the summer of 2001, in the northern Gulf of Mexico, an area of sea almost the size of Massachusetts was stripped of oxygen. Shrimp, fish, and other swimmers suffered a loss of habitat, but many were able to flee the dead zone. However, oysters, worms, and other bottom-dwellers couldn't get away and suffocated.

Many scientists blame these dead zones in the Gulf of Mexico primarily on nitrogen-rich fertilizers that are transported by the Mississippi River. The fertilizers come from farms and urban runoff many miles upstream. Nutrients in the fertilizers encourage the growth of tiny marine plants at the water's surface. Microscopic animals called *zooplankton* feast on the plant explosion, and their numbers skyrocket too. The rapidly increasing zooplankton population creates an imbalance between predators and prey: There aren't enough fish and other predators to eat the zooplankton and keep the populations in check. Billions of dead zooplankton fall to the seafloor, where bacteria digest them. In this process, the bacteria use all the oxygen in the deep water.

This may be one of the reasons that the population of brown shrimp in the area is declining, but no one is entirely sure. Some conservationists believe that one solution would be to cut back on the excessive use of fertilizer on farms, suburban lawns, and golf courses, as well as the excessive runoff of animal waste from intensive livestock operations.

Shrimp Farming

Not all of the shrimp you eat were caught in the sea. The total world supply of shrimp has increased because of aquaculture. Currently, aquaculture accounts for over 25 percent of world shrimp production and nearly 50 percent of internationally traded shrimp. Those are impressive numbers for an industry that got its start in the 1970s.

Shrimp farming seems like a great way to reduce pressure on wild shrimp populations and to limit the damage from bycatch. But aquaculture creates its own set of environmental headaches:

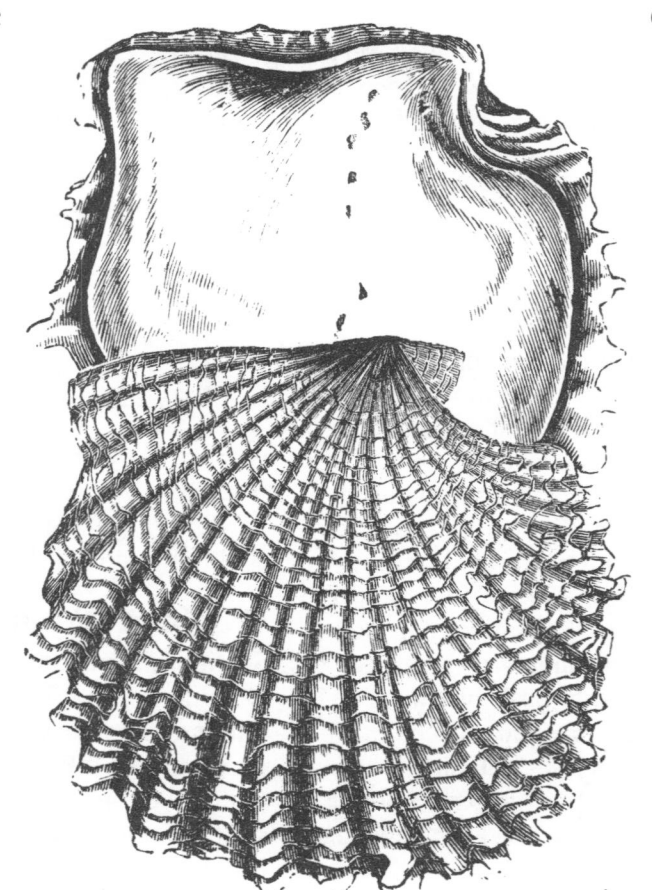

pearl oyster

BIOFACT

Typically, after three to ten years, shrimp farms must relocate as they use up the local supply of clean fresh water and salt water, and the buildup of sediments creates a habitat that is not conducive to hatching and raising additional shrimp.

- **Habitat Loss:** The location of shrimp ponds can be a big problem. In some places, the ponds are built on salt flats or other habitats where few other creatures live. But in many areas, shrimp farmers have destroyed mangrove forests and other coastal wetlands to make way for shrimp ponds. Ironically, these wetlands are the very nurseries where wild shrimp and many other animals start their lives.

- **Pollution:** Shrimp farming creates pollution. Shrimp food, antibiotics, and fertilizers used to raise shrimp can drain into fragile coastal areas, contaminating coastal mangroves and other wetlands. And, when consumed by humans, antibiotic residue found in shrimp may increase the risk of resistant diseases.

- **Impact on Wild Shrimp and Other Species:** Today, the majority of shrimp ponds are stocked with young shrimp that have been raised in hatcheries. But in some countries, such as Bangladesh, shrimp farmers use the young of wild shrimp because they are thought to be stronger and to survive better in ponds. The high prices farmers pay for young wild shrimp encourage collectors to catch them with fine-mesh nets.

  Some experts estimate that, for every young shrimp caught in the wild to be raised on a farm, 100 other marine creatures die in the collectors' nets. In addition, farmed shrimp are often fed fish meal made from ground-up wild-caught fish. As many as three to four pounds of wild-caught fish may be used to feed every pound of shrimp produced on a farm.

- **Disease:** Raising thousands of shrimp in a small area makes those shrimp more vulnerable to infection. In addition, they become even more prone to disease when they are stressed because of deteriorating water quality. And when diseased farms are drained or there's a flood, farmed shrimp can escape and infect wild shrimp and other species.

- **Problems for Local People:** Shrimp farming can have negative effects on local poor people who depend on fish as a source of protein. Expansive shrimp farms built on the coast restrict local people's access to the sea and other coastal resources. And, when mangrove and wetland areas are destroyed, people have fewer places to fish—and fewer fish to catch.

Shrimp Solutions

With the many pressures facing shrimp populations around the world, it's hard to find an easy fix for our shrimp problems. But people are searching for solutions. For example, conservationists, scientists, and fishers are working together to find shrimp-catching methods that don't damage the seafloor. They're finding, for example, that trawling at mid-depths, particularly at night, can do less harm to ocean-bottom communities. (Of course, it's a trade-off, because not as many shrimp are caught at mid-depth as on the bottom. But some fishers are willing to catch fewer shrimp in order to do less harm.) Scientists and fishers are working to educate policymakers about the destructive effects of bottom trawling and to encourage them to pass regulations that limit this practice.

Turtle Technology

A true breakthrough in trawling technology has come in the development of *turtle excluder devices,* or TEDs. In the 1980s, one of the most publicized controversies surrounding shrimp fishing occurred with the public outcry over the accidental killing of sea turtles. Turtles need to swim to the surface periodically to breathe air, but those that were caught in shrimp nets usually drowned before the nets could be hauled to the surface. During that decade, an estimated 200,000 turtles were killed worldwide each year. TEDs were developed to solve this problem. TEDs are escape hatches that can be sewn into the trawl nets to dramatically reduce the number of turtles killed.

In 1990, Congress passed a law requiring U.S. shrimp trawlers to use TEDs when fishing in the Atlantic Ocean and Gulf of Mexico. Although shrimpers were initially concerned about the increased costs related to TEDs, with use they realized that TEDs made fishing more efficient and reduced labor costs by keeping unwanted species out of the nets. And as for helping protect turtles, the devices seem to work. In fact, TEDs have been shown to reduce turtle bycatch by 97 percent with minimal loss to shrimp catch. In the 1990s, the United States banned the import of all shrimp taken in the waters of countries that didn't require shrimp trawlers to tow TEDs. This ban has helped turtles worldwide and has made the economics of shrimping more fair for U.S. fishers. The Kemp's ridley turtle population, for example, now appears to be recovering because of strict fishing regulations and the protection of nesting females and their eggs.

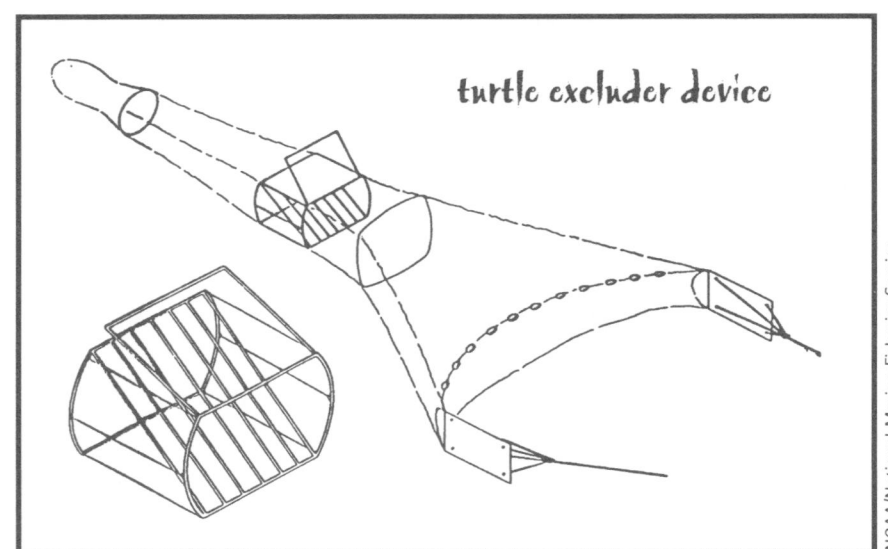

turtle excluder device

NOAA/National Marine Fisheries Service

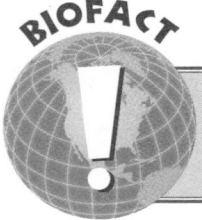

BIOFACT

The international value of trade for all shrimp products is approximately $7 billion annually.

The success of TEDs led to the development of other new technologies to deal with the overwhelming waste of bycatch. Bycatch reduction devices—or BRDs (pronounced "birds")—reduce the overall bycatch rate by guiding fish through escape panels in the nets.

Low-Harm Farms

And what about aquaculture? When farming is done sustainably, it can be a good alternative to catching wild shrimp. Farmers are working to make their operations cleaner and more efficient and to locate them in areas that won't destroy important coastal habitats.

In fact, some scientists have estimated that as much as 90 percent of the environmental impact of a shrimp farm results from where and how it is built. Building the farms above the high-tide mark usually reduces the harm to fragile coastal wetlands of all types. And building them correctly in the first place is cost effective and cuts down on repairs and the need to fix embankments. In addition, well-constructed low-harm shrimp farms can improve the feeding efficiency of the shrimp by 50 percent or more. This also reduces the amount of waste discharged, increases shrimp survival rates and size, and improves the overall profitability of the farm.

Live and Learn

Finally, educating consumers can be one of the biggest steps to conserving our seafood resources. Organizations such as the National Audubon Society and Monterey Bay Aquarium publish wallet-sized cards that help consumers with their seafood choices. (See **www.audubon.org/campaign/lo/seafood/cards.html** and **www.mbayaq.org/cr/seafoodwatch.asp.**) The types of seafood with abundant wild populations, low bycatch rates, and safe farming practices are at the top of the lists. With the cards in hand, seafood lovers can make better buying decisions. Recently, World Wildlife Fund and Unilever, the world's largest buyer of seafood, helped create the Marine Stewardship Council (MSC). The MSC certifies fisheries that have healthy populations and use sustainable fishing methods. Even though only a few fisheries have been certified so far, by looking for the MSC label and encouraging local restaurants and markets to carry sustainably caught fish, consumers can exercise their power of choice for marine biodiversity.

> *"When done sustainably, aquaculture has the potential to reduce pressure on fisheries while still producing seafood for consumption. If sustainable aquaculture can play a role in conserving marine life, then the practice can provide hope for the future of marine diversity."*
>
> **—Jason Clay, Vice President, World Wildlife Fund**

A lot more research needs to be done, and scientists, shrimp fishers, and lawmakers all have an important role to play in helping sustain shrimp populations. And those of us who simply like to eat shrimp can help ensure that they—and the myriad species they support—remain viable today and in the future. We can help by buying certified fish, supporting marine legislation based on sound scientific information, and educating ourselves and others about shrimp issues.

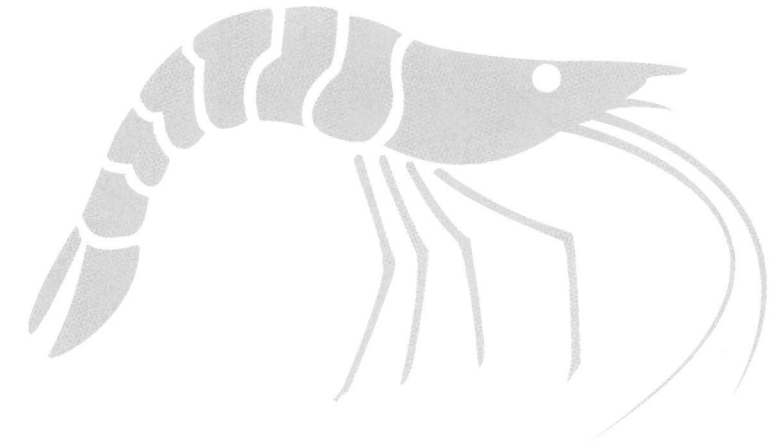

Sizing Up Shrimp

SUBJECTS
social studies, language arts, science

SKILLS
gathering (reading comprehension, identifying main ideas), interpreting (generalizing, relating, inferring, reasoning, elaborating), evaluating (assessing, critiquing, identifying bias)

FRAMEWORK LINKS
5, 30.1, 33.1, 34.1, 37, 40, 41, 42, 59, 60

VOCABULARY
benthic, **bycatch**, **crustacean**, decapod, filter feeders, **habitat**, scavengers, shrimp, values

TIME
one session

MATERIALS
copies of "Views of Shrimp" (pages 163-165), paper, pencils

CONNECTIONS
Use "The Spice of Life" in Biodiversity Basics to examine more general ways in which people value biodiversity. "Career Choices" (pages 176-184) can encourage students to consider economic aspects that may affect people's perspectives toward shrimp and fishing.

AT A GLANCE
Explore different perspectives on shrimp and identify the many ways that people value these sea creatures.

OBJECTIVES
Identify your own views and understanding of shrimp. Be able to explain the economic importance of shrimp. Explain how different ways of valuing shrimp can lead to conflict.

When someone says the word "shrimp" to you, what comes to mind? A small hard-shelled creature with lots of legs and long antennae? A tasty appetizer with cocktail sauce? Or a tiny person?

Many people know very little about the natural history of the sea creature in their shrimp dishes, the economies that depend upon shrimp, or the culture of the people who live near shrimp-filled waters. In this activity, your students will have a chance to explore different perspectives on shrimp and consider some of the values that underlie these perspectives. In the process, they'll identify—and broaden—some of their own views of these marine creatures.

THE FAR SIDE® BY GARY LARSON

"Listen, you want to come over to my place? I get great FM."

Make one copy of "Views of Shrimp" (pages 163-165) for each student.

1. Introduce the term "shrimp."

Have your students draw what they think of when they hear the word shrimp. After about five minutes, have them share their results. You may have a few jokesters in the class who drew small people. Rather than suppressing this response, ask your students why they think that the word "shrimp" has come to mean a small person? If the students drew a shrimp dish they've eaten, ask them if they can tell you what kind of animal a shrimp is. Do they know where the shrimp in the seafood dishes they like to eat come from? If they drew an animal in the wild, ask them where they learned about shrimp and what more they can tell you about them.

2. Hand out "Views of Shrimp."

Have the students read the views of shrimp, then discuss them as a class. How do these descriptions of shrimp relate to their own perceptions of shrimp? You might want to review the accounts one by one. What is the first one? *(a shrimp recipe from a cookbook)* Does this recipe tell you anything about what shrimp are or where they come from? *(no)* What is the second one about? *(the natural history of shrimp)* Did the students learn anything new about shrimp in this description? *(Answers will vary.)* What kinds of animals are shrimp? *(crustaceans)* Where do they live? *(oceans of the world)* What do they feed on? *(dead animals or tiny plants and animals in the water)* What benefits do they provide to people and other species? *(important part of food web, break down dead organic matter, and provide food for wild species and humans)*

Now turn to the third account. What is this poem about? *(a day spent catching and cleaning shrimp with a family)* In what ways are shrimp important to this young author and her family? *(Your students should pick up on the fact that shrimp aren't simply important as food. The poet conveys the value of shrimping as a family tradition, a ritual that brings people together over the course of an entire day.)*

James P. McVey, NOAA Sea Grant Program

How about the fourth account? What is this information? *(the annual income and expenses of a shrimper)* What is the shrimper's initial income? *($85,400)* How much of this is paid out in expenses? *($52,100)* What does this tell you about the livelihoods of people who shrimp? *(They have a lot of business expenses, which means their final income isn't necessarily very large.)*

Ask the students to characterize the values of the authors of the four accounts as best they can. Which of them seem to value shrimp as a food? *(recipe writer, poet, possibly shrimper)* As a wild creature? *(biologist)* As a source of income? *(shrimper)* As a source of recreation? *(poet, maybe shrimper)* Can the students see any ways in which the lives of shrimp are tied up with the cultures of the people who live near them? *(Poet's family traditions intertwined with shrimp fishing, cooking, and cleaning, and her poetry springs from these experiences, and so on. You may want to point out to students that human cultures are often influenced by resources around them, and, similarly, that sustaining the resource can help sustain these cultural traditions.)* What other ways of valuing shrimp do the accounts present? *(Answers will vary.)*

3. Discuss values.

Ask the students if they can think of any ways in which the values presented in the four accounts could come into conflict. If the students have trouble coming up with ideas, you might tell them that some people are concerned that the cheapest ways of catching shrimp are too destructive to the marine habitats where they live, and they kill too many other ocean species. Who might be concerned about this problem? *(Answers will vary, but presumably all of the authors could be* concerned.) Who might be concerned about adopting less destructive fishing techniques? *(Shrimper may think it would be too expensive.)* What if the result of this change in fishing techniques were higher shrimp prices—who might be concerned about that? *(People who buy shrimp, such as author of the recipe.)* Tell the students that they will explore in greater depth the problems and controversies surrounding shrimp fishing in the rest of this case study.

BIOFACT

Deep-sea shrimp have long antennae nearly four times the length of their bodies, which help them find food or mates by sensing chemicals of other animals.

WRAPPING IT UP

Assessment

Have students write a personal reflection on whether they value shrimp. They should explain their own values in terms of their personal experiences. Then ask them to use what they learned from the "Views of Shrimp" accounts to help discuss how their values may conflict with those held by other people.

Unsatisfactory—The reflection does not include personal values or does not relate personal values to the stories or potential conflicts.

Satisfactory—The reflection presents personal values related to shrimp and provides alternative values and potential conflicts that reflect the stories.

Excellent—Personal values are clarified using information learned in the activity. The reflection presents a balanced approach (critical thinking) to support the personal values while comparing them with differing, valid opinions and explaining how conflicts could arise among these views.

Portfolio

Include the students' personal reflections (see Assessments), along with their drawings, in the portfolios.

Writing Idea

Have your students create a mini field guide to shrimp. Each student can create one page for a specific species, including a picture of the shrimp and any relevant natural history information.

Extension

Have the students look for articles in newspapers, in magazines, and on the Web about problems with shrimp fishing. What are the views of the authors of the articles they've found? Of the people quoted in the articles? Of decision makers?

Shrimp Scampi

Shrimp are almost everyone's favorite. Your guests will want this recipe.

> 6 tablespoons unsalted butter, room temperature
> 1/4 cup olive oil
> 1 tablespoon minced shallots
> 2 tablespoons snipped fresh chives
> salt, garlic, and freshly ground black pepper, to taste
> 1/4 teaspoon paprika
> 2 pounds large shrimp, peeled and deveined

Preheat the broiler.

Combine the butter, olive oil, garlic, shallots, chives, salt and pepper, and paprika in a large bowl. Blend thoroughly.

Toss the shrimp in the mixture until thoroughly coated.

Cook the shrimp as close as possible to the heat source for 2 minutes on each side. Serve immediately.

4 portions

Excerpted from *The New Basic Cookbook.* © 1989 Julee Rosso & Sheila Lukins. Used by permission of Workman Publishing Co., Inc. New York.

Notes from a Shrimpy Ecologist

It used to be that whenever people called me a shrimp, it ticked me off. I thought, hey, if I'm a shrimp you're a bloated hippopotamus. So there.

Then I became a wildlife biologist, and I felt bad about the hippopotamus comment. It was a bit too insulting (to the hippos).

I also learned some very cool stuff about shrimp. Do you know that some shrimp make popping sounds as loud as a popgun? Others are called *opossum shrimp* because the babies are carried in a little pouch on their mother's underside. Cleaner shrimp linger at special underwater cleaning stations waiting for fish to arrive. When they do, the shrimp climb on board and eat parasites on the fish's body. Then they move inside the mouth for a little teeth cleaning. Hey, who needs dental floss?

Shrimp are *crustaceans.* More than 2,000 species cruise the world's oceans, ranging in size from less than an inch to more than a foot long! Some are scavengers, breaking down dead animals. This prevents the animals from piling up all over the ocean floor. Others are filter feeders, floating on their backs like people at a pool party, sipping tiny organisms from the water.

Fish, sea turtles, birds, and, let's face it, even many people like having these marvelous little crustaceans in their diets. All in all, shrimp are pretty important to the world. So tell me now, what's so bad about being called a shrimp?

Shrimping

Amelia Sides, age 18

Laughter on the water, at the dock, cast and pull,
music of water and voices.
Salt water in the mouth, taste the river mud.
Reach for the net, arm goes down, hold with your teeth, cast, spin,
and release.
Breathe.
Crash of water. Spray on the wind.

Hand over hand, cast and pull,
laughter at a caught fish, a squid. Stop to watch a heron.
Missed throw, the net twists.
Crash, pull it in, and throw again.
Laughter as a ten-year-old boy tries to throw a fifty-pound net.
Catch him before he goes in.

Sun goes down in the marsh. Light the lamps.
Crickets sing and moonlight reflects off the water.
Moths hum and bump at the lights, shrimp till the tide changes.

Orange fades to blue, night sky. Night on the marsh, the river.
Sit and watch the tide.
Birds cry, marsh smell of salt and water, marsh mud and wood smoke.
Pine bugs whir and scream in the dark. Lap of water on the dock.
Tired voices murmur, soft laughter.
A cool breeze whips wet and tired faces.
Cools the body and the mind.

Pack up the nets; blow out the lamps, head home.
Sit in the kitchen and clean shrimp.
Get kicked out of the kitchen and sit on the porch and clean shrimp.
Pick up by antenna, pinch off the head.

Old men drinking beer and telling stories.
Flash of cigarettes in the dark, sweet smoke.
Glow of charcoal, hamburgers on the grill.
Old women in the kitchen cracking jokes, laughter as they cook.
Crabs on the stove, coleslaw on the counter, peanuts on the boil.
Life, the river.

2002 Business Income and Expenses
Eddie Richardson, shrimper

Income from charter company and sale of shrimp	=	$85,400
EXPENSES		
Gear and vessel repair	=	$14,500
Fuel	=	$4,200
Crew salaries	=	$24,000
Insurance	=	$4,600
Licenses, dock fee	=	$2,500
Miscellaneous	=	$2,300
subtotal	=	$52,100
Total Earnings 2002	=	**$33,300**

SUBJECTS
science

SKILLS
gathering (reading comprehension, simulating, brainstorming), interpreting (relating, making models, identifying cause and effect), applying (experimenting, hypothesizing, proposing solutions, problem solving)

FRAMEWORK LINKS
12, 29, 30.1, 33.1, 35, 47.1, 50.1, 52.1, 55, 56, 65, 71, 71.2

VOCABULARY
aquaculture, bycatch, *bycatch reduction devices, efficiency, fertilizers, mangroves,* **sustainability, trawling,** *turtle excluder devices*

TIME
two sessions

MATERIALS
copies of the Muffy and Marcy comic strips (pages 173–175), "Shrimp Trawling" (page 171), and "Shrimp Farming" (page 172); materials for trawling simulations: 8 or more 7-inch embroidery hoops, 4 mesh bags or fabric with holes of different sizes (such as from bags for onions, bait, or hosiery), mixture of dry beans (1/4 pound each of lima, pinto, kidney, lentil, and black beans), 4 large (7-inch diameter) plastic containers

CONNECTIONS
To learn more about fishing practices and their impacts on the environment, conduct the hands-on simulations described in "Sharks in Decline" (pages 230–238). "The Many Sides of Cotton," in Biodiversity Basics, *offers tips on mediating conflict and addressing complex issues.*

 **AT A GLANCE**
Conduct a simulation to explore the problems of bycatch in shrimp trawling, then take a closer look at the pros and cons of shrimp farming.

 **OBJECTIVES**
Discuss some of the pros and cons of trawling and aquaculture in catching shrimp, including trade-offs in terms of efficiency in changing trawling net size, trawling at different depths, and so on. Describe the benefits of turtle excluder devices. Create designs for more effective trawling nets and more sustainable shrimp farms.

Have you ever dined at an all-you-can-eat seafood restaurant? Most likely the menu included shrimp in a seemingly endless supply. If so, you probably never thought about the controversy involved in collecting those shrimp. Worldwide, about three-quarters of the shrimp in restaurants and markets are taken from the sea with nets called *trawls*. The rest of them are raised on shrimp farms in a process called *aquaculture*. People disagree about which of these is the best (or worst!) way to obtain shrimp. Some people think it's better to catch wild shrimp in nets, while others think that it's more economical, efficient, and ecologically sound to farm them. Some people have decided not to eat shrimp until the Marine Stewardship Council certifies a shrimp fishery. And, of course, there is a range of opinions in between. (For more on the pros and cons of shrimping methods, see pages 155–157 in the introductory information.)

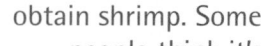

In this activity, your students will explore the complexities of shrimping. In the process, they'll learn about efficiency issues, pollution, and new technologies designed to improve shrimp fishing. They'll also find out that there are no easy solutions to the complex issues surrounding shrimp farming and fishing.

Before You Begin

Make one copy of the "Muffy and Marcy" comic strip, "Shrimp Trawling," and "Shrimp Farming" for each student or team of students. Before the first session, gather materials for trawling simulations (there should be enough for four teams of students, but you can adjust according to your needs). To make trawling nets, stretch a piece of mesh across each of the embroidery hoops. Try to have an equal number of hoops for each size of mesh. Mix the beans and divide them among the plastic containers.

What to Do

1. Discuss shrimp fishing methods.

Ask the students if any of them know how shrimp are caught. If they don't know, explain that there are two main ways that people obtain shrimp for market: trawling and shrimp farming (aquaculture). Worldwide, about three-quarters of the shrimp that people consume comes from trawling, and one-quarter comes from farming. Explain to the students that they're going to be taking a closer look at trawling in this session and a closer look at shrimp farming during another session.

2. Hand out the "Shrimp Trawling" sheet and "Muffy and Marcy Get Trawled" comic pages.

Discuss the description of shrimp trawling with your students. Then have them read the first of the three comic stories, "Muffy and Marcy Get Trawled." (You may want to explain that this comic strip features two characters from an animated film that was created to accompany World Wildlife Fund's *Biodiversity 911: Saving Life on Earth* traveling exhibit. These characters, Muffy and Marcy, are fictional "society shrimp.")

According to the comic strip, what are some of the benefits of shrimp trawling? *(catches lots of fish; relatively inexpensive)* What are some of the problems? *(can harm the seafloor, can catch many unwanted species called bycatch)* Be sure your students can define the word *bycatch*. What are some things that can be done to address the problems associated with shrimp trawling? *(trawling at different depths and using devices called "turtle excluder devices" on the fishing nets)*

Collect the comic pages for use in the next session.

BIOFACT

More than one-third of all shellfish and seafood that people consume start life in salt-marsh nurseries.

3. Introduce trawling simulation.*

Explain to the students that experts believe that trawl nets, which are used to catch shrimp and other fish, would be less destructive if they didn't drag along the seafloor, especially in sensitive areas such as beds of seagrass. But unfortunately the seafloor is where the concentrations of shrimp are greatest. Discuss the use of turtle excluder devices, or TEDs, which are pictured in the Muffy and Marcy comic. TEDs are a very helpful invention—turtles and many large fish can push open the bars on the device and escape. But trawl nets still accumulate a lot of bycatch—approximately four pounds for every pound of shrimp caught. Some fish that are smaller than shrimp will escape through the holes in the net, but that leaves a large number of medium-sized fish that get caught. When the net is hauled up, the shrimp are removed. Then the bycatch species are thrown overboard and generally don't survive. So there's a tradeoff when fishing at mid-depths. Tell the students that they'll be doing a simulation designed to give them a better sense of some of the dilemmas associated with trawling.

4. Organize class into four teams.

Give each team at least two embroidery hoops with different size mesh and a container filled with beans.

In this simulation, your students will play the part of shrimp fishers. Tell the students to assume that the different beans are different species of fish. Ask each team to choose one variety of bean to represent shrimp. Remember that shrimp are quite small relative to most other marine animals, including mammals, fish, and other crustaceans.

The team members should take turns selecting a net and dragging the net through the beans. After each turn, have that person count the results of his or her catch. How many shrimp did that person catch? How much of the catch was bycatch? Tell the students they should record their results. Encourage different team members to try out different size nets, which may mean trading nets with other teams until they've used all four net sizes.

5. Discuss results.

Have the students report back on their results. You may want to list the four kinds of nets on the board and write the results from each student who fished with that size net. And, if you want, you can have students average the results for each net size and create a bar graph comparing them.

What was the relationship between bycatch and the size of net used? *(presumably, the larger the mesh size, the less bycatch caught)* What was the relationship between the number of shrimp caught and the size of net used? *(the number of shrimp caught probably decreased with increased net size)* What sort of dilemma does this pose for fishers? *(They need to use a net with small enough mesh to catch the shrimp, which are very small, but that increases the number of larger, unwanted fish they'll catch.)*

 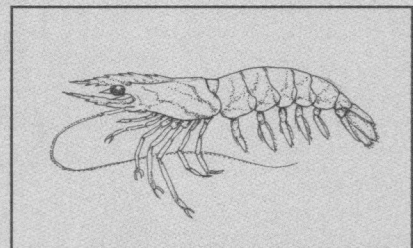 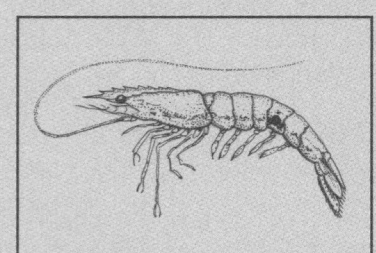

brown shrimp   white shrimp   pink shrimp

Chances are good that the shrimp you eat at a restaurant will be one of these species. Brown, white, and pink shrimp are abundant in U.S. waters, where fishers use trawl nets to catch these and other commercially valuable species.

6. Wrap up the assignment.

In the remaining minutes, divide students into groups of two or three and ask them to try to devise possible technological solutions to the problem of bycatch. Can they come up with a net size or sketch a device that reduces the amount of bycatch in trawl nets without significantly affecting shrimp catch? Then tell them they will be learning about a new device in the next lesson.

7. Begin the second session by discussing the wrap-up assignment.

Have the students share any ideas they came up with at the end of the last session. Did anyone think of a way to further reduce bycatch? Review the use of TEDs in trawl nets and point out that a newer development is the Bycatch Reduction Device, or BRD, which is similar to a TED. BRDs allow even more unwanted fish to escape the nets. Did anyone design something that resembled a TED or BRD? Anything more effective? (See page 158 for illustration of a TED.)

8. Hand out the "Shrimp Farming" sheet and the comic pages.

Read and discuss the techniques used in shrimp farming. Then have the students read the "Muffy and Marcy Down on the Farm" comic strip. Based on what they've learned, what are the advantages of shrimp farming, or aquaculture? *(Aquaculture is an efficient way to collect a lot of shrimp that doesn't result in lots of bycatch or seafloor destruction.)* What are the disadvantages? *(results in destruction of coastal habitats, especially mangroves and other wetlands; causes water pollution and spreads disease)*

9. Introduce the term "sustainability."

Ask the students if any of them can define the term "sustainability." *(It's the ability to use the Earth's resources—including land, forests, water, and wildlife—so that the needs of the present are met without diminishing the ability of people, other species, or future generations to survive. In other words, sustainable use of shrimp resources would mean that they are harvested or farmed at a rate at which they can be replenished without producing harmful pollutants or waste and at which there will be shrimp resources left for future generations.)* Ask the students if any of them think shrimp trawling is more sustainable than shrimp farming. Do any of them think the opposite? Which of these fishing practices has the potential to be more sustainable? *(Answers will vary to all of these questions. You might point out that even the experts disagree about which fishing approach is, or could be, more sustainable.)*

10. Read "Muffy and Marcy Get an Education."

Have the students read through the final Muffy and Marcy comic strip. Ask them to look for a definition of sustainable fishing in the comic and discuss the meaning as it applies to fishing in general. Then have them review the "Shrimp Farming" comic and make suggestions for ways that shrimp farming might be made more sustainable. Tell the students not to worry if they don't know the precise details about how shrimp are farmed. They should simply use their imaginations as they did in designing better trawl nets to think up a more sustainable design. In the remaining time, have the students share their design ideas, and conclude with the ideas discussed in "Low-Harm Farms" (page 159).

---

*The trawling simulation is adapted from *Taking Stock of Our Fisheries: The Ground Fish Crisis in the Gulf of Maine.* Manomet Center for Conservation Sciences, 1994.

WRAPPING IT UP

Assessment

Tell the students, "You be the judge." Have them prepare a ruling on whether trawling or farming, or a combination, should be allowed. The ruling should explain clearly why the judge has made this decision.

Unsatisfactory—The position is not clear or is not supported by facts as learned in the activity.

Satisfactory—The position of the judge is clear, and the judge has provided some solid, logical reasons to support the decision.

Excellent—The judge supports his or her position using information that shows careful thought regarding alternatives and includes pros and cons of the decision.

Portfolio

Students should include a drawing or short write-up on their proposed solutions for shrimp fishing (based on the first session's assignment).

Writing Idea

Because shrimp are such a popular food, many students (and their parents) may be interested in learning about some of the controversies surrounding the ways in which shrimp are caught. To help inform the school community about the "shrimp debates," have students write a short "pro/con" opinion piece on shrimp fishing, which can be published in the school newspaper. Students should present the topic in such a way that showcases the multiple angles and complexity of it, including economic, health, and environmental considerations.

Extensions

- Have the students create their own comic strips to illustrate either the "Shrimp Trawling" or "Shrimp Farming" handout. They can create the comic strips from the point of view of the fisher, the shrimp, or any other creature.

  - Have the students do "Experiment Two: Dissolved Oxygen" from "Much Ado about O_2" (On the Web at www.worldwildlife.org/windows/marine.) in the Salmon Case Study to get a better understanding of how shrimp farming can lead to a decrease in water quality.

  - Encourage students to take a historical look at fisheries. Have fish populations ever crashed? Where and why? How is the sustainability of a fishery affected by human population growth? Consumption? Market forces?

Ralph F. Kresge (NOAA)

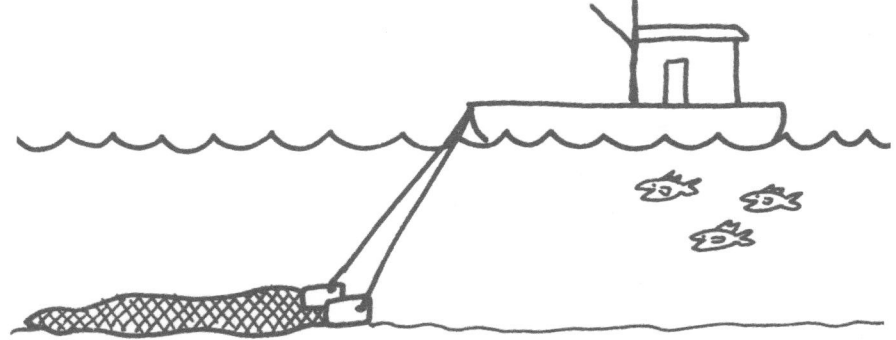

1. Fishers set out to sea on shrimp boats.	**4.** The trawl nets are hauled to the surface and emptied onto the deck of the boat.
2. Fishers drop large trawl nets into the ocean.	**5.** Fishers sort through the catch, throwing back the unwanted animals, known as bycatch. Many of the animals are killed when they are caught in nets.
3. The trawl nets are dragged just above or along the ocean bottom. Because many shrimp species slightly burrow into the ocean floor, boats tend to drag the nets through the top layer of sediment, often destroying habitats on the seafloor.	**6.** The shrimp are packed on ice or in large freezers for the trip back to port.

1.

Farmers build a shrimp farm, including ponds for raising shrimp. Sometimes they clear mangrove trees to make space for the farm or build it in a wetland area. This process can harm local wildlife and disrupt pollution filtering and other benefits of healthy wetlands.

2.

Most farmers buy young shrimp from hatcheries. In some cases, rural poor collect young shrimp from the wild, using very finely meshed nets, to sell to shrimp farmers. In the process of collecting the shrimp, other marine creatures are often accidentally caught and killed.

3.

Young shrimp are released into the ponds. Sometimes, wild-caught shrimp can introduce diseases into the ponds.

4.

Shrimp are fed a diet of commercial shrimp feed and are given medicines as needed to prevent diseases. Fertilizer is added to the ponds to stimulate the growth of plankton, which the shrimp also eat.

5.

Waste from the shrimp ponds (including excess food, medicines, fertilizer, fecal matter, and exoskeletons from shrimp), is sometimes released into the environment, often into coastal and marine ecosystems. (In "closed" farming operations, shrimp farmers have a better way of handling the waste. It is recycled to prevent polluting the surrounding environment and transmitting diseases to wild shrimp. But many farmers don't use this method because it is more expensive and not as well known.)

6.

Young shrimp are confined in a fenced area within the ponds where they can grow and adapt to their environment. The shrimp farmers build the ponds either right on the coast or in inland areas that are connected to the ocean. If there is a flood, young shrimp can be washed from the ponds and into the ocean. Eventually they are released from the fenced areas into the open ponds.

7.

When the shrimp reach an appropriate size for the market, the ponds are drained and the shrimp are harvested. Sometimes shrimp escape while the ponds are draining. These escapees can introduce diseases into the wild.

8.

When the ponds are drained, the waste material may be dumped into the environment. Or, if the waste is to be recycled, the water is put into other ponds or canals. After the waste settles to the bottom, the water is reused, and the solid waste can be used as organic fertilizer for salt-tolerant crops.

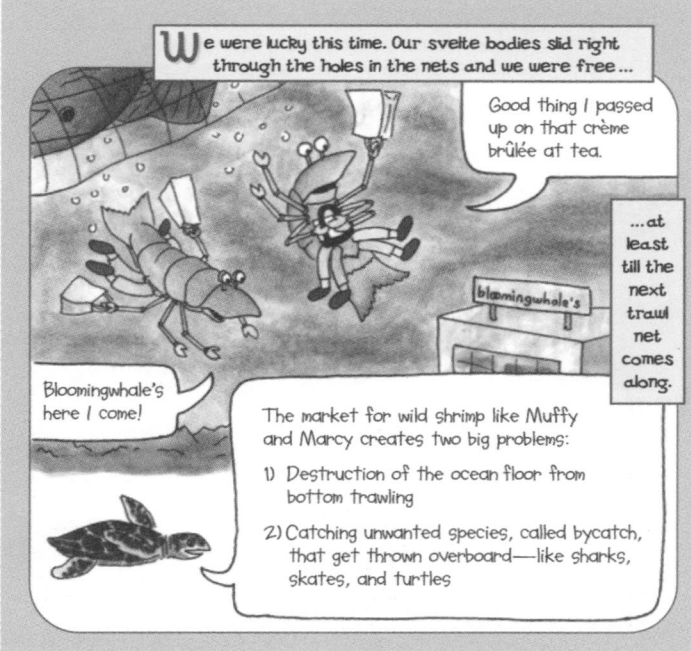

Muffy and Marcy Down on the Farm

Come on, pardner, let's get some chow.

Yee-haw!

SHRIMP FARM

After seeing one too many performances of "Oklahoma" on Scrodway, we decided to move to a shrimp farm.

Shrimp farms produce a lot of the shrimp that people eat. But they can also cause some big problems.

Why is there a "No Swimming" sign here?

Because the water's polluted from shrimp farming. Some of the waste and chemicals from the shrimp ponds and the shrimp feed wash into nearby waters and create big problems for sea life.

NO SWIMMING

BAG O' SHRIMP FEED

Yuck! This stuff tastes worse than sea urchin poop!

And in many places, shrimp farmers have cut down coastal mangroves to make room for more shrimp ponds.

Where have all the egrets gone?

NO SWIMMING

SHRIMP FARM

This isn't the wild adventure I was hoping for.

After a week, we'd had it. We packed our bags and headed back to Ocean Boulevard ...

Do you think we can get back in time for a sea cucumber facial?

And boy, am I craving a latte!

Raising shrimp on farms is called aquaculture. When it's done sustainably, it can be a good alternative to catching wild shrimp. But in many places, shrimp farms destroy coastal mangrove forests. And they cause overfishing and pollution because growing shrimp eat lots of fish meal and produce lots of waste.

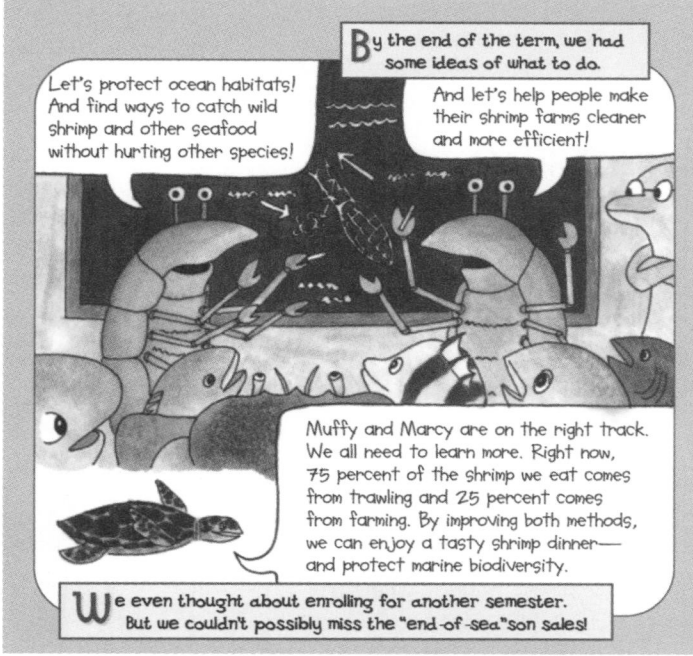

3 Career Choices

SUBJECTS
social studies, language arts

SKILLS
gathering (researching, listening), interpreting (inferring, drawing conclusions, defining problems, reasoning), applying (predicting, proposing solutions, problem solving, developing and implementing investigations), citizenship skills (working in a group, evaluating a position, evaluating the need for citizen action)

FRAMEWORK LINKS
13, 29, 30.1, 33.1, 34.1, 35, 37, 40, 46, 47.1, 50.1, 52.1, 59, 60, 62, 63, 67, 68, 72

VOCABULARY
bacteria, **bycatch, dead zone, dissolved oxygen,** fertilizers, **fish farms, plankton, trawl**

TIME
one session, optional second and third sessions

MATERIALS
copies of "Job Fair Key" (page 181), "Job Fair Roles" (page 182), and "Shrimp Struggles" (pages 183-184); paper, tape, markers, and other art supplies (optional)

CONNECTIONS
"Decisions! Decisions!" (pages 284–289) makes a good follow-up to this activity, as it asks students to consider various perspectives on the controversial topic of alien species. To focus on career development, have students try "Calculate Your Wildlife Career Profile" in Wildlife for Sale and "Career Moves" in Biodiversity Basics.

AT A GLANCE
Play the role of someone whose job relates to shrimp or shrimp issues, and think through your rights and responsibilities. Analyze shrimp-related jobs in your community and elsewhere.

OBJECTIVES
Describe the problems associated with fish farms, bycatch, and dead zones. Name some of the ways that people in their work lives can have an effect on these shrimp-related issues. Present and defend a position on one of these issues from the point of view of someone with a career that interests you.

Although marine issues such as bycatch and overfishing may sometimes seem far removed from our daily lives, that's hardly the case. Even if we never see an ocean, much less a wild fish or fishing boat, our lives are connected to marine habitats and species through the food we eat, the air we breathe, the products we use, the waste we produce, and much more. Each of us as consumers, and simply as inhabitants of planet Earth, makes decisions every day that affect the health and viability of marine resources. Even a surprising number of jobs connect to shrimp and other marine species in some way.

In this activity, your students will take on the role of people whose careers relate to shrimp and shrimp issues. Then they'll explore the rights and responsibilities of those workers when it comes to certain problems affecting shrimp. By doing this, the students will grapple with some difficult ethical and professional dilemmas and think about the role that they themselves might play in using and sustaining natural resources. Afterward, they'll have a chance to meet people in their own community with these and other shrimp-related careers and find out more about what they do.

Before You Begin

For each student, make one copy of "Job Fair Key," "Job Fair Roles," and "Shrimp Struggles." Also collect art supplies for the students to use to make "props" for their roles (optional).

If you plan to do the optional part of this activity, arrange for some speakers to come to the classroom for the second session. Depending on where you live, there may be local fishers, government officials, marine biologists, seafood restaurant owners, and fisheries managers whose work relates directly to shrimp and who could engage your group in a lively and informative discussion. (If your community doesn't have such people, you will probably have to skip this option.) Try to select people who have a background in the topic and who would enjoy talking to middle school students. You may want to suggest to the visitors that they cover some of the following questions during their presentations.

■ What is your job? What kind of schooling/training is required for someone with your position?

■ In what ways does your position relate directly or indirectly to shrimp?

■ Did you learn about shrimp as part of your training? If so, what kinds of things did you learn?

■ Do you have ongoing access to information on shrimp issues?

■ Are you aware of any issues affecting the availability of shrimp?

■ Are you concerned about the health and viability of shrimp populations? Why or why not? If so, are you taking any actions to help protect shrimp?

■ Do you feel a sense of responsibility for the health of marine ecosystems where shrimp live? Why or why not? If so, are you taking any actions to help protect marine ecosystems?

Note: See "Resources" pages 360-369 for Web sites on marine-related careers.

What to Do

1. Talk about shrimp and related terms.

This activity works best when students already have some basic familiarity with shrimp and shrimp issues. If you have not done any of the other shrimp activities in the module, begin by introducing shrimp, describing what they look like, and explaining their life cycle (see page 153). Also, you may want to review any relevant vocabulary terms the students do not know, such as dead zone, bycatch, trawl, and so on.

2. Hand out "Job Fair Key" and "Job Fair Roles" to each student.

Tell the students that this simplified key is designed to lead them to a job that they might be interested in. More specifically, the key leads them to a role that they will be playing as they continue with this activity. Tell the students not to worry if the job they end up with is not something they would really like to do. The point of the exercise is to think about how a person with that position might think or act when it comes to certain shrimp-related problems. More than one student will likely end up with each job.

Note: Most students can only assume their role to a limited extent and the career key is, of necessity, general and simplistic. This activity is intended to get them thinking about how people in different occupations have an effect on shrimp-related issues.

3. Organize class for mini-discussions.

Once everyone has had a chance to select a job/role, hand out copies of "Shrimp Struggles." Tell the students that they will be meeting in small groups to discuss one of the problems from the perspective of their chosen role. Explain that the letters at the end of each description correspond to the jobs on the "Job Fair Roles" handout. People with those jobs will meet to discuss that particular problem. In other words, people with jobs A, H, and J will get together to discuss "Dealing with a Dead Zone." Ideally, each of the mini-discussion groups will have one person from each job. If necessary, students can double up in a group, or you can form two discussion groups that are dealing with the same problem. But if you have *no* students for a particular job, see if you can convince someone to switch from another job that has too many students.

krill

4. Have students create props for their roles (optional).

You might want to have students create a single prop that helps represent the role they are playing to other students. For example, the chef could make a paper chef hat or could draw, color, and cut out a food item to hold. The journalist could carry a notepad and pen.

5. Explain assignment.

Now tell the students to individually read through their "Shrimp Struggle" and think about how someone with their assigned role might respond to the problem described. Do they perceive it as a problem? If so, have they contributed to this problem? How? Have they been hurt by this problem? In what ways? What actions might they want to take? What actions might they not want to take?

Then organize the discussion groups. Assign each group to a particular part of the classroom where they can sit together and talk about their "Shrimp Struggle."

Now the students should take turns introducing themselves to the other people in their discussion group and explaining their positions. They should follow their introductions with a group discussion of the problem. Do they agree about who is responsible for the problem? Do they agree about what should be done? What is the best solution they can come up with, or is it impossible to reach a solution? What other information did they need?

6. Discuss scenarios.

Give each small group a chance to present briefly its "Shrimp Struggle" to the rest of the class. Have them explain the major points of discussion and whether they came up with a solution. Were your students surprised by how many careers could have a tie to something as specific as shrimp? In what cases were solutions driven by a sense of responsibility? In what cases were people reluctant to take action? Why? Do they think their responses to these issues were realistic?

BIOFACT

Female mantis shrimp are protective mothers. They carry 50,000 eggs on their front legs for several weeks.

WWF-Canon/Jürgen FREUND

sustainable crab and shrimp farming in the Philippines

7. Discuss upcoming visit of community members (optional).

Tell your students that, for the next session, there will be some visitors from the community whose jobs relate directly or indirectly to shrimp. Ask your students to think about relevant questions they might want to ask the visitors.

8. Before the discussion begins, introduce the visitors.

Seat the visitors at the front of the room. Then moderate the discussion so that it is informative and friendly, even when speakers hold differing views. Ask the speakers the questions that you have given to them beforehand as well as any others that seem appropriate. Allow your students to ask questions that occur to them during the discussion.

9. After the visitors leave, discuss the experience.

How much did each panel member seem to know or think about shrimp? Were they all concerned about shrimp issues? Marine issues? Why or why not? Were any of them taking any direct actions to protect shrimp or other marine species?

Ask the students if they think any of the people on the panel or other people they know might taken different actions if they knew more about shrimp. If so, you might encourage your students to create public education materials (such as a library bulletin-board display) to help consumers and professionals make better-informed decisions about things that relate to shrimp.

BIOFACT

Noise pollution from human activity—such as shipping, oil exploration, and military maneuvers—may be harming marine species. Loud noises may disrupt feeding, mating, and migration, drive species from their habitat, and cause deafness.

WRAPPING IT UP

Assessment

List occupations related to shrimp and shrimping on the board, such as *shrimp farmer, shrimp trawler, marine biologist, fisheries manager, seafood seller, seafood restaurant worker*, and *seaside resort owner*. Ask each student to choose one occupation and write three short summary paragraphs describing that person's views on problems associated with fish farms, bycatch, and dead zones.

Unsatisfactory—The student shows understanding of problems associated with none or only one of the three topics (fish farms, bycatch, and dead zones).

Satisfactory—The student shows an adequate understanding of all three problems.

Excellent—The student has a better-than-average grasp of the three problems and has written creative and interesting paragraphs.

Portfolio

Students should include their Assessment assignments in their portfolios.

Writing Idea

Ask students to continue role-playing in their chosen occupation. In their career roles, have them write an essay about what their jobs would be like without shrimp. Once students have completed their essays, ask them to get back into their original groups and compare how the absence of shrimp would impact their livelihoods.

Extension

Have students conduct research to find out about what professional associations and organizations exist that deal with shrimping, shrimp-related issues, and people who work in this field. Students should look for Web sites and publications related to these groups. Students can create a proposed "career plan" by learning more about people who are involved in jobs that are interesting to them. Using careers in shrimp-related industries as an example, have students research the background of people who are working in this field (such as what kind of degrees they have, what other positions they have held, and what kind of professional organizations they belong to) and develop a career plan based on what they find.

JOB FAIR KEY

Follow the tree below to find a job that might interest you.

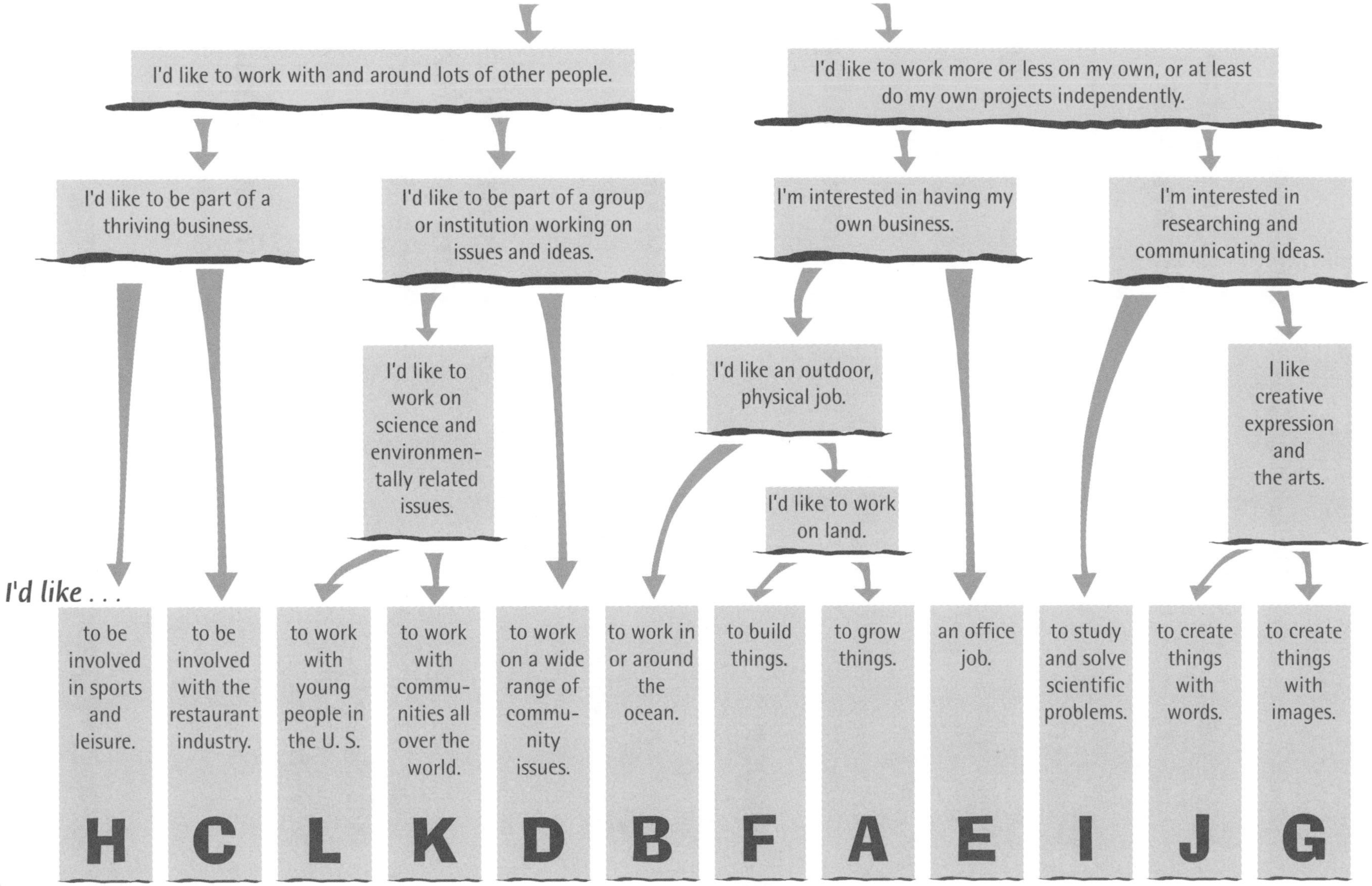

I'd like . . .

I'd like to work with and around lots of other people.

- I'd like to be part of a group or institution working on issues and ideas.
  - I'd like to work on science and environmentally related issues.
    - to work with communities all over the world. — **K**
    - to work with young people in the U. S. — **L**
  - to work on a wide range of community issues. — **D**
- I'd like to be part of a thriving business.
  - to be involved with the restaurant industry. — **C**
  - to be involved in sports and leisure. — **H**

I'd like to work more or less on my own, or at least do my own projects independently.

- I'm interested in researching and communicating ideas.
  - I like creative expression and the arts.
    - to create things with images. — **G**
    - to create things with words. — **J**
  - to study and solve scientific problems. — **I**
- I'm interested in having my own business.
  - an office job. — **E**
  - I'd like an outdoor, physical job.
    - I'd like to work on land.
      - to grow things. — **A**
      - to build things. — **F**
    - to work in or around the ocean. — **B**

A. Farmer

You grow crops on land bordering the Mississippi River.

B. Fisher

You own a medium-sized boat and catch shrimp and other fish in the Gulf of Mexico.

C. Chef

You're the head chef at a large seafood restaurant in Atlanta.

D. Mayor

You're the mayor of a small seaside town that hopes to develop its tourist industry.

E. Seafood importer

You own a business in Boston importing seafood for restaurants and grocery stores. You buy lots of farm-raised shrimp from on-line vendors.

F. Construction worker

You operate large machinery at new construction sites.

G. Documentary filmmaker

You are currently making a documentary film describing life in mangrove forests.

H. Resort owner

You run a large seaside resort in Texas that's popular for swimming, boating, and recreational fishing.

I. Marine biologist

You study the ecology of marine animals and have a special interest in sea turtles and sharks.

J. Journalist

You write news stories for magazines and newspapers.

K. Director of a conservation organization

You design and run projects that protect wildlife and wild places around the world.

L. Science teacher

You teach life science to sixth through ninth graders.

1. Dealing with a Dead Zone

In recent years, scientists have detected large areas of oxygen-depleted water in many parts of the world's oceans. Because most living things need oxygen and thus cannot live there, scientists call these areas "dead zones." In the summer of 2001, in the northern Gulf of Mexico, an area of sea almost as large as Massachusetts was found to be stripped of oxygen and devoid of life. Most experts believe that dead zones in the Gulf of Mexico are caused by fertilizers that wash off of farmland and yards into the Mississippi River. The fertilizers then spill from the Mississippi into the northern Gulf of Mexico. All those fertilizers cause a sudden surge of aquatic plant growth—and a rapid growth of the creatures that eat the plants. When these creatures die, they are broken down by bacteria, which consume huge amounts of oxygen. The resulting dead zone is inhospitable to life. As the oxygen decreases, some fish can swim away, but coral, shrimp, and many other marine species suffocate and die.

What will you do about this dead zone? (A, H, J)

2. Bycatch Blues

Thanks to the development of Turtle Excluder Devices, far fewer sea turtles are now caught in shrimpers' trawl nets in the Gulf of Mexico. But the amount of bycatch in trawl nets is still high—approximately four pounds for every pound of shrimp harvested. Scientists suggest that shrimpers use bycatch reduction devices (BRDs), change their net size, or change fishing methods altogether. But many shrimpers argue that these changes are impossible. They believe they can't make the changes without increasing their costs, and then they may not be able to compete with shrimpers from other countries that have fewer restrictions.

What will you do about bycatch? (B,C, I)

3. Down on the Shrimp Farm

In many parts of the world, people have cut down native mangrove trees to create room for shrimp farms. Mangrove trees grow along the shores of bays, estuaries, and other saltwater areas, and the water that flows among their roots provides sheltered, nutrient-rich habitat for young shrimp and other creatures. But some shrimp farmers reap short-term gains from cutting down mangroves to build artificial ponds. When they do this, all the native species that rely on mangroves lose their habitat. In addition to the habitat loss, waste from the shrimp farms sometimes creates pollution that further disrupts the coastal environment and local fisheries. And eventually the highly acidic soils where the mangroves had grown prove to be unfit for long-term aquaculture operations. While shrimp farming creates jobs for some people, farming can also hurt local people because they no longer have a place to fish for wild shrimp and can't afford the farm-grown shrimp.

What will you do about mangrove clearing? (E, G, K)

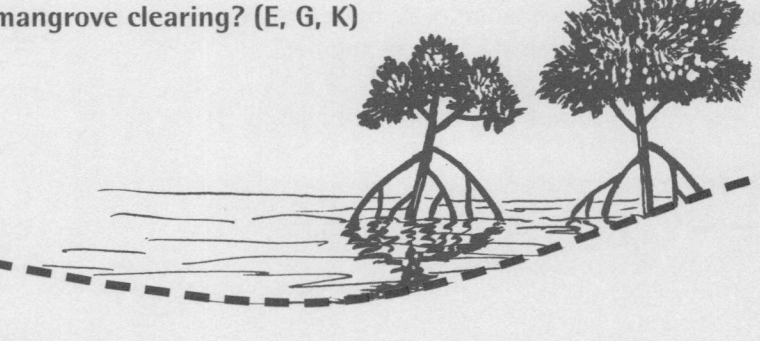

4. Wetland Woes

Oceanfront development such as new hotels and restaurants can often help draw tourists to an area and strengthen the local economy. But constructing new buildings in these communities can lead to the degradation and even destruction of coastal wetlands—nutrient-rich habitats that young shrimp and many other marine species depend on for survival.

What will you do about coastal development? (D, F, L)

"The diversity of life in the oceans provides a natural 'hope chest' for this and future generations. . . . Living marine resources provide essential economic, environmental, aesthetic, and cultural benefits to humanity."

—**Marine Biodiversity Values,**
National Marine Fisheries Service (NMFS)

4 | Imagining Oceans

SUBJECTS
language arts, science, social studies

SKILLS
applying (proposing solutions), presenting (writing, articulating, explaining, clarifying), citizenship (evaluating the need for citizen action, evaluating results of action)

FRAMEWORK LINKS
40, 68, 69, 71.1

VOCABULARY
crustacean, **climate change,** mangrove, marine protected areas (MPAs), sea bed

TIME
two sessions

MATERIALS
copies of "Story Starters" (page 190), paper, pens or pencils

CONNECTIONS
To help students better envision a positive and sustainable future, try "Thinking About Tomorrow" and "Future Worlds," both in Biodiversity Basics. For other creative writing exercises, use "Going Under" (pages 94-98), and "Drawing Conclusions" (pages 270-279), as well as "The Nature of Poetry" in Biodiversity Basics.

 **AT A GLANCE**

Finish a "story starter" to share your visions for the future of the world's oceans.

 **OBJECTIVES**

Use vivid descriptions to create a portrait of the oceans of the future. Demonstrate an awareness of current marine issues and the kinds of actions and innovations that could help address these issues in coming years. Develop creative ideas and images to articulate a personal vision.

When teaching your students about environmental topics, one of the most important things to avoid portraying is a feeling of hopelessness. That's why this lesson is an important conclusion to the problems your students have learned about. After all, as the topic of shrimp demonstrates, most environmental issues aren't *all* bad news. In many cases, there's a long history of innovations and actions that people have taken to solve problems once they've been detected.

In this activity, your students will have a chance to think about all the possible technologies, innovations, and actions that may play out in the coming decades, and use them to envision oceans of the future. What will oceans of the future look like? Which species will still live in them? Will we be eating seafood from the wild, from farms, or from other sources entirely? How will the way we use and enjoy oceans change? If your students want, this can be a madcap science-fiction writing adventure. Whatever the approach, this is a helpful exercise to get them thinking past the problems of today to the promise of a future that they'll be helping to shape.

Before You Begin

Make one copy of "Story Starters" for each student.

What to Do

1. Discuss people's treatment of the oceans in the past.

Ask the students to think about whether people are taking better or worse care of oceans than they did many years ago. To get a discussion started read aloud the information below.

Americans used to dump all kinds of trash and other waste into oceans. During the 1800s, for example, the beaches of Cape Cod, Massachusetts, were covered with everything from dead horses and wood planks to household garbage. Even as recently as the 1970s, barges full of trash were hauled from New York City and dumped at sea. Regulations about ocean dumping didn't exist in those days, and marine protected areas were rare. Many people thought the ocean was so big that no harm could come to it.

People began passing laws about dumping trash in the ocean, but then other problems surfaced. We learned that dolphins were being killed in large numbers by tuna fishers and sea turtles were being killed by shrimpers. Eventually, laws were passed to help protect both dolphins and sea turtles. And now there are even bigger problems to worry about. Fisheries worldwide are in decline, and vital habitats, from mangrove forests to fragile sea beds, are being destroyed in an effort to raise or catch more fish. Whales and polar bears are showing signs of having poisons in their body fat—poisons that come from pollution produced all around the world. And coral reefs are dying because ocean temperatures are rising due to global climate change, which is also caused—at least in part—by worldwide pollution.

Tell your students that they should take several minutes to consider the marine problems they've learned about while studying shrimp and other ocean topics, as well as the ways that people are responding to them—changes in policies, new inventions, public education, and so on. Based on this information, what do they think the oceans will look like in another 200 years?

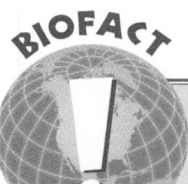

BIOFACT

Cleaner shrimp select a specific part of the reef as their "cleaning station," and fish "customers" can tell cleaner shrimp apart from other species by the cleaners' vivid coloration.

2. Hand out "Story Starters."

Tell the students to use one of the story starters, or another sentence of their own choosing, to write a two- to three-page story about the world's oceans circa 2200. Encourage them to develop their ideas from real issues they've learned about. But also tell them to be as imaginative as they can. Explain that the grading will be based on the vividness of their descriptions, originality of their ideas, and attentiveness to real species and places. Give the students one class period plus homework time to complete their stories, and give them the option of illustrating the stories.

3. Share stories.

After the students have completed their stories, have them form pairs and swap stories with their partners. Then have them compare their vision of the future with their partner's. Are they similar? Very different? Are they equally optimistic or pessimistic? Why?

4. Display stories.

Have students display their stories around the room so others can read them.

5. Create action lists.

As a wrap-up activity, have the students walk around the room and read the stories. Ask them to generate a list of the five most important things they can think to do in their lives to help create a more hopeful future for oceans. Then have the entire class make a composite list from the individual ones, avoiding duplication. (You might also want to check the back of the "Joy to the Fishes and the Deep Blue Sea" marine biodiversity poster for a list of ways to help the oceans.) Post the composite list in the classroom. By sharing as a class and then posting the list, the group will develop a sense of ownership.

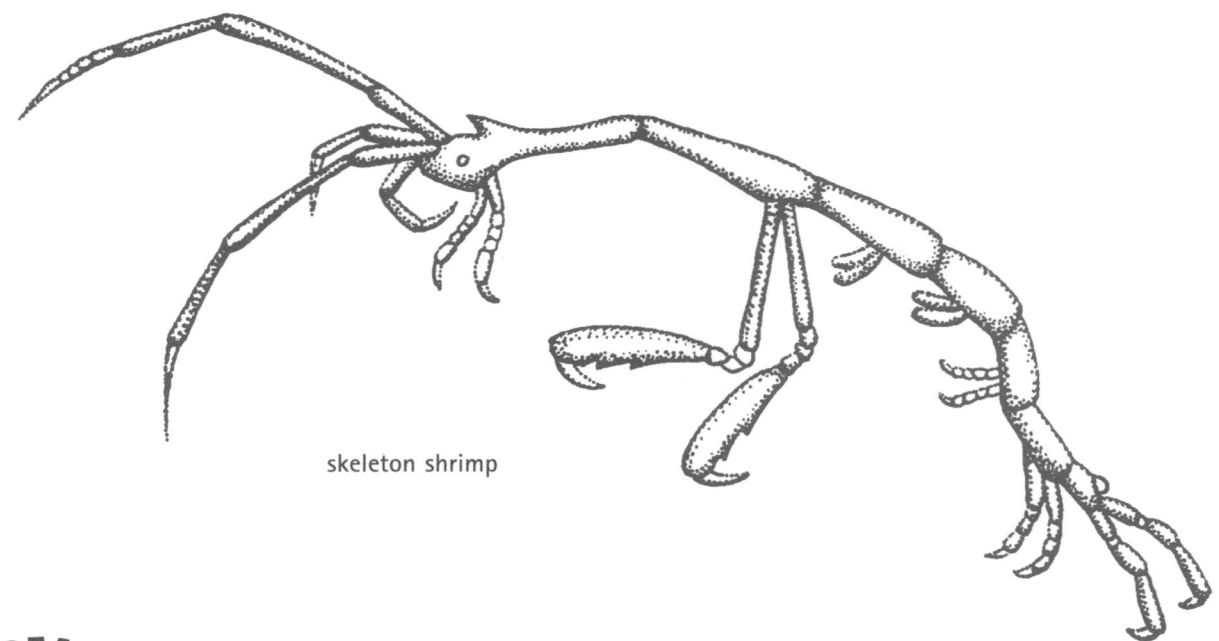

skeleton shrimp

BIOFACT

The skeleton shrimp is very hard to spot in its habitat of eelgrass. These tiny shrimp, only 1.6 inches long, have clear, stick-like bodies that blend into their surroundings.

WRAPPING IT UP

Assessment

Use the assignment as presented in the activity for grading. Create a rubric that includes:

- Evocative descriptions
- Originality
- Real species and places
- Language arts elements being covered in class (if desired)

Unsatisfactory—Description is incomplete or is complete but not provocative. The story fails to display any sense of creativity. Student fails to include any real species or places in the writing.

Satisfactory—Description is complete and provocative. The personality of the student is evident in the writing. Named species and places exist.

Excellent—Draws the reader into the story using descriptive language. Unique. Shows critical thinking. Uses real data to create the scenes provided in the writing.

Portfolio

Have students include their stories and action lists in their portfolios.

Writing Idea

Ask students to think about what kind of world they would like to see in 50 years and what kinds of changes would need to be made today to reach that desired future. Have students write short letters that specifically focus on marine-related issues, and send the letters to local newspaper editors. The letters should discuss their concerns about marine diversity and the actions that they and others may want to take now to create the best future 50 years from now. If possible, have some of the letters published in the school newspaper.

Extensions

- Have students research predictions that were made about the twenty-first century. After looking at predictions from the mid- and late-1800s, as well as the early 1900s, how well do students think scientists and others were able to visualize the future? What kinds of marine-related events were occurring at that time? Was the focus on conservation and protection, or was it on exploration? Why and how has that changed today?

- Have your students participate in "River of Words," an international environmental poetry and art contest that nurtures respect for and understanding of the natural world. For more information see **www.riverofwords.org** or contact International Rivers Network, 1847 Berkeley Way, Berkeley, CA 94703. (510) 848-1155.

STORY STARTERS

1.

We returned to Earth in the year 2200 after 50 years at the lunar substation. Our capsule landed in the ocean, and that's the first place we noticed how much our old world had changed. . . .

2.

Mosquito-powered cars, edible computer monitors, and school-based pizzerias—sure, those were all nice inventions of last century. But what really made me proud to be an Earthling as we entered the twenty-third century was something going on in the oceans. . . .

3.

The year was 2200. The place, Planet Earth. Jerrison called me on my cell-ear and said, "Meet me at the beach. Pronto." . . .

4.

Touring the world's oceans in a plastic submarine isn't normally recommended, especially when you're planning to spend a lot of time in shark habitat. But Lydia said this was the speediest way to go, and we had a lot of ground . . . that is, water . . . to cover before we wrote our "State of the Seas Report 2200." . . .

5.

My band, "Crustacean Frustration," was scheduled to perform in one week on Miami Beach for Earth Day 2200, and I still hadn't written the lyrics for our ocean songs. Why did I have such a bad case of writer's block? . . .

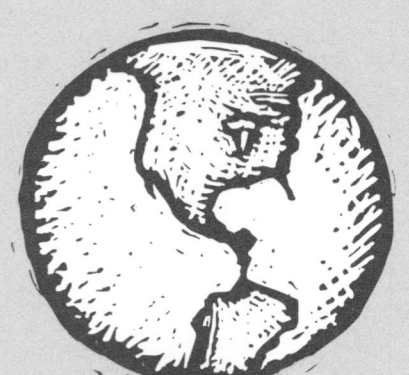

Shrimp Resources

Here are some additional resources to help you design and enhance your Shrimp Case Study. Keep in mind that this resource list includes materials we have found or used; however, there are many other resources available on shrimp. For a list of general marine biodiversity resources, see the Resources section on pages 360-369.

Organizations

Global Aquaculture Alliance (GAA) is an international, nonprofit trade association that advocates environmentally friendly aquaculture practices. The GAA created the Responsible Aquaculture Program, which outlines production standards for fisheries. 5661 Telegraph Rd., Suite 3A, St. Louis, MO 63129. (314) 293-5500. **www.gaalliance.org**

Industrial Shrimp Action Network aids nongovernmental organizations in addressing environmental threats related to shrimp farming. The network supports organizations from Europe, Asia, North America, Latin America, and Africa. 14420 Duryea Ln., Tacoma, WA 98444. (253) 539-5272. **www.shrimpaction.com**

National Marine Fisheries Service (NMFS) is the federal agency within the National Oceanic and Atmospheric Administration (NOAA) in the Department of Commerce responsible for the stewardship of the nation's living marine resources and their habitats. (301) 713-2370. **www.nmfs.noaa.gov**

National Sea Grant College Program has an Exotic Species Resource Center with educational materials. Sea Grant, which is managed by NOAA in the Department of Commerce, is a partnership among government, academia, industry, scientists, and private citizens that helps Americans understand and sustainably use our precious Great Lakes and ocean waters for long-term economic growth. **www.nsgo.seagrant.org**

Sea Grant News Media Center (Connecticut), Shrimp Aquaculture. See "National Sea Grant College Program" listing above for more about Sea Grant. **www.seagrantnews.org/news/ctshrimp**

Shrimp Farming and the Environment Consortium is a joint initiative of the World Bank, the Network of Aquaculture Centres in Asia-Pacific, World Wildlife Fund, and the Food and Agriculture Organization of the United Nations. The consortium investigates effective ways to manage shrimp aquaculture in coastal areas and releases regular online reports on current projects and issues worldwide. **www.enaca.org/shrimp**

Shrimp Resources (Cont'd.)

Curriculum Resources, Books, and Web Sites

Audubon Society's Seafood Lover's Guide ranks seafood species on a scale ranging from abundant to severely depleted. Shrimp falls into the red category of severely depleted.
www.audubon.org/campaign/lo/seafood/cards.html

The Bridge—Ocean Sciences Education Teacher Resource Center is a growing collection of the best marine education resources available on the Web. It provides educators with a convenient source of accurate and useful information on global, national, and regional marine science topics and gives researchers a contact point for educational outreach. The Bridge is supported by the National Oceanographic Partnership Program and is sponsored by the National Marine Educators Association and the national network of NOAA/Sea Grant educators. Sea Grant Marine Advisory Services, Virginia Institute of Marine Science, College of William and Mary, Gloucester Point, VA 23062. **www.vims.edu/bridge**

Faces of Fishing: People, Food, and the Sea at the Beginning of the Twenty-First Century (Adult), by Brad Matsen, is an engaging photo essay with compelling imagery. Highlights the people who are changing the face of fishing for future generations. (Monterey Bay Aquarium Press, 1998). $19.95

Greenpeace Oceans Campaign: Shrimp Aquaculture includes facts about shrimp farming and current world news relating to aquaculture practices. **www.greenpeace.org/~oceans**

Marinecareers.net, a Web site of NOAA's Sea Grant Program, introduces young adults to careers in marine-related occupations such as oceanography, marine biology, and ocean engineering. The site includes profiles of various marine professionals and a resource page with links to information on other aspects of working in marine-related fields. **www.marinecareers.net**

Monterey Bay Aquarium's Seafood Watch offers information on how consumers can make smart choices when buying seafood. The Web site includes a shrimp fact sheet on conservation issues and links to more shrimp resources. **www.mbayaq.org/cr/seafoodwatch.asp**

Shrimp News publishes an annual report on world shrimp farming. Published in January 2002, "World Shrimp Farming 2001" features two special reports: one on a new shrimp-farming technology that promises to revolutionize the industry and one on the development of shrimp aquaculture in the United States. **www.shrimpnews.com**

Shrimp Sentinel Online: Shrimp Links is a governmental and non-governmental organizational dialogue organized by Earth Summit Watch on the sustainability of shrimp trawling and farming. The dialogue represents citizens' initiatives to monitor and encourage actions by governments to implement the promises made at the Earth Summit in Rio and thereafter to move toward sustainable development. **www.earthsummitwatch.org/shrimp**

"We have overfished the seas systematically everywhere we have gone. We must act now, not 20 years from now . . . if we are to prevent further degradation of the marine environment."

–Elliot Norse, President, Marine Conservation Biology Institute

Case Study
Sharks

Read a recent article about sharks and you'll probably come away convinced that your next ocean swim will end with a deadly attack. But what those articles *don't* always say is that sharks rarely attack people. What's more, the frightening reports obscure a much bigger issue: Shark populations are actually declining rapidly worldwide. In this case study, your students will learn more about the diversity of shark species, the causes and costs of their decline, and the way attitudes and values of people around the world influence shark conservation.

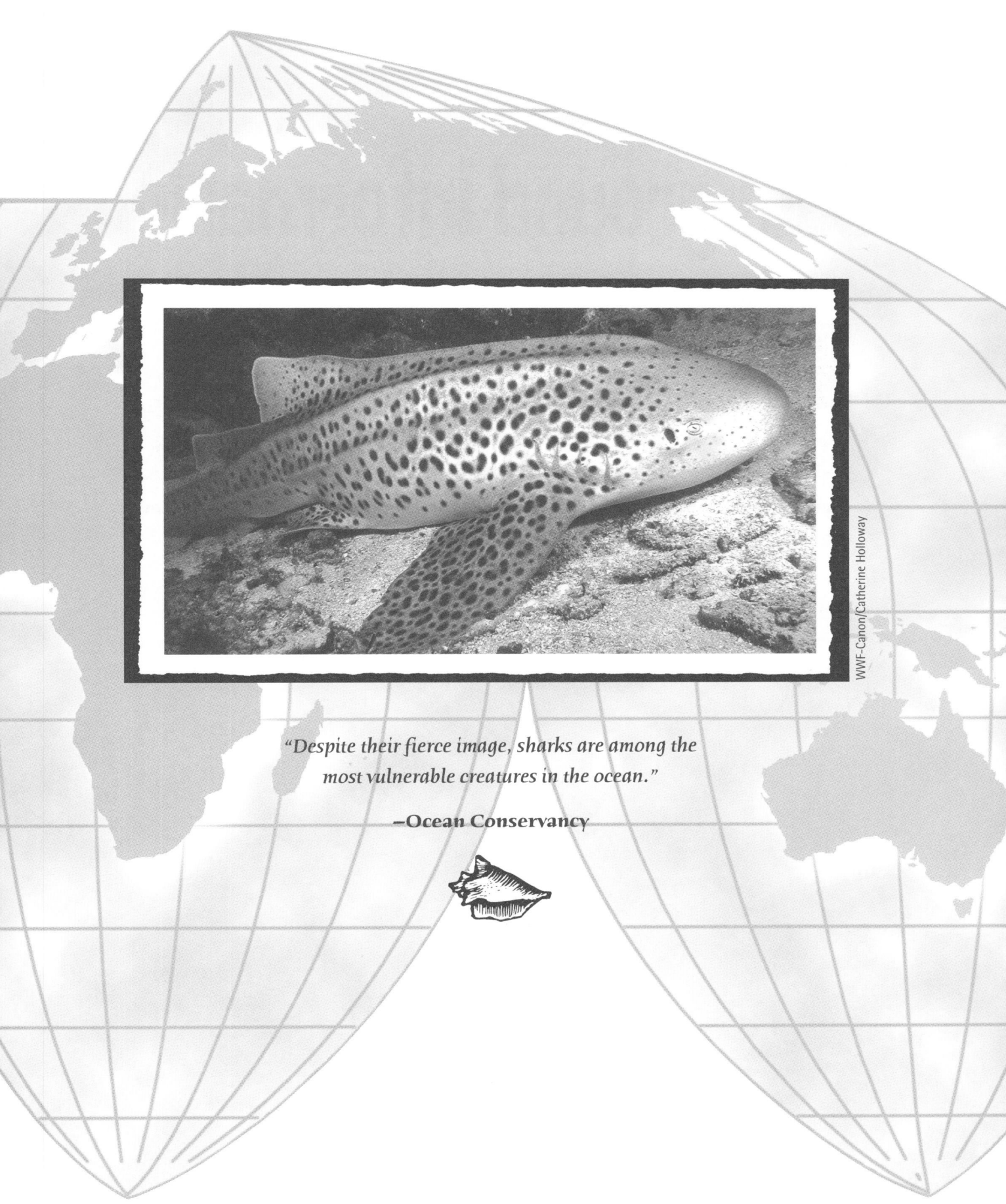

WWF-Canon/Catherine Holloway

"Despite their fierce image, sharks are among the most vulnerable creatures in the ocean."

—Ocean Conservancy

Background Information

It's a hot summer day at the beach. Ready to cool off, you wade into the ocean and dip under the waves. You swim five, ten, twenty feet. And then it happens. For a split second, you think about sharks. What if a giant killer is lurking in the dark waters beneath you? What if you see a triangular fin heading your way? Your heart thumps and your spine tingles. Even though you know you shouldn't be so afraid, there's no fighting your imagination now. Spooked to the core, you turn in for shore. So much for an ocean swim. You hope the shower feels safe.

If you're like a lot of people, you may find sharks terrifying. But you might be surprised to learn that sharks don't really deserve their horror-movie reputation. They aren't prowling the ocean waters hunting for the next person to attack. In fact, these days it's the other way around: Humans are prowling the ocean waters hunting for sharks!

BIOFACT

Who's really more dangerous? For every person killed by sharks, an estimated 10 million sharks are killed by people.

Fishing for Sharks

Catching Sharks Catches On

The annual capture of sharks has been rising steadily in recent years. That may come as a surprise to those of us who have never seen a shark product, much less a shark. But three activities—recreational (sport) fishing, commercial fishing, and accidental catch—are contributing to the decline of shark populations.

Recreational Fishing: In growing numbers, people on both the Atlantic and Pacific coasts are paying for the thrill of trying to catch a shark. The sharks' tremendous fighting spirit gives people all the excitement they've been seeking in this extreme sport (see "The Magnificent Mako"). White sharks, commonly referred to as "great whites," histori-cally have been considered a prize catch for sport fishers. But the National Marine Fisheries Service has banned targeted fishing of great whites because they have declined so dramatically. They are naturally rare, so they're easily overfished in the few areas where they are known to live.

In terms of numbers, recreational fishing isn't the biggest threat to shark populations. But it does present certain problems. One is that offshore fishers aim for big trophy sharks, such as makos. Makos grow faster than some other sharks and mature early, so large individuals have high reproductive potential. Capturing them means a great loss in terms of their potential offspring. In addition to offshore fishing, a lot of recreational fishing takes place near the shore, which is often a location for shark nurseries. That means that fishers may be taking pregnant females or juvenile sharks that haven't had a chance to reproduce at all. One good option for sport fishers is to practice "catch and release" when fishing for sharks.

Commercial Fishing: Many markets for shark products have expanded in recent years, driving commercial shark fishing to an all-time high. For example, the meat of thresher, porbeagle, and other

The Magnificent Mako

One of the most sought-after shark species is the shortfin mako. Mako sharks dwell in tropical and warm-temperate seas around the world. In North America, makos live off the coasts of southern California and Baja California as well as in the western Atlantic Ocean, including the

shortfin mako

Gulf of Mexico. Makos are powerful and fast, and they're known for leaping, thrashing, and attacking boats to resist capture. Fishers even tell stories of makos leaping into boats, jaws gnashing, scaring the people aboard into the water! But ultimately, even makos aren't fierce enough to avoid being killed by people.

sharks is now on the menu at many restaurants, replacing more traditional seafood that has become too rare or expensive. Restaurants in California have even been known to try to pass off the tasty meat of mako sharks as swordfish—a popular ocean fish that is depleted in some regions.

shark Fins—Past and Future

An expensive bowl of shark-fin soup may not sound like a delicacy to you, but in Asia, it's a traditional dish that has been around for many years. In fact, sharks and shark parts are the basis of numerous foods, as well as health and beauty treatments, that are important in ancient Asian cultures. Hundreds of years ago, when shark products first became popular, shark populations were most likely larger and healthier. At the same time, human populations involved in exploiting these creatures were smaller, which meant that harvesting the sharks wasn't as destructive to the animals' populations.

©Mako Hirose/Seapics.com

But now, with growing human populations and the accompanying increase in demand for shark products, shark-fishing practices are causing serious conservation concerns. Reducing unsustainable shark fishing, while working to decrease the demand for shark products, may offer hope for reestablishing healthy populations in the future. It's important for conservation and cultural groups to work together to protect the diversity of life in the seas while respecting well-established cultural traditions.

Interest in other shark products is spurring commercial shark fishing too. Shark livers are used as a source of lubricants, vitamins, and cosmetics. Shark skin—once a popular material for cowboy boots—is still made into leather products. And powdered shark cartilage, considered by some people to be a powerful cure-all for everything from sore eyes to cancer, sells for as much as $100 per bottle, even though there is no reliable evidence that the powder is effective in fighting disease.

What's more, the market has skyrocketed for one small part of the shark: its fins. In Hong Kong and other places around the world, diners pay up to $90 for a bowl of shark-fin soup. When catching fish for their fins, many fishers will simply slice off the shark's fins and throw the shark back into the water alive. These injured sharks soon drown or die of starvation, infection, or predation.

Accidental catch: Fishers don't always catch sharks on purpose. Many times their nets and hooks accidentally snare sharks, as well as other fish, sea turtles, and marine mammals. When the species that are caught are unwanted, they are called *bycatch.* (Blue sharks, for example, are considered bycatch because their meat has very little commercial value.)

One of the biggest causes of accidental bycatch of sharks is *longline* fishing for tuna and swordfish. Longlines are thin cables or monofilament lines that may stretch as far as 40 miles across ocean waters. They have a float every few hundred feet and a baited hook every few feet. Unfortunately, longlines aren't very discriminating: Approximately 25 percent of all animals caught with longlines are discarded, and of those, up to 75 percent are sharks.

In 1989, about 80 percent of the sharks caught in the northwestern Atlantic as bycatch were killed and dumped back into the ocean. Today, as the value of shark meat and fins has grown, fishers keep more of the sharks they catch. But, of course, that doesn't increase the sharks' survival rates—it just reduces waste. Even worse, many sharks that are caught are immature and have not lived long enough to produce young.

"Sharks around the world are facing a bleak future. In order to turn this tide, people will need to not only stop fearing sharks, but care enough to take action on their behalf."

–Sonja Fordham, Shark Fisheries Specialist, The Ocean Conservancy

The species that fishers catch and *keep,* even though they aren't the targeted species, are called *incidental catches.* For example, people fishing for tuna keep the makos and thresher sharks that get caught in their longlines or nets because the sharks are just as valuable at market as swordfish and tuna.

As you might have guessed, all these activities spell bad news for sharks. Scientists estimate that some species of coastal sharks in the Atlantic waters of the United States have declined by as much as 80 percent over the last 20 years. These numbers would create concern about any fish population. But in the case of sharks, they're especially troublesome. While sharks may be some of the top predators of the sea, they're not good at bouncing back when their numbers get low. To understand why, it's important to know more about sharks and their life cycles.

great white shark

Shark Natural History

A Shark's Life

Sharks, along with skates and rays, belong to a group of fish called *elasmobranchs*. Elasmobranchs are distinguished from other fish mainly by their skeletons, which are made of cartilage rather than dense bone.

Most sharks have a few key characteristics in common. They have five or more gill slits on each side, unlike most fish, which have a gill cover, or *operculum.* They have leathery skin covered with tiny, sharp scales. Most sharks have tails that are asymmetrical—the upper lobe extends out over the lower lobe. (Lammid sharks, white sharks, and makos have symmetrical tails.) And sharks don't have swim bladders to keep them buoyant, which means they have to swim to keep from sinking. (*Pelagic* sharks, such as makos, also have to keep swimming in order to breathe, but not all sharks swim constantly. Angel sharks, like other *demersal* species, live at or near the bottom and can rest on the ocean floor.)

Variety of Lifestyles: Beyond those few simple facts, it's hard to generalize about sharks. That's because there are so many species of sharks—nearly 500 by last count—and the variation among those species is tremendous. (See pages 202-203 for examples of different shark species.) Some sharks live in fresh water, some live in coastal areas, and some live only in the deep sea. One of the smallest sharks, the spined pygmy shark, is only 8 to 10 inches long; the largest, the whale shark, can grow to be more than 40 feet!

Interestingly, not all sharks reproduce in the same way. Some, including horn sharks and cat sharks, release their fertilized eggs into the sea, leaving them completely unguarded. (The eggs are protected inside tough, leathery egg cases.) Others, such as lemon sharks and hammerheads, retain the fertilized eggs, hatch them internally, and nourish them through placentas until they are old enough to be born. Still others, such as cookie-cutter sharks, retain the fertilized eggs, which hatch internally but receive no nourishment from the mother. Instead they must survive by eating unfertilized eggs and their smaller siblings! Despite these dramatic differences in reproductive strategies, all sharks have relatively long, slow life cycles.

All animal species go through their life cycles at different rates. At one end of the spectrum are insects such as fruit flies that can hatch, mature, reproduce, and die in a matter of days. At the other end are species such as elephants and people that take many years to mature, reproduce, and die. Sharks are more like elephants than insects. Most shark species grow slowly and mature late. In fact, dusky sharks don't reach their breeding age until they are more than 20 years old. Many shark species reproduce only every other year. Some sharks carry their young for two years. And many produce only a small number of young at a time.

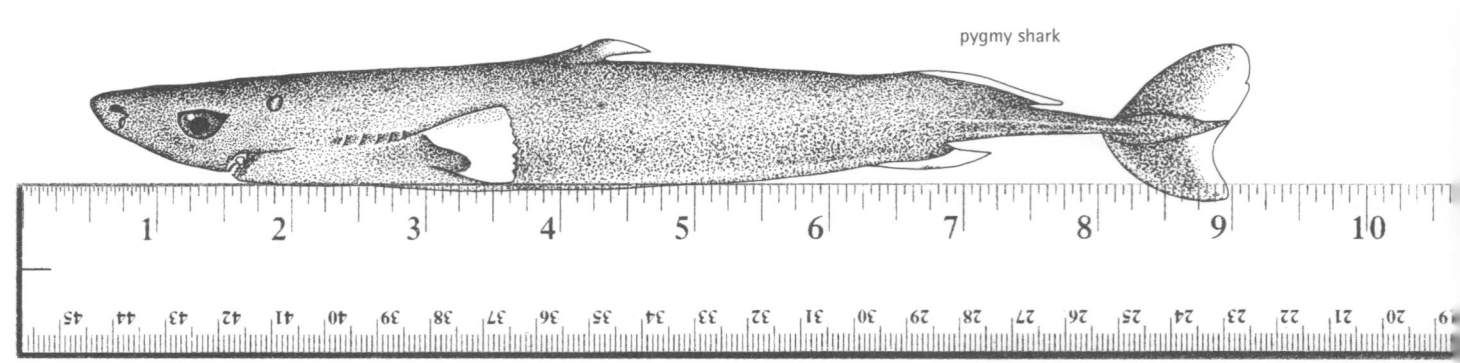

pygmy shark

Long Life Cycle Is No Longer an Advantage: When species live in unstable environments, it's advantageous for them to have a short life, mature early, and produce many young. This helps populations bounce back quickly when, for natural reasons, many individuals die. Sharks, on the other hand, evolved in a relatively stable environment, so they can invest more of their energy in living longer, fertilizing internally, and producing fewer and larger offspring. One advantage is that their young are born large enough to avoid their few predators and start feeding immediately on the fish, crustaceans, and cephalopods that form the bulk of their diet.

Unfortunately, sharks no longer have a stable, relatively predator-free environment. Overfishing by humans has reduced the numbers of many fish that sharks depend on for survival, making it harder for them to find food. And humans have become the predator that many sharks never had.

If sharks matured quickly and reproduced more rapidly, they might have a better chance of surviving the impact of these human activities. But sharks, like other species, cannot suddenly change their life cycles. Moreover, even swordfish—which release millions of young at a time—have been rapidly depleted by humans because females are being caught before they are sexually mature. So it's easy to see why late-maturing, slow-growing, small-litter-producing sharks are especially vulnerable. In the late 1990s, scientists estimated that sharks off the Atlantic Coast were being killed twice as quickly as they were reproducing.

What Good Are Sharks?

Many people would agree that sharks are remarkable to observe. Their sleek bodies, sculptured fins, and gaping jaws inspire not just fear, but also awe. Still, their real value is the role they play within their ecosystem.

Some sharks, such as basking sharks, feed by opening their mouths wide and straining small fish and invertebrates from the water. Others, such as makos, chase down tuna, swordfish, and other large fish. Great white sharks and tiger sharks seek out larger prey, such as seals and sea lions. But adult sharks have few predators other than humans. For that reason, some are top predators within their ecosystem.

Some sharks eat the same kinds of fish that people do. But when it comes to selecting which individual to catch, sharks and people have different approaches. Sport fishers aim for the biggest, heaviest fish, and commercial fishers often capture fish at random. But scientists believe that sharks tend to catch sick, injured, older, or less agile animals—in other words, those individuals that are less capable of escaping an attack. In this way, sharks may help ensure that the fittest animals survive to reproduce, boosting the overall health of ocean populations.

Scientists are concerned about the effects that shark depletion could have on marine biodiversity—the overall species diversity of the sea. In addition to weeding out less healthy individuals, sharks take advantage of big population booms in their prey populations. In so doing, they keep any one species from becoming dominant and overwhelming other species that share their home.

And, of course, sharks themselves are part of the overall diversity of the ocean. If they decline, the richness and variety of the ocean will be diminished too.

BIOFACT

Shark's teeth, which are actually modified scales, grow in rows, and sharks can sometimes have as many as five rows of teeth at a time! Rather than growing along with the shark, sets of teeth are replaced with other sets of teeth as the animal matures.

Name That Shark!

Few people know that there are nearly 500 species of sharks around the world. Check out these pictures to get a glimpse of the incredible diversity of shark species.

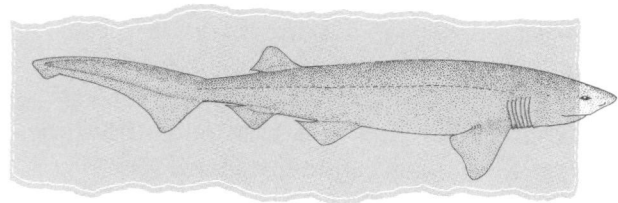 **Bluntnose Sixgill Shark**

Prickly Shark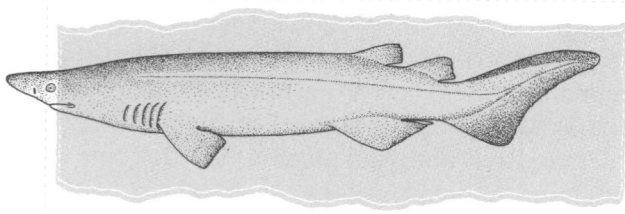

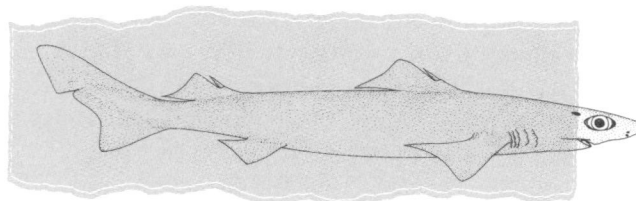 **Gulper Shark**

Mandarin Dogfish Shark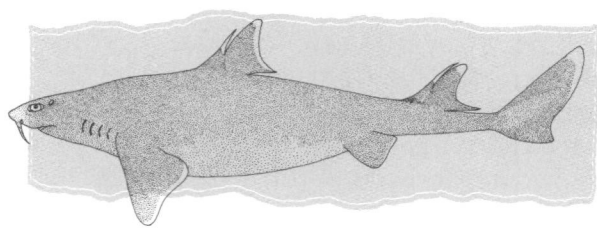

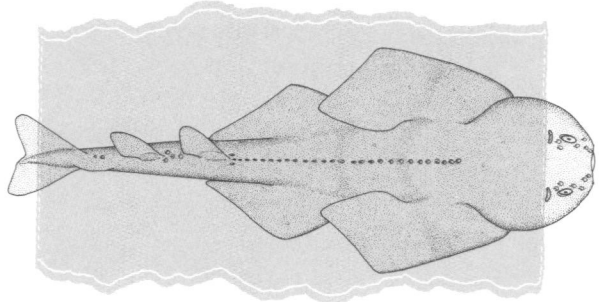 **Sawback Angelshark**

Zebra Bullhead Shark

Name That Shark! (Cont'd.)

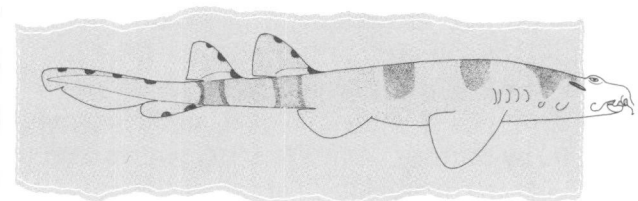

Northern Wobbegong Shark

Zebra Shark

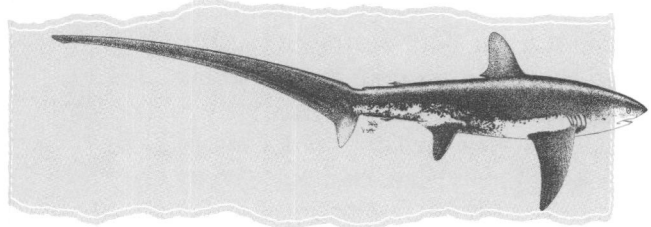

Thresher Shark

Barbeled Catshark

Hooktooth Shark

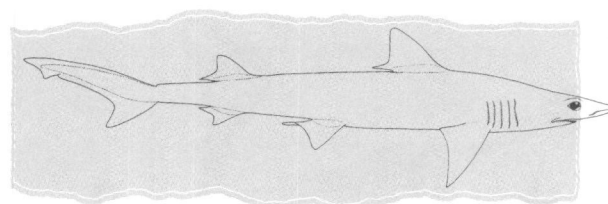

Sandbar Shark

Shark Solutions

Laws for Jaws

The U.S. government and some state governments have taken a number of steps to try to protect sharks from overfishing. In 1976, the United States declared exclusive control over fishing within 200 miles of its coastline. The federal government has now made it illegal to kill sharks in the Atlantic and Pacific Oceans just for their fins. It also established shark catch quotas, which set specific annual limits on shark takes in the Atlantic. Concern about shark depletion has led to even greater restrictions on shark fishing since 1997, when the government reduced the quota on some sharks in the Atlantic by 50 percent. In addition to these fishing controls, the establishment of marine protected areas in certain parts of the world is helping to protect shark nurseries and habitat.

All of these laws are good for sharks, but problems remain. After all, the United States is only one of 125 countries actively involved in trading shark products. Not all shark populations enter into international waters, so the United States can protect some species of sharks, but if other countries don't set limits on shark takes, laws that exist in only a few countries will not be enough to keep all shark populations healthy.

Also, shark quotas don't address the number of sharks killed as bycatch and thrown back into the sea. As it is now, the bycatch numbers are very large, and many marine conservationists think they need to be better controlled. (For more about marine legislation, see pages 341-344.)

Changing Views of Sharks

People's attitudes toward sharks vary widely from place to place and from culture to culture. For example, traditional Hawaiian cultures treat all sharks with respect in their religion, mythology, and daily life. By contrast, some Western cultures tend to view sharks as frightening creatures of the deep, that can pose a serious threat to human life. This attitude has done a great deal to fuel shark hunting, while making it very difficult for shark conservation to gain public support.

But there are signs that things are changing. Peter Benchley, the author of *Jaws,* now writes articles and essays explaining his new understanding of the creatures he once portrayed as ruthless man-eaters. He believes people need to respect and protect sharks. And perhaps they are beginning to do just that: New laws for shark conservation suggest that people are coming to recognize the importance of these fascinating fish. Perhaps as we learn more about sharks, we can get past some of our fears. We may still get spooked when we take a swim at an ocean beach, but that doesn't mean we can't recognize the vital role sharks play in the ocean environment and take action to ensure that they survive into the future.

BIOFACT

Scientists believe that ancestors of sharks swam through Earth's seas more than 400 million years ago—about 200 million years before dinosaurs.

"Jaws" movie poster

"What I definitely have become—to the best of my ability—is a shark protector, a shark advocate, a shark appreciator, and above all, a shark respecter. Sharks have an extremely important place in the natural order . . . and we're just beginning to learn how complex and wonderful they are. I know so much more about sharks than I did when I wrote **Jaws** *that I couldn't possibly write the same story today."*

–Peter Benchley, author

SUBJECTS
language arts, social studies, science

SKILLS
interpreting (inferring, identifying cause and effect, reasoning, elaborating), evaluating (critiquing, identifying bias)

FRAMEWORK LINKS
37, 40, 59, 60

VOCABULARY
attitude, fact

TIME
one session

MATERIALS
copies of "A Short Shark Story" (page 211), "Shark Survey" (page 212), and "Shark Meters" (page 213)

CONNECTIONS
For other values-clarification activities, try "Sizing Up Shrimp" (pages 160-165), as well as "The Spice of Life" in Biodiversity Basics and "Perspectives" in Wildlife for Sale.

AT A GLANCE
Explore your knowledge of and attitudes toward sharks by reading a short story.

OBJECTIVES
Identify statements that are facts versus those that are attitudes. Explore personal attitudes toward and knowledge of sharks. Describe some of the ways that knowledge and attitudes are related. Name several ways that attitudes about sharks can influence our actions toward them.

Man-eater. Predator. Monster of the deep. Read a popular account of sharks, and you'll likely come across these or similar phrases. In contemporary Western culture, sensational articles and spine-tingling movies such as *Jaws*, *The Deep*, and *Deep Blue Sea* have largely shaped our views of sharks. As a result, most people view sharks with fear and animosity.

But these views of sharks are not universal. Cultures that have long-standing ties to the ocean often view sharks with respect, if not reverence. New research is also helping people to see that sharks are not the indomitable people-killers we make them out to be. In fact, they are much more vulnerable to the actions of humans than we are to them.

In this activity, your students will get a chance to gauge their own knowledge of and attitudes toward sharks. And they'll review their classmates' results to see if there are any connections between what we know about sharks and what we *think* about them.

Before You Begin

Make one copy of "A Short Shark Story," "Shark Survey," and "Shark Meters" for each student.

What to Do

1. Give each student a copy of "A Short Shark Story."

Have the students read the story to themselves or ask for volunteers to read different paragraphs aloud.

2. Identify statements of fact and statements that are personal feelings or attitudes.

Have the students review the statements made by the characters in the story. Then have them:
(1) Underline at least three statements that reflect the personal feelings or attitudes of the speakers, and (2) put a circle around at least three statements of fact.

If there is any confusion, ask one of the students to explain the difference between a personal feeling or attitude and a fact. Then have the students share some of the fact statements and attitude statements they selected.

Afterward, ask the students if a person can have wrong feelings or attitudes toward sharks. *(No, because a feeling or attitude is just a personal belief or view.)* Are statements of fact ever wrong? *(Yes, because people may be making something up, incorrectly quoting an information source, or quoting a source that is unreliable.)* You might mention that not all the statements of "fact" in this story are accurate. For example, the story states that there is nothing you can do to prevent shark attacks, but see "Be Shark Smart" on page 210 for some ideas.

3. Hand out the "Shark Survey."

Have the students complete the "Shark Survey." In the first part of the survey, students will record their attitudes toward sharks. You might remind them that there are no wrong answers in this section. The second part of the survey poses questions of fact. There are correct answers to these questions (see page 214), but students should simply answer according to their knowledge and best guesses.

4. Hand out "Shark Meter" pages.

Now tell the students that they're going to tabulate the results of their "Shark Surveys." They should follow the instructions on the handout, filling in as many circles on each of the two sharks as is appropriate. The top shark measures attitudes: The more circles that are filled in, the more positively that student feels about sharks. The bottom shark measures knowledge: The more circles that are filled in, the more that student knows about sharks.

5. Review and discuss results.

When students have finished calculating their attitudes and knowledge about sharks, encourage them to share their results. You might want to have the students post their "Shark Meters" in the classroom. Or, for a more active approach, have the students line up across the front of the room in five groups: those who scored a 1 or 2 in their attitude measurement, those who scored a 3 or 4, those who scored a 5 or 6, those who scored a 7 or 8, and those who scored a 9 or 10. You should end up with five groups separated just slightly in the order described. Into which category did most of the students fall?

Now ask for a show of hands from all the students: Who got 1 or 2 of the factual questions right? Who got 3 or 4 right? Who got 5 or 6? Who got 7 or 8? Did anyone get 9 or a perfect 10? As you survey the students, encourage everyone to keep an eye on where the hands are going up. Can they detect any relation-ship between knowledge and attitudes? For example, did people who knew the least about sharks generally have a better attitude or a worse attitude toward sharks than those who knew the most?

Ask some of the students to share examples of where they picked up their knowledge about sharks. Based on the sources mentioned, is this information likely to be reliable? Why or why not?

Finally, ask the group for a show of hands to vote for one or two possible answers to the following: If they had $100 to designate for wildlife conserva-tion, and they could choose to give all of it to protect pandas or half of it to protect pandas and half of it to protect sharks, which would they choose? Tell

them to assume that sharks are suffering major population declines (which they are). After students have voted, ask the students to again reflect on any patterns they observed. Did people's attitudes toward sharks correspond with their willingness to help protect them?

Explain to the students that, as this unit continues, they'll be learning more about sharks and the problems they face. It might be interesting to see if anyone's attitude toward sharks changes along the way.

THE FAR SIDE® By GARY LARSON

"Well, somehow they knew we were—whoa! Our dorsal fins are sticking out! I wonder how many times *that's* screwed things up?"

WRAPPING IT UP

Assessment

The students are going to be the teacher! It's time for a test, but the students are going to create their own tests. In the center of a page, have the students write some statements about sharks—either facts or personal attitudes. On both the left side and the right side of the statements there should be blank lines. The lines on the left will be marked as "Fact" or "Attitude." The lines on the right side will be marked "Positive" or "Negative" to show how the student thinks each statement would shape a person's attitude.

Have the students put in the answers they believe are correct in their own tests. For fun, you could use statements from different students' assessments and do a "fun quiz" with the class. Grades won't be necessary, but see how well the statements worked.

Unsatisfactory—Elements of the test are missing, facts and attitudes are not clearly distinguished, and positive/negative responses are not all reasonable.

Satisfactory—Five to seven statements are presented with correctly distinguished facts and attitudes as well as reasonable positive and negative answers.

Excellent—Eight or more statements are clearly written with correct factual and attitudinal responses as well as reasonable positive and negative answers.

Portfolio

In their portfolios, students should include their "Shark Meters," as well as a few sentences explaining how people's attitudes about sharks may be linked to their level of knowledge about sharks. They can also include the tests they wrote for their Assessments.

Writing Idea

Each student can write a short piece that informs community members about sharks and the roles sharks play in marine ecosystems. In their pieces, students should address several common shark myths and provide factual information to help readers better understand how those myths came to be popularized and in what ways the myths may affect people's willingness to protect sharks.

Extensions

- Look for references to sharks in the media. Your students can compile a list of the representations of sharks they found and collect images of sharks in newspapers and magazines.

- Create a shark bulletin board. Have your students post their shark meters so people can compare the results. Or, have the students post any shark articles, poems, photographs, and drawings they find over the course of the unit.

- Show a short film or video about sharks. What views of sharks does this film convey? Did it affect the students' attitudes in any way?

- Have the students do some research to come up with a list of tips for being safe in coastal areas where sharks might be active. (Share the "Be Shark Smart" tips on page 210 with your students.)

1. Always swim in a group. Sharks are more likely to attack someone who's alone.

2. Never swim at night or at dusk or dawn. Sharks are most active at those times.

3. Swim in clear water. In murky water, a shark may mistake you for its usual prey.

4. Stay far from places where people are fishing or cleaning fish. Fish blood and guts can attract sharks and put them in a feeding mood.

5. Stay away from places where lots of small fish are leaping from the water. This could be a sign that a shark is chasing them. Also stay away from places where lots of seabirds are diving. That's a signal that small fish are nearby, potentially accompanied by sharks that like to feed on them.

6. Don't stay in the water if you are bleeding. Blood can attract sharks.

7. Don't wear shiny jewelry. Jewelry can look like flashing fish scales.

8. If you see a large shark, don't panic and start splashing around. That can make the shark think you're injured prey—an easy target. Just warn others and calmly leave the water.

9. If a shark ever does attack you, fight back. Hit its eye or gill areas with your fists or feet.

And here's one more tip: Don't worry, swim smart, and have fun!

Adapted with permission from *Ranger Rick*, June 2002, published by the National Wildlife Federation, © 2002.

BIOFACT

Each year lightning kills more than 70 times as many people as sharks do!

A SHORT SHARK STORY

Carlos, Katie, and Katie's twin brother, Nick, stood on the beach and stared out at the blue waters of the Pacific.

"It looks so welcoming today," Katie said.

"No it doesn't. Look at those little waves," Nick said. "The more you look at them, the more they look like hundreds of shark fins popping out of the water."

Carlos shook his head. "You can't think about yesterday, guys, or you'll never get in the water. Do you want to learn the joys of surfing, or don't you?"

Katie nodded, hesitantly. Nick shrugged.

The day before had been a terrifying one at this same beach. A surfer had been lying on his board, waiting to catch the next wave, when a large shark came up and grabbed onto his board. Without even thinking, the surfer whacked the shark on the nose. The shark took a large bite of board and disappeared. The frightened surfer paddled to shore, too scared to even look behind him. Now everyone in Santa Cruz was talking about sharks—surfers, swimmers, you name it. Nick thought it was a pretty lousy time to be visiting California and getting his first surfing lesson.

"Sharks freak me out," Nick said. "I don't know if I can do this."

"Sure you can," Carlos said. "Did you know that you have a better chance of being killed by lightning than by a shark?"

"That may be true," Katie said. "But I can do stuff to avoid being struck by lightning. There's nothing a person can do to avoid being eaten by a shark."

"Except to stay out of the water," Nick said quickly.

"I'll admit it, every surfer I know has thought about sharks at one time or another," Carlos said. "How can you not? Scientists say a surfer lying on a board looks a lot like a seal or sea lion from a shark's perspective. And great white sharks eat lots of seals and sea lions."

"You're making me feel much better, Carlos," Nick said sarcastically, taking a step back from the waves.

"OK, so there is a risk," Carlos said. "But we take risks every day. It's risky to drive a car on a highway. At least out here I can catch a big, high wave and ride it in, with the sun beating on my shoulders and the water sparkling like a sapphire. That's worth a little risk!"

"It does sound pretty great," Katie said. "Should we go for it?"

"I think I'd rather surf in a swimming pool," Nick said.

"It's up to you," Carlos said. "But I can't stand wearing a wet suit without getting wet any longer." He waded into the water. "C'mon, Katie. Let's teach you how to catch a wave!"

"Are you coming, Nick?" Katie asked, turning to her brother.

"I don't think so," he said. "But I'll keep an eye on the two of you, just in case you get into trouble out there."

"You might regret this for the rest of your life, Nick," she said, wading in after Carlos.

"At least I'll have the rest of my life!" Nick answered.

What Do You Believe?

1. Sharks are scary. .... agree/disagree
2. People who swim in oceans where sharks live are crazy. .... agree/disagree
3. Sharks are a little frightening, but they're not bad. .... agree/disagree
4. All sharks that swim near the shore should be killed. .... agree/disagree
5. Some kinds of sharks don't seem scary to me at all. .... agree/disagree
6. Sharks are interesting. .... agree/disagree
7. The world would be better off if there were no sharks. .... agree/disagree
8. Sharks are mean. .... agree/disagree
9. I worry about people killing too many sharks. .... agree/disagree
10. The ocean is a better place with sharks in it. .... agree/disagree

What Do You Know?

1. Sharks are a kind of fish. .... true/false
2. Almost all sharks live near the coast where people swim, snorkel, and surf. .. true/false
3. If you see a shark while you're in the water, it will probably attack you. .... true/false
4. If a shark bites you, you will probably die. .... true/false
5. There are fewer than 100 species of sharks in the world. .... true/false
6. Some of our medicines are derived from shark products. .... true/false
7. Some adult sharks are less than a foot long. .... true/false
8. Some people eat sharks. .... true/false
9. Because people are taking special precautions, there are
   fewer shark attacks now than ever before. .... true/false
10. Sharks have few natural predators, so their populations are stable. .... true/false

SHARK METERS

What Do You Believe?

Compare the answers below with those you gave on your survey. Then fill in a dot on the shark meter for every answer you gave that matches the numbered answers here. (For example, if you had four matching answers, fill in four dots in a row, starting at the shark's tail.)

(1) disagree, (2) disagree, (3) agree, (4) disagree, (5) agree, (6) agree, (7) disagree, (8) disagree, (9) agree, (10) agree.

What Do You Know?

Fill in a dot in the shark for every one of the questions you answered as follows:
(1) true, (2) false, (3) false, (4) false, (5) false, (6) true, (7) true, (8) true, (9) false, (10) false.

ANSWERS TO THE "SHARK SURVEY"

1. **True.** Sharks are a special kind of fish, though, because their skeletons are not made of dense bone. Instead, their skeletons are made of cartilage, like your ears or the end of your nose. So they're called *cartilaginous* fish.

2. **False.** Sharks live in a variety of ocean settings—from coastal areas to the deep ocean.

3. **False.** Sharks rarely attack humans, and when they do it's likely to be a case of mistaken identity: The sharks mistake the person for a sea lion, another marine mammal, or an injured fish. Sharks also may attack divers who bother them in some way.

4. **False.** Scientists think sharks often give people a bump or bite to investigate what they are (or to make them go away if they're bothering the sharks) and will not necessarily continue to attack.

5. **False.** Scientists have identified nearly 500 species of sharks worldwide.

6. **True.** Parts of sharks have been used for everything from artificial skin for burn patients to anticoagulants for people with heart problems.

7. **True.** There are many species of small sharks. For example, adult pygmy sharks grow to be only 10 inches in length.

8. **True.** Shark is a popular dish at many restaurants, and shark-fin soup is a delicacy in some Asian countries as well as in the "Chinatown" areas of many large, U.S. cities.

9. **False.** Despite increased understanding of shark behavior, shark attacks have increased over the past several decades. Scientists believe that human population growth—simply having more people in the water than ever before—explains most of this increase.

10. **False.** While it is true that most adult sharks have few natural predators, humans now kill sharks intentionally or accidentally at extremely high rates. For that reason, scientists believe that some coastal shark populations in U.S. Atlantic waters have declined by 50 to 75 percent over the last 20 years.

blue shark

"We now know that the best shark is not a dead shark; that these oft maligned fish play critical roles in preserving balance in the marine ecosystem."

—Mike Hayden, President/CEO, American Sportfishing Association

Where the Wild Sharks Are

SUBJECTS
science, language arts

SKILLS
organizing (classifying, categorizing, arranging), interpreting (translating, relating, reasoning), applying (restructuring, composing)

FRAMEWORK LINKS
1, 1.1, 2, 3, 17.1, 18, 21, 23.1

VOCABULARY
barbels, bathypelagic, **biodiversity,** *bottom trawls,* **coral reefs,** *epipelagic,* **estuaries,** *kelp beds, luminescence, mesopelagic, mollusks, pelagic,* **plankton,** *rays,* **trawl**

TIME
one to two sessions

MATERIALS
copies of "Meet the Sharks" (pages 220-223) and Clues—"Where the Wild Sharks Are" (page 224); large drawing of "Ocean Zones" (page 226); scissors, string, tape, crayons or markers (optional)

CONNECTIONS
Use "Sea for Yourself" (pages 72-81) to get students interested in learning about different marine species as well as the importance and uniqueness of their habitats.

AT A GLANCE
Identify shark species and determine where in the ocean each species lives.

OBJECTIVES
Describe some of the different ocean zones. Describe several species of sharks and the parts of the ocean in which they live.

The vast expanses of water in the oceans seem so much alike. That's why students are often surprised to learn that the ocean has different zones of life, defined primarily by the depth of water and distance from shore, as well as by geographic distribution—from polar to temperate to tropical regions. In this activity, students will learn about some of the different zones in the ocean as they meet a range of shark species and piece together information to determine where each one lives.

Anja G. Burns

Before You Begin

Make one copy of "Meet the Sharks" and "Clues—Where the Wild Sharks Are" for each student or pair of students. Redraw the "Ocean Zones" diagram (from page 226) on a large piece of paper. You can label the ocean zones yourself or leave them blank and have your students label them, using the clues on the handout. Do not put the numbers on your drawing. Students will be asked to place their sharks in the correct zones, but they do not have to have the exact placement as shown on page 226.

What to Do

1. Discuss shark attack scenario.

Begin by describing this scenario to your students:

"A 12-year-old girl is attacked by what witnesses call a 'big shark' in waist-deep water on the Florida coast. After her mother pulls her to safety, the girl receives 72 stitches to her leg and survives."

Ask the students if they think it would be difficult to identify the kind of shark that attacked the girl. Are there many different kinds of sharks? Are they similar or quite different? Explain to your students that sharks vary a lot from one species to the next. In this activity, each student or pair of students will determine the habitat of an assigned shark species and assess whether their species might have been the shark that attacked the girl.

2. Hand out copies of "Meet the Sharks" and "Clues—Where the Wild Sharks Are."

Tell the students that the information on "Meet the Sharks" will introduce them to 25 species of sharks, all of which can be found in North American waters. Can someone define the term species? *(A species is a group of organisms that have a unique set of characteristics [such as body shape and behavior] that distinguish them from other organisms. If they reproduce, individuals within the same species can produce fertile offspring.)* You might explain to the students that there are nearly 500 shark species worldwide. Assign one species to each student. (Some may need to work in pairs or work on two sharks, depending on your class size.) Tell the students that their job is to read the information and clues to determine where their particular shark species lives. Have the students cut out the shark with its description and, if you wish, tape one end of a piece of string to the back of it. They can also color the shark if they'd like.

3. Discuss ocean zones.

While the students are preparing their sharks, hang up the ocean zone diagram. When the students are ready, ask them to think about conditions in the ocean. If they walk into the water right off the beach, what is the ocean like? *(Shallow, cold in winter but warmer in the summer or in the tropics.)* If they were able to keep walking into deeper and deeper water, how would the ocean change? *(It would get darker and colder. Fewer or no plants would be growing on the ocean floor, there would be fewer or no coral reefs, and so on.)*

Remind the students that, because of variations in light, temperature, and other conditions, species of fish and marine mammals are usually better suited to one part of the ocean than the other, just as some land animals are suited to different climates, different parts of a forest, and so on.

Discuss the four oceans zones (epipelagic, mesopelagic, bathypelagic, and coastal), explaining the characteristics of each (see page 224). Point out to your students that, while most sharks spend the majority of their time in one particular zone, they do travel throughout various ocean zones especially to feed.

At this point, if you haven't already labeled the ocean zone diagram, ask for volunteers to use clues from the handout to fill in the blanks.

4. Affix sharks to chart.

Now invite the students to come forward and place their sharks in the appropriate zones of the ocean, explaining what clues led them to this conclusion. (Mention that some students may need to wait until other sharks have been placed before they can place their own species accurately.) They can either tape the sharks directly in the zones or tape them nearby using a piece of string to connect them to the proper zone. Also, have the students say whether they think their species of shark could have been responsible for the shark attack described at the beginning of the activity. Why or why not? (See page 226 for appropriate zones.)

You may want to remind the students that, although this diagram will indicate the particular ocean zone each shark species prefers, it doesn't reflect their geographical distribution. For example, blue sharks are found all around the world, but Atlantic angel sharks are found only in the ocean waters from Massachusetts to the Caribbean. Yet both appear in this diagram together.

5. Discuss results.

After students have completed their handouts, discuss the following questions.

- What are some of the reasons that different species of sharks favor particular zones of the ocean? *(Some follow preferred food sources, some are better adapted than others to cold water and dark water, some are too large for shallow water, and so on.)*

- Were you surprised that some of the larger sharks, such as great white sharks and tiger sharks, often venture into coastal areas? Why does this make sense? *(These sharks feed heavily on marine mammals, which are concentrated close to land.)*

- Biodiversity is a word that means the variety of life on Earth—including genes, species, and habitats. What aspects of biodiversity did this activity cover? *(The activity illustrated the great variety of shark species and provided examples of some of the different types of habitats where they live.)*

- Do the students notice any pattern in the distribution of sharks on this diagram? *(Although the diagram is not comprehensive, it does show more sharks in coastal areas than in other parts of the ocean.)* You might take this opportunity to explain to your students that marine biodiversity is not evenly distributed across the oceans. Certain areas—such as coral reefs and other coastal habitats—contain disproportionately more species than other areas. And these areas are often more vulnerable to harm from pollution, destructive fishing practices, and other problems caused by people.

whale shark

BIOFACT

Whale sharks are the largest fish in the sea, reaching lengths of over 40 feet and weights of up to 20 tons.

WRAPPING IT UP

Assessments

Ask students to draw a cross-section (side view) of the ocean floor. They should identify some of the different zones in their diagrams and label what changes there would be between each zone. Have the students name a shark that would live in each zone.

Unsatisfactory—Only one or two zones are identified with differences clearly noted, or several zones are identified but no differences are explained.

Satisfactory—Three zones are identified with differences named. At least one shark is assigned correctly to each zone.

Excellent—Four zones are identified with differences named. At least one shark is assigned correctly to each zone.

Portfolio

If students complete one of the shark mystery challenges (page 227), have them include a short piece on their mystery species in their portfolios. Or, if students pursue research as outlined under the first Extension activity, have them include the reports in their portfolios.

Writing Idea

Have students create field guide entries for each of the species highlighted in the "Meet the Sharks" diagram. Field guide descriptions include information on the species, such as its common and scientific names, length, coloration, habitat, range, and preferred diet. Or ask students to write up their answers to the "Shark Mystery Challenges."

Extensions

- Have each student choose one species of shark and put together a short report on it. The report should describe its preferred food, geographic distribution, method of reproduction, and other interesting aspects of its life and behavior.

- Give each student a copy of "Shark Mystery Challenges" on page 227. Explain to the students that our current understanding of sharks lags well behind our knowledge of many other kinds of animals. Can they guess why? *(For many years, information about sharks was obtained primarily through brief glimpses of live sharks and the study of dead specimens. As diving equipment became more sophisticated, scientist began to slowly accumulate data on shark behavior. Year-round, close-up observations were first possible when people began to keep sharks in captivity. Another boost has come quite recently with the development of advanced scuba equipment, which enables divers to stay underwater for long periods of time. Now scientists are able to observe sharks in their natural environment and tag sharks for long-term monitoring.)* Why might this information be helpful? *(Corrects misconceptions about sharks. Guides decisions about how best to protect sharks over the long term.)*

Have each of your students investigate what scientists do and do not know about sharks by researching one or more of the mysteries on page 227. Turn to page 228 for help with evaluating their answers.

MEET THE SHARKS

Note: These pictures are not to scale.

1. Frilled Shark

Frilled sharks are about six feet long and have soft, eel-like bodies. This has led some people to observe that they look more like sea snakes than sharks. Frilled sharks feed on octopuses and bioluminescent squids and are sometimes caught in bottom trawls.

2. Bramble Shark

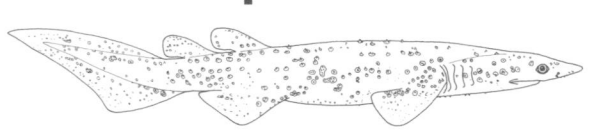

The bodies of bramble sharks are covered with thornlike spines that help them glide through the water. The large, slow-moving shark will sometimes float motionless, perhaps looking for hidden octopuses and rockfish.

3. Spined Pygmy Shark

The spined pygmy shark is one of the smallest sharks in the world, reaching lengths of only about eight to ten inches. It is blackish-brown on most of its body and has a luminescent underside.

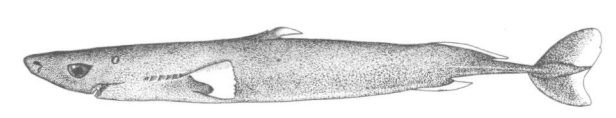

4. Cookie-Cutter Shark

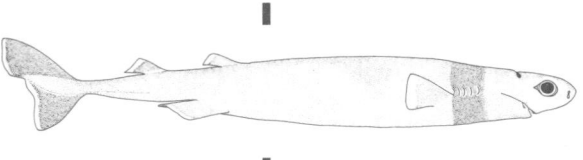

Scientists finally discovered why small, round holes covered the bodies of some tuna, porpoises, whales, and sharks when they observed how the cookie-cutter shark feeds: The shark holds its mouth against the skin of its prey, sucks in a cylinder (or plug) of flesh, tears it off with its sharp teeth, and leaves a hole. These strange sharks emit a greenish glow.

5. Atlantic Angel Shark

Also known as a sand devil, the Atlantic angel shark looks a lot like a ray. Angel sharks feed on mollusks, skates, flounders, and other creatures. They are found in the same zone as lemon sharks, but tend to burrow under the sand.

6. Horn Shark

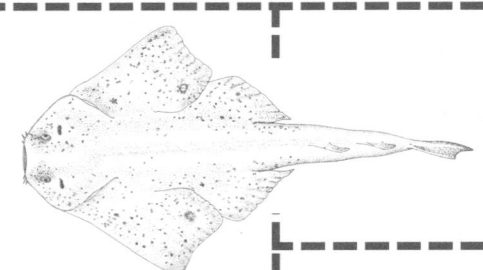

Horn sharks have a short, blunt head with high ridges above the eyes. By day, they often hide in kelp beds, emerging at night to feed on small fish, mollusks, and other creatures.

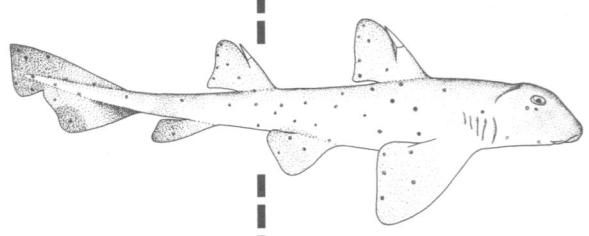

7. Nurse Shark

Sluggish and generally docile, nurse sharks often lie on the ocean floor without moving. But watch out! Some divers have made the mistake of touching the quiet sharks, only to receive a sudden, severe bite in response. You can recognize nurse sharks by long fleshy appendages, called barbells that hang below their snouts. These sharks are commonly found in the vicinity of reefs.

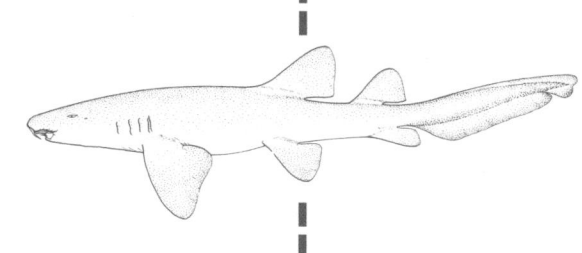

8. Whale Shark

Whale sharks are gentle giants that are covered with spots and stripes. They're the largest fish in the world, averaging about 32 feet in length. Certain individuals may reach lengths of more than 40 feet! But whale sharks are very docile, feeding on plankton and small crustaceans on the surface, usually well off shore.

9. Sand Tiger Shark

Sand tigers are voracious eaters, consuming a steady diet of fish and squids, and they've even been known to feed on sea lions and other mammals. Nonetheless, these large sharks are essentially gentle and rarely dangerous. Groups of sand tiger sharks often gather around rocky reefs.

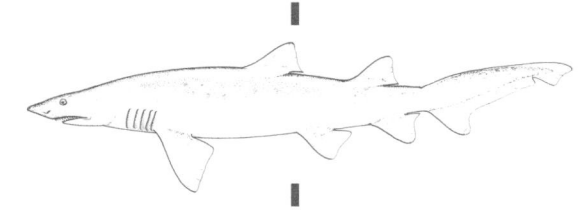

10. Goblin Shark

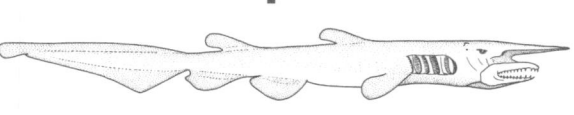

The unusual snout of the goblin shark extends out from its head in a long, flat blade. It is one of the rarest of shark species. Little is known about the goblin shark's feeding habits, but it is thought to spend most of its time swimming at depths of more than 3,500 feet.

11. Megamouth Shark

In 1976, a U.S. Navy crew stationed off Oahu, Hawaii, found an enormous shark entangled in a large sea anchor about 500 feet below the surface. Apparently the shark had tried to swallow the anchor and died. The shark, new to science, was given the name megamouth shark because of its huge, wide mouth.

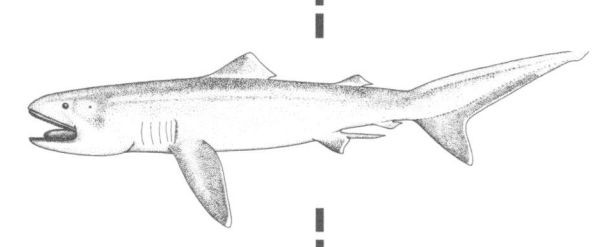

12. Bigeye Thresher Shark

Bigeye thresher sharks feed primarily on squids and small tuna, and often get caught in tuna fisheries longlines. Like other thresher sharks, they have extremely long tails, which scientists think they may use to round up or even stun fish.

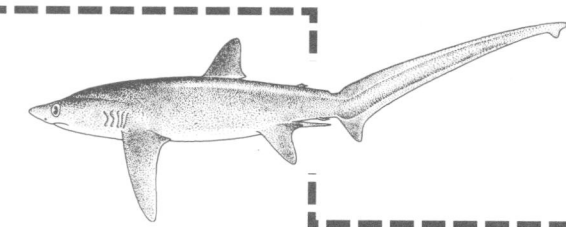

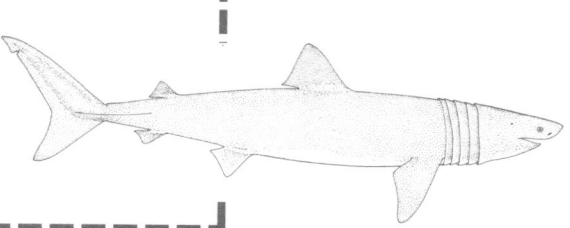

13. Basking Shark

Basking sharks swim with their mouths open, gulping water and plankton and then straining out the water. The second largest species of shark, basking sharks are reported to attain lengths of 40 to 45 feet. In summer, they are often seen basking near the surface of the water.

14. Great White Shark

Few animals cause as much terror as great white sharks. They are strong swimmers, and they prey on seals, sea lions, porpoises, tuna, sea turtles, and other sharks. They often congregate around seal and sea lion rookeries and are responsible for one-third to one-half of all human shark-attack fatalities each year. But that doesn't mean that a lot of people are killed by them: The total number of fatalities internationally from great white shark attacks was only ten for the entire 1990s.

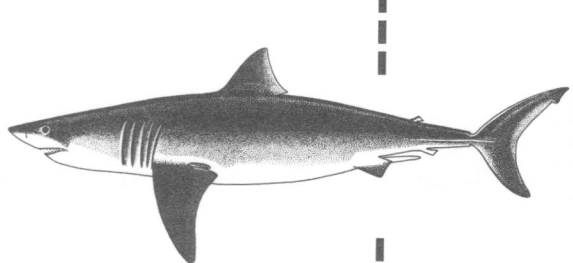

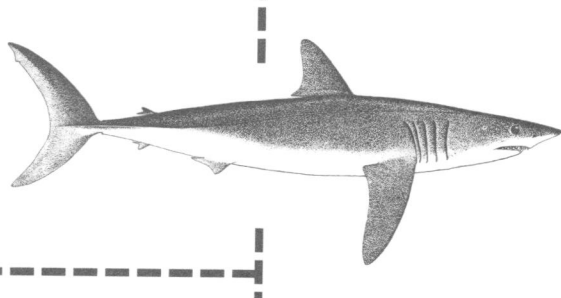

15. Shortfin Mako

The fastest of all sharks, shortfin makos prey on other sharks, swordfish, and tuna. They are among the most beautiful and powerful fish in the sea. But they're probably best known for their fierce antics when caught on a fishhook. They can leap high in the air to try to shake out the hook and, in an attempt to escape, may even attack fishing boats and the people in them.

16. Porbeagle Shark

In the nineteenth century, porbeagle sharks were heavily fished for their liver oil, which was used to tan leather. These sharks feed on squids, and fish such as mackerel, cod, and flounder.

17. Brown Cat shark

Only about two to three feet long, the brown cat shark feeds primarily on shrimp and small fish. It has a chocolate brown body and green eyes.

18. Leopard Shark

This shark is spotted like a leopard and eats small fish, crabs, sea worms, and other organisms. Leopard sharks are sometimes seen in large schools by divers and kayakers, but they are harmless to humans.

19. Silky Shark

A fast-moving shark with unusually smooth skin, the silky shark feeds on organisms such as squids, mackerel, tuna, and pelagic crabs. Sometimes they swim in the same ocean area as blue sharks.

20. Bull Shark

Bull sharks are sizable predators—they can grow to the length of 11 feet. They have been known to attack swimmers in estuaries, rivers, and freshwater creeks that flow directly into the ocean.

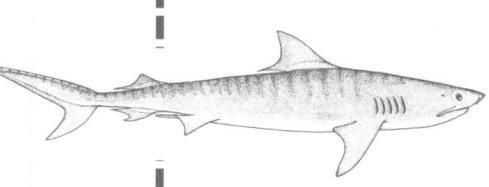

21. Tiger Shark

The tiger shark is striped like a tiger and weighs as much as 2,000 pounds. Although they usually feed on marine birds and seals, a variety of items have been found in the stomachs of tiger sharks, including: squids, lobsters, smaller sharks, turtles, canned peas, lumps of coal, the leg of a sheep, and human remains.

22. Night Shark

Night sharks search for fish and shrimp with their large, green eyes. They give live birth, and their litters usually range from 12 to 18 pups.

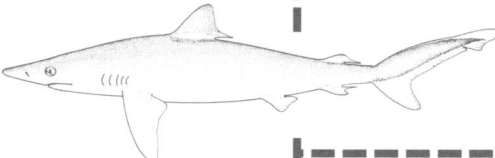

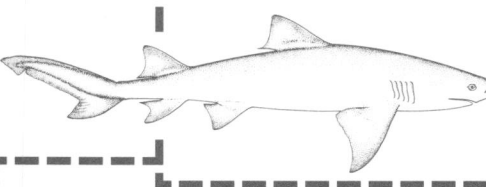

23. Lemon Shark

Lemon sharks are active around docks and estuaries, and they feed on fish, crabs, seabirds, and more. They look somewhat like bull sharks, but they have distinctive yellowish undersides.

24. Blue Shark

One of the widest ranging sharks, blue sharks have been known to swim more than 40 miles a day! Their migrations keep them in cool waters, where they feed largely on schooling fish and squids.

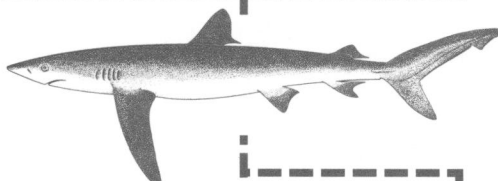

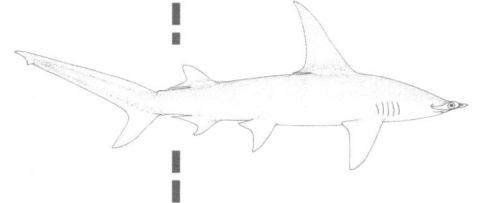

25. Great Hammerhead Shark

You won't have any trouble identifying a hammerhead: Its head is shaped like a wide rectangle with eyes at either end. It is a voracious eater, making a meal of rays, smaller sharks, and other fish. Hammerheads are very common around tropical reefs.

1.
The two largest sharks in the world reside in the same ocean zone.

2.
Kelp beds grow just off the coast in many temperate climates.

3.
Sharks with large eyes and/or green eyes tend to live in the deepest parts of the open ocean.

4.
Swordfish, tuna, and squids are often found in the epipelagic zone.

5.
Sharks that dwell in the bathy-pelagic zone often have luminescence (the ability to glow), which may allow them to communicate and capture prey in the ocean's darkest, deepest waters.

6.
An estuary is a place where fresh-water creeks or rivers empty into the sea.

7.
Seals and sea lions tend to congregate on islands and rocky coasts.

8.
Cookie-cutter sharks and green dogfish are found in the same ocean zone.

9.
Bottom trawling is a fishing method in which a large net, or trawl, is dragged along the seafloor bottom to catch shrimp and pelagic fish.

10.
Many coral reefs are found in coastal zones.

11.
Porbeagles and shortfin makos are found in similar habitats.

12.
Leopard sharks and horn sharks live in the same zone, but they're found in different parts of the world.

OCEAN ZONES

Coastal: Located near the shore, and stretching from the ocean's surface to a depth of 650 feet, this sunlit zone is home to a wide variety of marine species—from squid to sea lions.

*__Epipelagic:__ This zone, similar to the coastal zone, is located in the open ocean rather than near the shore. Phytoplankton flourish in the abundant natural light, providing nutrients for a wide array of marine animals.

*__Mesopelagic:__ Extending from about 450 feet to 3,300 feet, light penetrates the upper areas of the zone, but the lower reaches are almost completely dark.

*__Bathypelagic:__ No sunlight touches this region (3,300 feet to about 13,200 feet, not including the sea floor), but bioluminescent animals thrive in the dark waters, producing their own light to lure prey.

*Pelagic means "of the open ocean."

(1) **bathypelagic**

(2) **bathypelagic**

(3) **bathypelagic**

(4) **bathypelagic**

(5) **coastal**

(6) **coastal**

(7) **coastal**

(8) **epipelagic**

(9) **coastal**

(10) **bathypelagic**

(11) **mesopelagic**

(12) **bathypelagic**

(13) **epipelagic**

(14) **coastal**

(15) **epipelagic**

(16) **epipelagic**

(17) **bathypelagic**

(18) **coastal**

(19) **epipelagic**

(20) **coastal**

(21) **coastal**

(22) **bathypelagic**

(23) **coastal**

(24) **epipelagic**

(25) **coastal**

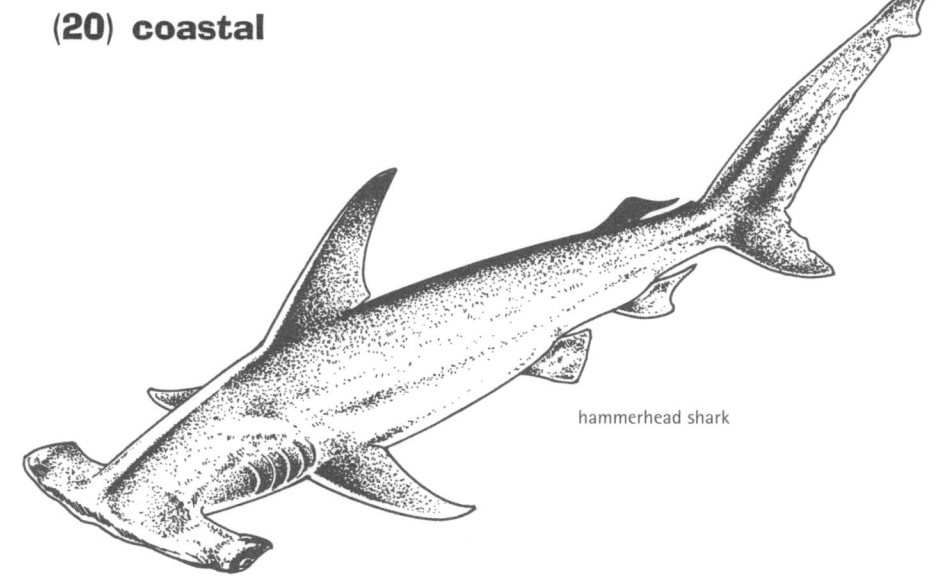

hammerhead shark

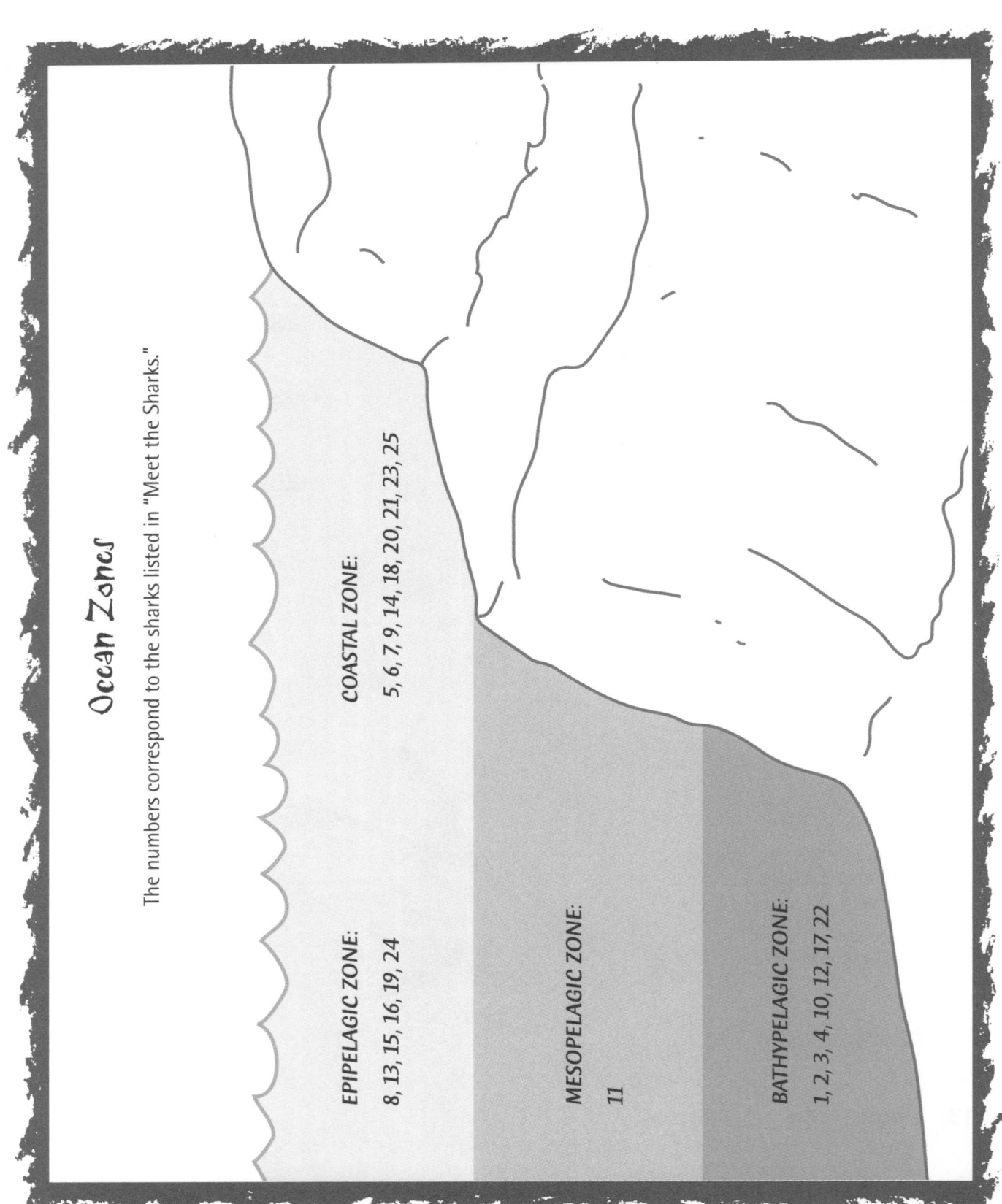

Ocean Zones

The numbers correspond to the sharks listed in "Meet the Sharks."

COASTAL ZONE:
5, 6, 7, 9, 14, 18, 20, 21, 23, 25

EPIPELAGIC ZONE:
8, 13, 15, 16, 19, 24

MESOPELAGIC ZONE:
11

BATHYPELAGIC ZONE:
1, 2, 3, 4, 10, 12, 17, 22

1.

Your Uncle Hughie is the director of a large aquarium. For years, his visitors have been begging him to display a live great white shark. Now Uncle Hughie has asked you to look into the matter. Do great white sharks survive in captivity? If so, what are their basic living requirements? If not, can you recommend another shark that would be more suitable . . . and still please the public?

2.

You've learned how to scuba dive and can't wait to take a dive off the coast of California. But there's one problem: You're terrified of sharks. You've decided you're willing to buy the best equipment to keep you safe. What kinds of inventions have been designed to help protect people from sharks? Which ones work? How? Is there anything else you can do to reduce your risk of attack?

3.

A friend tells you, "Sharks aren't good for anything." You disagree, so you propose a bet. He'll pay you $5 for every benefit sharks provide for people and the planet. How many benefits can you find? What are they? How much money does your friend owe you?

4.

You've learned that a tour boat leader has started "chumming" the waters about three miles from your favorite beach. Chumming means spreading bait, such as animal blood and oil, to attract sharks for viewing and filming. You think this sounds bad. Can you justify your concern? Why or why not?

5.

One day you're walking through a fish market in Southeast Asia, and you come across a shark you can't identify. You buy the shark—not to eat it, but to compare it with pictures you have of sharks in your favorite shark book. After a bit of searching, you tell your travel companions that you think you've found a new species of shark. They think you're crazy. How could you justify your position? How could you find out if you really do have a new species of shark?

6.

You've been hired to help create the set design for a new *Jurassic Park* movie. The producers want to have an ocean scene, but they aren't sure which animals were around during the Jurassic period. Can you find out which sea animals existed at that time? Were sharks present? When did the first shark relatives appear on Earth?

7.

Your mother tells you that she's found a delicious-looking recipe for shark-fin soup, which she hopes to serve at her next dinner party. You think the recipe must be a joke, and you're determined to talk her out of it. What can you find out about shark-fin soup?

ANSWERS TO "SHARK MYSTERY CHALLENGES"

1.

There are about 100 species of sharks on display in aquariums around the world. The sand tiger shark, bull shark, sandbar shark, blacktip reef shark, and whitetip shark are some of the sharks that have adapted most easily to aquarium life. By contrast, the great white shark has never done well in captivity. In general, sharks do well if they have enough space and if scientists know enough about their feeding habits to keep them well fed.

2.

One important thing you can do to lower your risk of a shark attack is to avoid diving, swimming, or snorkeling when sharks are most active—at dusk or dawn. Some wetsuits have been designed to mimic the striped coloration of pilot fish, which sharks don't eat. But these suits don't work: Sharks don't avoid pilot fish because they're striped, but because healthy and strong pilot fish are too hard to catch. More successful suits have been made with tiny interlocking stainless steel rings, which are able to protect a person from a shark's bite. Unfortunately, these suits are very expensive. (For more tips on avoiding shark attacks, see "Be Shark Smart" on page 210.)

3.

Many very large sharks, such as great white sharks and tiger sharks, feed on seals, sea lions, and other marine mammals that eat large amounts of mollusks and other small organisms. The sharks' predation helps keep populations of those mammals from depleting their own prey populations. Smaller sharks provide food for other species. Sharks have also been useful as medicines: In the 1930s, shark liver was used widely as a source of vitamin A, but now vitamin A can be synthesized, so this use is no longer common. Shark corneas have been used successfully as transplants for human corneas. Shark cartilage yields a kind of artificial skin used for victims of burns. Anticoagulants from sharks are used for treatment of cardiac problems. And an extract from shark bile has been shown to be useful in treating acne.

4.

Scientists are still investigating the effects of chumming on sharks and people, but in general people are concerned that chumming makes sharks unnecessarily dependent on people and may put nearby swimmers and divers at risk.

5.

So far, scientists have identified nearly 500 species of sharks. Most scientists believe that we have not yet discovered all the shark species that exist in the world. For example, in a study conducted in 1998 and 1999, scientists discovered 14 potentially new species of sharks at a fish market in the Philippines. So finding a new species of shark at a market isn't as crazy as it may seem. You could consult a shark field guide or an expert to see if your species has previously been recorded.

6.

Sharks were beginning to dominate the sea by the Jurassic period, some 208-155 million years ago. Other animals present in the ocean at that time were turtles, clams, snails, and corals. The first primitive sharks appeared at least 400 million years ago. That means they preceded dinosaurs, trees, mammals, and flying insects!

7.

There are indeed many people who believe that shark-fin soup is a delicacy. Unfortunately, the food spurred a practice called finning—catching a shark, removing its fins, and throwing it back into the water. These sharks die after being finned and, since so little of the shark has been used, the practice is both wasteful and a contributing factor in the decline of sharks worldwide. Shark finning has been banned in U.S. waters, and many shark-fin products worldwide now come from fisheries that make use of the entire fish.

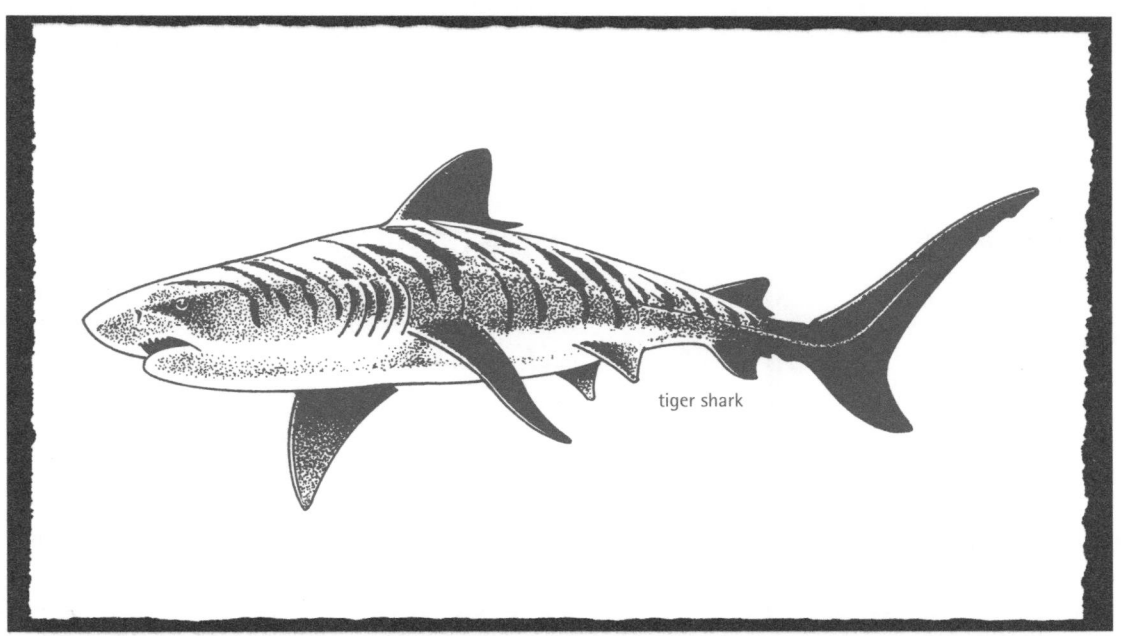

tiger shark

"In all our lives there are milestones, important moments we remember long after. This was one of them. For the brief time of his appearance I drank in every detail of the shark—his eyes, black as night; the magnificent body; the long gills slightly flaring; the wicked white teeth; the pectoral fins like the wings of a large aeroplane; and above all the poise and balance in the water and the feeling conveyed of strength, power, and intelligence."

—Hugh Edwards, naturalist

3 | Sharks in Decline

SUBJECTS
mathematics, science

SKILLS
interpreting (drawing conclusions, inferring, defining problems, reasoning, elaborating), applying (predicting, hypothesizing)

FRAMEWORK LINKS
30.1, 40, 47.1, 50, 50.1

VOCABULARY
bycatch, gill nets, longlines

TIME
two sessions, one night for homework

MATERIALS
bandanas or other strips of cloth, three Nerf balls or other soft objects that can be thrown safely, two 12-foot ropes/clothesline, 8 to 12 clothespins, notebook paper, copies of "Fishing Worksheets A and B" (pages 236-237)

CONNECTIONS
"Where the Wild Sharks Are" (pages 216-228) can set the stage for helping students understand how certain fishing methods may affect some shark species more than others, based on where in the ocean the sharks live. For more on fishing methods that can affect shark populations, try "Catch of the Day" (pages 166-175).

AT A GLANCE
Carry out group simulations of common fishing methods and assess why these methods and sharks' reproductive biology are together contributing to a rapid decline in shark populations.

OBJECTIVES
Describe several methods by which sharks are captured. Discuss some of the advantages and disadvantages of each method.

When shark-attack stories make the news day after day, people start to think that sharks are becoming more aggressive or that their populations are growing. However, sharks aren't increasing in numbers or ferocity. In fact, sharks are suffering significant population declines. Scientists estimate that some species of coastal sharks have declined by between 50 and 75 percent in just the last 20 years.

One reason that shark populations have declined so rapidly is that many common fishing methods accidentally capture sharks in addition to the targeted fish. Another reason is that a growing market for shark meat, shark fins, and other shark products has made sharks a direct target of fishers who previously didn't capture sharks, or at least didn't keep the sharks if they were caught. (For more on shark-fishing methods, see pages 197-199.)

But these practices might not take such a dramatic toll on sharks if it weren't for some basic aspects of sharks' reproductive biology. Sharks are slow-growing, late-maturing animals that don't reproduce very quickly. And they are extremely susceptible to population declines if large numbers of them are killed.

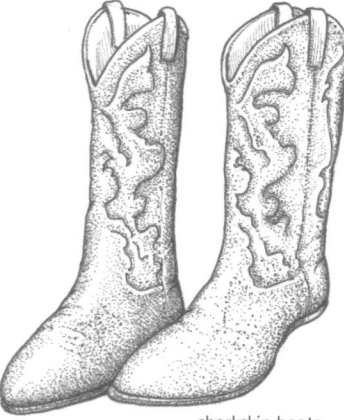

sharkskin boots

This activity contains a series of simulations that explore different fishing methods and how they intentionally or unintentionally lead to the capture of sharks. Then the activity highlights why some fishing methods are so disruptive to shark populations, particularly in light of sharks' reproductive biology.

Before You Begin

You should decide in advance whether you want to conduct the trawling simulation with your students. It is part of the Shrimp Case Study on pages 166-175, but can also be done as part of this activity. If you decide not to do it, strike out all references to trawling on the handouts and do not include the trawling chart.

Before beginning any of the simulations, push desks to the sides of the room, leaving a large open space in the middle. Gather the materials for the simulations. Copy Charts A, B, and C from "Fishing Worksheet A" onto the board, and make a copy of "Fishing Worksheets A and B" for each student.

What to Do

1. Discuss fishing.
Ask students if they have any idea how people catch fish in the open ocean. Have a few students share what they know about the topic, then tell them that you're going to conduct a series of classroom exercises to show different fishing methods and their effectiveness in catching targeted species. Write the following list on the board:
   a) Hook and line
   b) Gill nets and drift gill nets
   c) Longlines
   d) Trawling (optional)

hook and line

2. Simulation A: Hook and Line
The hook-and-line fishing method is used by sport fishers as well as by some commercial fishers. In this simulation, some of your students are going to be fishing for yellowfin tuna using a hook and line. The other students are going to be the tuna, sharks, and other sea creatures.

Ask for three volunteers to be fishers. Have the fishers stand aside while you divide the remaining members of the class as follows:
   1) 3 to 4 pairs of students (with arms linked) = adult tuna
   2) 3 to 4 individual students = juvenile tuna
   3) 3 to 4 pairs of students (with arms linked) = adult sharks
   4) 3 to 4 individual students = juvenile sharks
   5) remaining students = other fish

Tie a bandana or strip of cloth around the arm of every tuna. You need not label the other students, but they should remember what identity they've been assigned.

Now present the rules of the game. The fishers will have one minute to "fish" for a tuna from the group. Since it wouldn't be safe to throw a hook and line at their classmates, they'll "fish" by throwing the Nerf ball or other soft object. To make things harder for the fishers, they have to be touching a desk with a part of their body when they throw the ball. None of the fish may run. Any fish the fishers hit is considered "caught," but if it's not an adult tuna, the fishers should "throw" the fish back into the group and toss the ball again. Have the adult tuna that are caught stand next to the fishers who caught them. Whichever fisher has caught the most adult tuna when the minute is over wins the game. To begin the game, group the fish in the middle of the room. Then tell the fishers to begin. As the fishers catch their fish, record the results on the board on Chart A. (Be sure to count every fish caught, even if the fish is thrown back.) You might want to do another round of fishing if time permits. (To do this, "restock" the waters and select new fishers.)

Afterward, have the students copy the results from the board onto Fishing Worksheet A and analyze the results. How many fish were caught that were not adult tuna? Tell the students that sharks are generally able to survive when they are caught using a hook and line and then thrown back. That being the case, what was the expected total shark mortality in these simulations? *(Answers will vary, but it's unlikely that many would die.)*

3. Simulation B: Gill Nets

Explain to the students that some commercial fishers use gill nets to catch fish in the open ocean. Gill nets allow a fish to fit its head and gill covers, but not its fins or other parts of its body, through the net holes. The gill covers get caught in the net and prevent the fish from wriggling loose. So any fish that are larger at the gills than the holes in the net will get stuck. Once pulled onto the deck of a fishing boat, the fish will quickly die. You might point out that, in addition to being directly targeted by commercial fishers, a lot of sharks are accidentally caught in gill nets by fishers that are targeting tuna.

Some gill nets are fixed in one place and collect fish until they're hauled in. Others are allowed to float through the open water. (These floating gill nets are called drift nets.) Sometimes drift nets get lost; they can float for years gathering fish and other sea creatures in them.

To simulate gill net fishing, select one student to be the fisher. Have that person place the two ropes down on the floor to create three equal-sized "lanes." Then have that person secretly designate one lane to be where the gill net will be. (Be sure the

person tells you which lane he or she has selected before the other students start "swimming.")

Meanwhile, divide the rest of the students as follows (you need not label them, but they should remember the identity they've been assigned):

1) $1/4$ of the students = adult tuna
2) $1/4$ of the students = juvenile tuna
3) 1 student = sea turtle
4) 1 student = dolphin
5) 2 to 4 students = small fish
6) $1/2$ of remaining students = adult sharks
7) other $1/2$ of remaining students = juvenile sharks

Now gather the students at one end of the classroom, and tell them they have to walk to the other end. When they reach the ropes, they should continue down one of the three lanes. Tell them that the fisher has placed a gill net across one of these lanes, but since fish cannot see gill nets, neither can the students. Tell them that they cannot change their lane once they have selected it.

The marine creatures should "swim" from one end of the room to the other, and they should stay in their lanes at the other end of the room. Then have the fisher announce which lane had the gill net, and have him or her count up the catch. All the small fish would have been able to swim through the netting in the gill net. The remaining creatures should be considered caught.

Run through the simulation again if time permits, recording both simulations on Chart B. Have the students copy the figures onto Fishing Worksheet A, Chart B and compare with the results logged on Chart A.

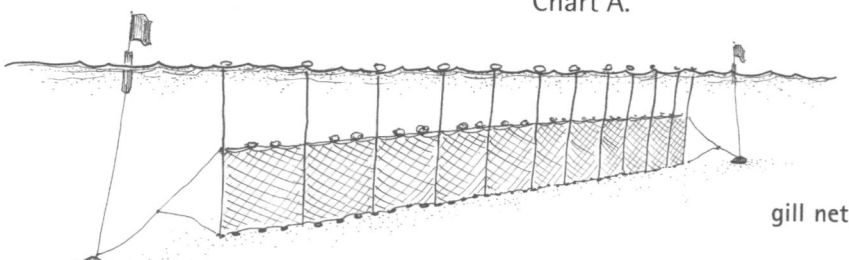

gill net

BIOFACT

In Hong Kong, a bowl of shark-fin soup can sell for as much as $90!

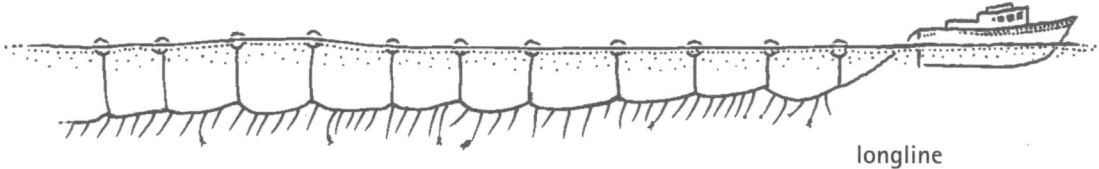

longline

4. Simulation C: Longlines

Explain to the group that longlines are just what they sound like: long, thin cables or monofilament strands that stretch as far as 40 miles across the ocean. (Help your students understand this distance by comparing the distance to a place about 40 miles away from your classroom.) Tell the students that on a longline, there is a float attached to the cable every few hundred feet and a baited hook every few feet. Longlines are often used to capture tuna and billfish such as swordfish. But they also unintentionally catch many sharks.

Choose two people to be longline fishers. Give them one rope, the clothespins, and 10 or more pieces of paper. Then have them go out into the hall and clip the paper on the rope in whatever distribution they want. Tell them that they'll learn how to "fish" with their longline when they get back into the room.

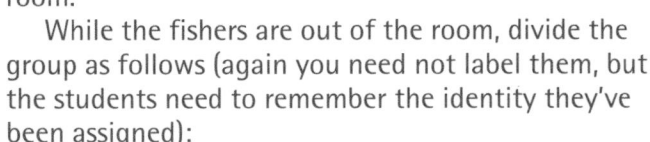

basking shark

While the fishers are out of the room, divide the group as follows (again you need not label them, but the students need to remember the identity they've been assigned):

1. $1/4$ of students = adult tuna
2. $1/4$ of students = juvenile tuna
3. 2 students = sea turtles
4. 1 student = dolphin
5. $1/2$ of remaining students = adult sharks
6. other $1/2$ of remaining students = juvenile sharks

Tell the fish to stand around the room in any configuration they want. The only thing they may not do is stand directly behind another fish. Tell the fish you haven't yet decided which side of the room (front or back) the fishers will start from, so there's no point in bunching up at the back of the class.

Bring the two fishers in and have them stand at the front or back of the room with their rope stretched out across the classroom. Explain that the papers on their longline are meant to represent their baited hooks. They should hold the rope so that the papers pass over the heads of some fish and brush against others. Then have them walk slowly down the length of the classroom, being sure not to shift their longline just to hit a particular fish. The fish may not duck or shift their bodies to avoid one of the "hooks." Every time a fish is brushed by a piece of paper, that student should remove the paper. (In real life, once a hook has caught a fish, no other fish can be caught on it.) Then the fish that are caught should go to the front of the room and identify themselves. Repeat the simulation if time permits.

Discuss the outcome of the fishing, record it on Chart C (with students copying the figures to Fishing Worksheet A, Chart C), and compare the results with those recorded on Charts A and B.

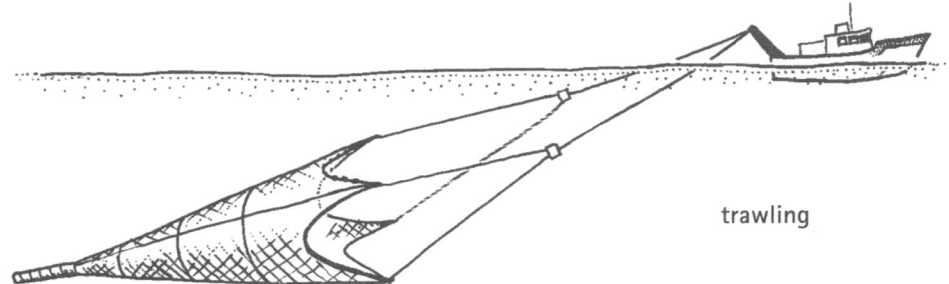

trawling

5. Simulation D: Trawling
(See "Catch of the Day" on pages 167-169 of the Shrimp Case Study.)

6. Discuss simulations.
Ask the students if they have any questions about the simulations. In each simulation, were they surprised by how many sharks and other fish were caught, even though they weren't the targeted species? Explain that this unwanted catch is called *bycatch*. Some students may express dismay that fishers are responsible for killing so many marine mammals and fish that they don't use. You might explain that people are working to minimize this bycatch, but that it is difficult and expensive to change common and ingrained practices.

7. Assign homework.
Assign "Fishing Worksheet B" for homework. Use the worksheets as a means of assessing each student's understanding of the concepts (see Assessments). Then return the sheets to the students and set aside a class period to review and discuss the answers. (An answer sheet is provided on page 238.)

8. Discuss status of sharks.
Tell your students that because of current fishing practices, many kinds of sharks are experiencing huge population declines. In fact, scientists estimate that humans kill at least 100 million sharks every year. What are some ways that people could try to reduce this number? *(Set limits on shark catches, set limits on the size of sharks that fishers may catch, reduce consumer demand for shark fins, or change fishing methods.)* Why might these changes be difficult to implement? *(It's hard to rally public concern for sharks; many sharks move from one country's waters to another's, so fishing limits set by one or two countries won't guarantee that sharks are protected; current fishing methods are profitable to the commercial fishing industry, so any changes are likely to be resisted.)*

9. Research shark conservation.
As a wrap-up to the activity, have your students research current efforts in shark conservation. They should search the Web, contact environmental organizations, check the newspaper for articles, and so on. Allow students to share their findings with the rest of the class.

BIOFACT

The fish used in England's famous "fish and chips" dish is sometimes from a shark—the dogfish shark.

WRAPPING IT UP

Assessments

For assessment, use "Fishing Worksheet B" as a homework assignment.

Unsatisfactory—Provides incomplete or insufficient answers.

Satisfactory—Adequately answers each question.

Excellent—Provides thoughtful responses to each question.

Portfolio

Students should include "Fishing Worksheets A and B" in their portfolios.

Writing Idea

Have students write a letter to the editor of a local newspaper from a fisher's perspective, explaining the pros and cons of the fishing methods explored in this activity. The letter should also discuss the need for further research on new ways to catch fish that cause minimal marine habitat destruction and reduce bycatch.

Extension

Have your students look into the reproductive biology of several shark species. Do sharks reproduce in the same way that other species of fish do? Or are sharks' reproductive habits closer to those of large mammals? Explain.

leopard sharks

Record the fishing results for each of the following methods.
Circle the types of fish that were ultimately kept.

Chart A: Hook and Line

	Adult tuna	Juvenile tuna	Adult shark	Juvenile shark	Other fish
Round One					
Round Two					

Chart B: Gill Nets

	Adult tuna	Juvenile tuna	Adult shark	Juvenile shark	Sea turtle	Dolphin
Round One						
Round Two						

Chart C: Longlines

	Adult tuna	Juvenile tuna	Adult shark	Juvenile shark	Sea turtle	Dolphin
Round One						
Round Two						

Chart D: Trawling

	Kind of netting	Targeted species	Bycatch
Round One			
Round Two			

FISHING WORKSHEET B

1. Explain what you think might be the *advantages* of each of the following fishing methods.

a. Hook and line

b. Gill nets

c. Longlines

d. Trawling

2. What do you see as the *disadvantages* of each of the following methods?

a. Hook and line

b. Gill nets

c. Longlines

d. Trawling

3. Most sharks reproduce slowly, producing small numbers of young at a time and maturing quite late in life—more like elephants or humans than cockroaches or rabbits. Why might some fishing methods, such as using drift nets and longlines, present particular problems for many shark species, whether they are intentionally or accidentally caught?

4. In recent years, many sharks that were caught accidentally were dumped back into the sea—dead or alive. Now, because of the rising popularity of shark-fin soup, many fishers are cutting off the sharks' fins and then dumping the sharks back into the ocean. What do you think of this practice?

ANSWERS TO "FISHING WORKSHEET B"

Note: Answers will vary according to simulation results.

1. **Advantages of hook and line:** Doesn't require extremely expensive equipment, better able to catch target species. **Advantages of gill nets, longlines, and trawls:** They catch many more fish and require less labor and precision.

2. **Disadvantages of hook and line:** Requires a lot of human labor and is time intensive. **Disadvantages of nets, longlines, and trawls:** They catch species indiscriminately so they accidentally kill many unwanted species, although the size of openings in the nets' mesh and the types of hooks used can help make these methods more selective. In some areas, up to half of the bycatch consists of sharks. Also, gill nets and longlines can get lost at sea and will continue catching and killing fish and other marine creatures. When trawling is done along the ocean floor, it destroys ocean habitat.

3. **Because nets and longlines catch sharks of all ages,** they catch many immature females—females that have never had a chance to reproduce. For example, if a female dusky shark is caught before she's 22 years old, she will not have had the opportunity to produce any off-spring. What's more, those sharks that do reach reproductive age do not produce many young at a time, so populations cannot easily bounce back from heavy fishing tolls.

4. **Answers will vary.** Some students may point out that cutting off the fins can give fishers some monetary reward for catching a shark and that it's less wasteful than throwing back a dead shark without using any part of it. Others may think that it's still wasteful to use so little of a shark and may point out that finning only increases the total number of shark deaths and encourages the market in illegal trade. Some students may point to the cruelty of throwing a mortally wounded animal back into the sea to suffer and die.

". . . Or can man not only learn how to live in harmony with his fellow men, but also contrive to co-exist amicably with sharks, avoiding unnecessary conflict and recognizing that planet Earth is the home of other species, as well as Homo sapiens, and that these species have an equal claim on the resources of the world."

—Rodney Steel, science journalist

SUBJECTS
language arts, social studies, art

SKILLS
applying (creating, synthesizing, composing), presenting (writing, illustrating)

FRAMEWORK LINKS
5, 37, 41, 42, 58

VOCABULARY
attitudes, **culture**, nenue, Seri, Tlingit

TIME
one to two sessions, depending on projects chosen

MATERIALS
copies of "Sharks in Culture" (pages 243-246), lined and unlined paper, pencil, paints, colored paper, scissors, glue, clay, musical instruments, and other items as needed

CONNECTIONS
To further explore cultural connections with biodiversity, use "Salmon People" (on the Web) and "The Culture/Nature Connection" in Biodiversity Basics, and "A Wild Pharmacy" in Wildlife for Sale.

 **AT A GLANCE**

Read a traditional Hawaiian story about sharks and then write a poem, make a poster, draw a comic strip, or create a piece of art that portrays your views toward sharks.

 OBJECTIVES

Describe different cultural views of sharks. Articulate your attitude toward sharks.

If you could travel around the world, you'd find signs of sharks everywhere. After all, they don't just inhabit the world's oceans—they also are found in the culture and imaginations of people. For example, sharks play a role in many Hawaiian and Polynesian stories. Sharks decorate the bark paintings of Australian aboriginal peoples and the hats and carved poles of the Haida and Tlingit peoples of Alaska and British Columbia. Pacific Islanders and the Seri Indians of Mexico carve shark sculptures.

Take a close look at those representations and you'll see that negative attitudes toward sharks aren't universal. Far from being the evil villains depicted in most Hollywood movies, sharks are often portrayed in other cultures as powerful guardians, even deities, of the ocean realm.

In this activity, your students will read a Hawaiian story about sharks and look at pictures of shark art from around the world. Then they'll make their own art to portray their perspective on sharks.

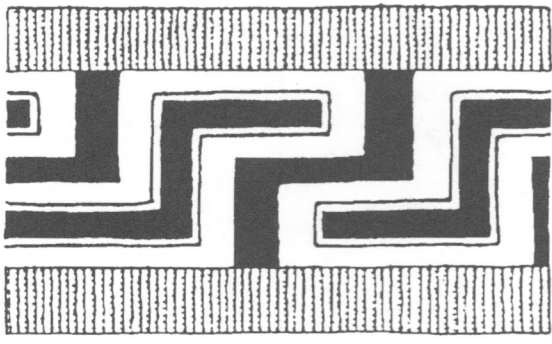

Tlingit design

Before You Begin

Make one copy of "Sharks in Culture" for each student. Gather whatever writing or art supplies you'd like to provide for the students' art projects.

What to Do

1. Hand out copies of "Sharks in Culture."

Have the students look over the handout. Explain that the story and artwork are examples of how sharks are portrayed by people from different cultures around the world. Read "The Shark Guardian," pages 243-244, to the class. Afterward, spend some time discussing the story and the artwork presented. Were the students surprised that the people in the story acted as guardians of sharks? What is the tone of the shark painting? *(Emphasizes the horror of the scene, grotesqueness of the shark.)* What sense of the artists' view of sharks comes through the other shark artwork? *(All three reflect an integration of sharks with the culture—Seri sculpture comes from direct observation of sharks in their habitat, and Australian and Tlingit pieces show spiritual connections to sharks.)* Why might people from different cultures have different attitudes toward and relationships with sharks? *(Answers will vary.)*

2. Assign shark project.

Tell the students that their assignment is to create their own written or artistic representation of their views of sharks. They can model their art piece after one of the examples on their handout (for example, by writing a story about sharks, painting a shark, or carving a shark figurine). Or, they may want to try one of the following:

- Write a poem about sharks, with the word "SHARKS" running down the left side and each line beginning with one of these letters.

- Make a poster for an imaginary movie about sharks. Would the shark be the villain? The hero?

- Draw a comic strip about a superhero who does or does not like sharks.

- Make a shark puppet out of fabric.

- Make a shark piñata.

Whatever format the students choose, their piece should reflect their personal view of sharks.

3. Share results.

Give the students an opportunity to share their art pieces informally or in a class art exhibit. They may want display their pieces in a public space and encourage other people to rethink their attitudes toward sharks.

> *"If there is poetry in my book about the sea, it is not because I deliberately put it there, but because no one could write truthfully about the sea and leave out the poetry."*
>
> **– Rachel Carson, ecologist**

WRAPPING IT UP

Assessment

Use the poem, poster, comic strip or art piece and have each student write an explanation of how their work reveals their personal attitude toward sharks. Students should also describe the reasoning behind their attitudes.

Unsatisfactory—Explains either attitude or reason (but not both) or fails to explain how the art shows the attitude and reason.

Satisfactory—Describes the attitude and reasons.

Excellent—Relates the reasons with the attitude and reveals how the work includes this.

Portfolio

Include students' poetry and stories in their portfolios. If artwork doesn't fit in the portfolio, have students make a sketch of it or take a photo of it.

Writing Idea

Have students interview neighbors and family members to find out about their perspectives on sharks. Based on these interviews, students should write short "Shark Stories" that explore their community's attitudes toward sharks and highlight how those attitudes might be linked to local culture.

Extensions

- The story you read to the students was about shark guardians. Do the students know any other shark guardians? Have them investigate careers that are devoted to protecting sharks.

- Have the students visit Web sites of organizations working on shark conservation. Then ask them to generate a list of what ordinary citizens can do to help sharks.

THE SHARK GUARDIAN

This is a story of the days when Mary Kawena Pūku'i was a little girl in Ka'ū on Hawai'i. One very rainy day she got to thinking of a certain kind of fish. "I want *nenue* fish," she said.

"Hush, child," her mother answered. "We have none."

"But I am hungry for *nenue* fish!" the little girl repeated and began to cry.

"Stop your crying!" said another woman crossly. "Don't you see we can't go fishing today? Just look out at the pouring rain. No one can get you *nenue* fish. Keep still!"

The little girl went off into a corner and cried softly so that no one should hear, "I do want *nenue* fish! Why can't someone get it for me?"

Her aunt came in out of the rain. It was Kawena's merry young aunt who was always ready for adventure. "What is the matter with the child?" she was asking. "The skies are shedding tears enough, Kawena. Why do you add more?"

"I want *nenue* fish," the little girl whispered.

"Then you shall have some. The rain is growing less. We will go to my uncle."

In a moment the little girl had put on her raincoat, and the two were walking through the lessening rain. It was fun to be out with this merry aunt, fun to slip on wet rock and shake the drops from dripping bushes.

At last they reached the uncle's cave. "Aloha!" the old man called. "What brings you two this rainy morning?"

"The grandchild is hungry for *nenue* fish," Kawena's aunt replied.

"And *nenue* fish she shall have," said the old man. Net in hand, he climbed the rocks above his cave home. Kawena and her aunt watched him as he stood looking out over the bay. He stood there like a man of wood until the little girl grew tired watching. The rain had stopped and sunlight touched the silent figure. Why didn't he do something? Why didn't he get her fish? Why did he stand there so long—so long?

Suddenly he moved. With quick leaps he made his way to the beach and waded out. Kawena and her aunt hurried after him and saw him draw his net about some fish and lift them from the water. Just as the girl and woman reached the beach the old man held up a fish. "The first for you,

old one," he said and threw the fish into the bay. A shark rose from the water to seize it. "These for the grand-child," the old man added. He was still speaking to the shark as he gave four fish to Kawena.

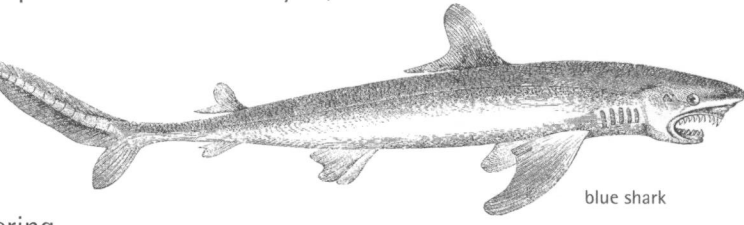

blue shark

The little girl took her fish, but her wondering eyes were following the shark as he swam away.

"That is our guardian," the uncle said. He too was watching the shark until it disappeared.

"Tell her about our guardian," said the aunt. "Kawena ought to know that story."

The uncle led them back to his cave. There, dry and comfortable, they sat looking down at the beach and the bay. "It was from those rocks that I first saw him." The uncle began, his eyes on rocks below.

"One day, many years ago, I found my older brother lying on the sand. For a moment I thought that he was dead. Then he opened his eyes and saw me. 'Bring *'awa* and bananas,' he whispered. I stood looking at him, not understanding his strange words. After a bit he opened his eyes again and saw me still beside him. ''Awa and bananas!' he repeated. 'Get them quickly.'

"As I started away I saw him pull himself to his feet, holding onto a rock. He looked out over the bay and called, 'Wait, O my guardian! The boy has gone for food.' Then he sank back upon the sand. I looked out into the bay, but saw no one.

"I got 'awa drink and ripe bananas and brought them to my brother. He pulled himself weakly to his feet once more and moved out onto those rocks, motioning me to bring the food. He called again and his voice was stronger. 'O my guardian, come! Here is 'awa drink! Here are bananas! Come and eat.'

"Suddenly a large shark appeared just below the rocks on which we stood. As my brother raised the wooden bowl of 'awa, the great fish opened his mouth. Carefully my brother poured the drink into that open mouth till all was gone. Then he peeled the bananas one by one and tossed them to the shark, until the great fish was satisfied. 'I thank you, O my guardian!' Brother said. 'Today you saved my life. Come here when you are hungry.' The shark turned and swam away.

"While my brother rested on the sand he told me his adventure. His canoe had been caught in a squall and overturned. He was blinded by rain and waves and could not find the canoe. It must have drifted away. The waves broke over him and he thought the end had come.

"Then he felt himself on something firm. 'A rock!' he thought, and clung to it. Suddenly he felt himself moving through the waves and knew that he was riding on the back of a great shark and clinging to his fin. He was frightened, but kept his hold.

"The storm passed on, and my brother saw the beach. The shark swam into shallow water, and Brother stumbled up the sand. It was there I found him.

"He never forgot that shark. Often I have seen him standing on the rocks above this cave with 'awa and bananas ready. Sometimes he called. Sometimes he waited quietly until the shark saw him and came. Sometimes the shark drove a small school of fish into the bay as you saw just now. My brother caught some and shared them with the shark.

"The time came when my brother was very sick. Before he died he beckoned to me. 'My guardian,' he whispered. 'You must give food to the one that saved my life.'

"I have not forgotten, and the shark does not forget. I feed him 'awa and bananas, and he sometimes drives fish into my net. Today he wanted *nenue* fish and put the thought of them into your mind. Always remember our guardian, Kawena."

Kawena Pūku'i is a woman now, but she has never forgotten the shark guardian.

Told by Mary Kawena Pūku'i

Reprinted with permission from *Tales of the Menehune* (Revised Edition), compiled by Mary Kawena Pūku'i, retold by Caroline Curtis, and illustrated by Robin Burningham, Copyright © 1960, Revised Edition Copyright © 1985 (reprinted 1996) by Kamehameha Schools Press.

BIOFACT

Remoras are fish that hitch rides on the backs of sharks and other fish and eat tiny shellfish that accumulate on the sharks' fins and gills.

1) "Watson and the Shark" by John Singleton Copley, 1778.

In 1749, a 14-year-old orphan was swimming in the harbor in Havana, Cuba, when he was attacked by a shark. This painting by American painter John Singleton Copley depicts the orphan's shipmates' desperate attempts to rescue him.

Watson and the Shark, Ferdinand Lammot Belin Fund, ©2002 Board of Trustees, National Gallery of Art, Washington, 1778.

Al Abrams

2) Seri Indian ironwood carvings of sharks.

The Seri Indians live on the coast of the Sea of Cortez in Sonora, Mexico, where they hunt and fish for much of their food. Some Seris carve these sculptures to depict the animals they see in their region.

Reprinted with permission of Mainz Didgeridoos. Admin@mainzdidgeridoos.com.

3) Australian bark painting with shark on it.

Australia's aboriginal people have been making bark paintings for thousands of years. They grind colored rocks to make paint and use eucalyptus bark as the canvas. The scenes on the bark paintings reflect the spiritual "dream time" stories of the aboriginal peoples.

4) Tlingit carved pole.

Creatures from the natural world are depicted on most of the traditional poles of the Tlingit people of the Northwest Coast of North America. The animals on these poles are totem animals who help shape the lives of the families connected to them.

U.S. Forest Service

Shark Resources

Here are additional resources to help you design and enhance your Sharks Case Study. Keep in mind that this resource list includes some of the materials we have found or used; however, there are many other resources available on sharks. For a list of general marine biodiversity resources, see the Resources section on pages 360–369.

Organizations

American Elasmobranch Society is a nonprofit science organization that conducts research on sharks, skates, rays, and chimaeras, and promotes public awareness of natural resources. **www.flmnh.ufl.edu/fish**

Mote Marine Laboratory's Center for Shark Research is an international center for research, scientific collaboration, consulting, education, and public information on sharks and their relatives (skates and rays). Their Web site includes shark facts and statistics. **www.mote.org**

Curriculum Resources, Books, and Web Sites

The Bridge—Ocean Sciences Education Teacher Resource Center is a growing collection of online marine education resources. Use the Search feature, or under "Ocean Sciences Topics," click on "Biology," then "Sharks." Sea Grant Marine Advisory Services, Virginia Institute of Marine Science College of William and Mary, Gloucester Point, VA 23062. **www.vims.edu/bridge**

Great White Sharks (Adult) by Richard Ellis and John E. McCosker centers on one of the most feared ocean creatures. Extensively illustrated, the book is the first-published compilation of information and research about great whites. (HarperCollins, 1991). $35.95

NOVA Online: Shark Attack! is an "online adventure" resource on the biology of sharks, their distribution, and the people who interact with them. **www.pbs.org/wgbh/nova/sharks**

The Shark Almanac (Adult) by Thomas Allen tries to dispel myths of sharks as man-eaters and delves into the unknown world of these marine creatures. The book provides recent scientific research as well as updates on continuing shark conservation efforts. (Lyons & Burford Publishers, 1999). $35.00

Shark Research Program of the Florida Museum of Natural History provides links relating to various aspects of sharks. Viewers can visit an image gallery featuring a range of species or click on "Education" for a discussion of shark natural history. **www.flmnh.ufl.edu/fish/sharks/sharks.html**

Sharks (Elementary) by Niki Walker and Bobbie Kalman is a general shark resource for children. Complemented by color photographs, the book outlines various species of sharks, from tiny cookie-cutters to giant makos. (Crabtree Publishing Group, 1997). $19.96

Sharks! (Elementary) by Irene Trimble and Mike Maydak, a work of the "Know-It-All" series, is a die-cut book featuring shark species, their diet, and physical abilities. (McClanahan Book Company, 1999). $2.79

Sharks in Question: The Smithsonian Answer Book (Adult) by Victor G. Springer and Joy P. Gold addresses commonly asked questions about sharks in the first half of the book. The second half covers the biology of various shark species and the reasons behind attacks on humans. (Smithsonian Institution Press, 1989). $24.95

Case Study
Alien Species

Voracious eels, killer weeds, and choking mussels may sound like the stuff of science fiction. But in fact they're real-life organisms that have unintentionally invaded marine habitats and created a host of problems for native species. In this case study, your students will meet a motley assortment of alien marine species, analyze the characteristics that make them so troublesome, and tally their ecological and economic impacts. Your students will also learn what people are doing to combat alien species and will debate the pros and cons of these efforts.

moray eel

Gary Buckingham, USDA ARS, Image 4723002, www.invasive.org, March 18, 2002

"Aquatic invasive species are devastating our native plants and animals, and can undo much of the progress we have made in conservation."

—Steve McCormick, President, The Nature Conservancy

Background Information

In 1999, divers exploring a marina in Darwin, Australia, made a startling discovery. Since their previous dive just six months earlier, hundreds of millions of tiny mussels had invaded the marina and taken up residence on the bottom of boats, on ropes, and on piers. Scientists later identified the mussel as a Central American cousin of the zebra mussel, a creature that has wreaked havoc in the Great Lakes in recent decades. This striped mussel, they determined, must have reached Australia by hitching a ride on the hull of a yacht.

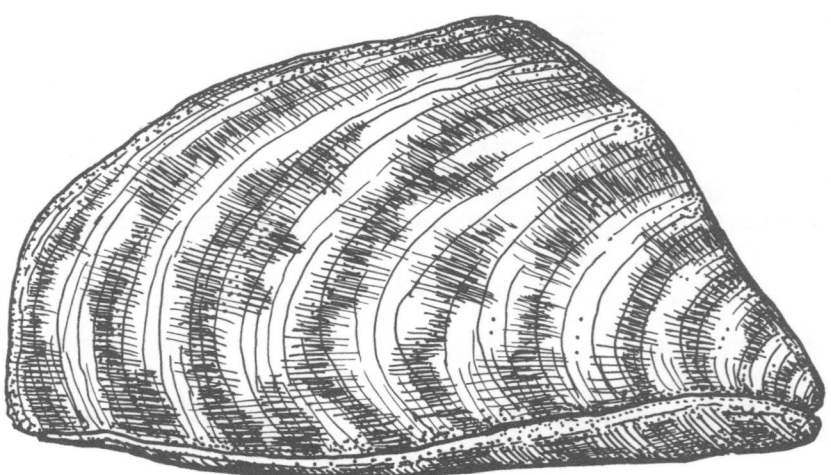

zebra mussel

All around the world, alien species—from mussels to microbes to monkeys—are popping up in unexpected places. Alien species, also called *nonnative species*, are any species that have been taken from their native habitat and transplanted to a new environment. Alien species don't always disrupt their newfound habitat, but, when they do, the ecological and economic costs can be enormous. (Alien species that actually take over habitat and displace native species are called *invasive species.*) That's why more and more people are starting to pay attention to invasive alien species— looking into where they are, what effect they are having, and what might be done to get rid of them. Invasive aliens can be found in almost every ecosystem on Earth, including oceans, marshes, forests, grasslands, and lakes.

All About Aliens

Place Invaders

If you live near an ocean or any large body of water, you've probably heard about one or more aquatic invaders in your area. Among the most notorious nonnative species are green crabs, which have munched their way up the East Coast of the United States, devastating the region's soft shell clam industry; zebra mussels, which have clogged water pipes, sunk buoys, and smothered native clams, mussels, and crayfish in the Great Lakes and Mississippi River; and *Spartina*, or cordgrass, an invasive species that has displaced oyster beds off the Pacific Northwest coast. These are just a few of the thousands of alien aquatic species making their presence felt in places all around the world, and new ones turn up frequently. For example, if biologists can't stop their spread, Asian carp could turn out to be an even greater threat to Great Lakes species than the zebra mussel is.

Of course, alien species have invaded countries all around the world, and many have originated in U.S. waters. In other words, our native species have become someone else's aliens. And in some cases, native species become aliens by spreading from one region to another within the same country.

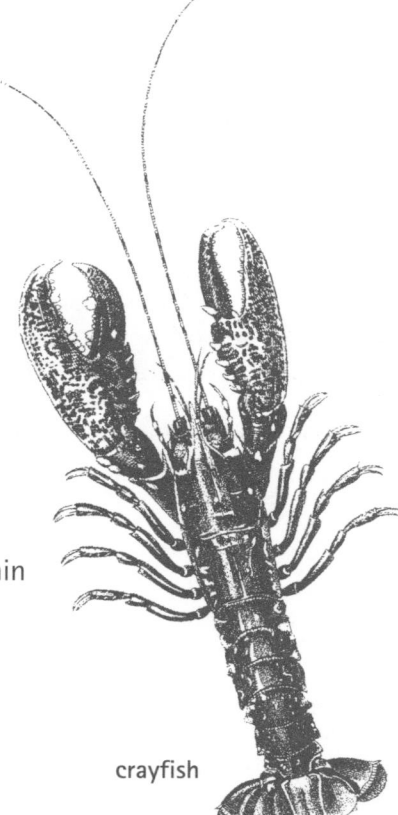

crayfish

purple loosestrife

Norman E. Rees, USDA ARS, Image 0022074, www.invasive.org, February 6, 2003

Most invasive alien species spread uncontrollably because there's nothing to keep their populations in check. Like the green crab, alien species can be voracious eaters that decimate local plant or animal populations because they have few if any predators. That's bad news not only for the species being consumed, but also for those that depend upon the same food supply. Zebra mussels, for example, have caused native freshwater species to decline by filtering out tremendous amounts of the phytoplankton upon which all species in the local habitat depend.

Alien species can also be aggressive, rapidly growing plant species that elbow native species out of their natural habitat. For example, purple loosestrife has displaced cattails and other native wetland species. As cattails have declined, so have birds such as bitterns and rails, mammals such as muskrats, and other animals that depend on cattails for shelter or food.

A Few Words of Introduction

All of the following words describe species that enter new territory. Just remember that not all such species are considered to be invasive or a nuisance.

alien species	introduced species	nonnative species
bioinvaders	invasive species	nuisance species
exotic species	non-indigenous species	weedy species

Alien species don't just affect diversity at the species level, they also disrupt it at the ecological level, changing the entire nature of ecosystems and biological communities. In San Francisco Bay, for example, the Asian clam is thought to be responsible for a large reduction in phytoplankton, an important element of the bay's food web. The Chinese mitten crab, another bay invader, burrows into banks and levees, accelerating erosion and siltation rates. The silt impairs the feeding and breathing processes of aquatic insects that the native fish feed on and damages insect and fish eggs.

Note: This case study places a special emphasis on the impact of alien species on saltwater environments. But as you'll see, the information and activities that follow also contain references to freshwater and terrestrial environments. Why cover freshwater and terrestrial species in a marine module? Many students using this case study live far from an ocean, but they can learn about the problems caused by aquatic alien species by studying these species in nearby freshwater and terrestrial ecosystems. And many students are likely to be familiar with invaders in backyards and local recreational areas. That's why local species offer opportunities for student action and restoration projects.

BIOFACT

Fur farmers first introduced nutria (a South American rodent) to the southern United States in the 1930s. Since then, they have escaped from fur farms and have been released in lakes to control vegetation. Unfortunately, they quickly spread to other areas, eating vegetation and damaging habitat.

> *"Aquatic nuisance species may constitute the largest single threat to the biological diversity of the world's coastal waters."*
>
> **—Marsha Gear, California Sea Grant Program**

Scientists now believe that alien species are among the most significant threats to biological diversity at every level. While habitat loss and pollution have gained more attention as causes of biodiversity decline, alien species have methodically been pushing hundreds of native species into threatened or endangered status and have driven many endangered species to the point of extinction. Scientists have calculated that alien species are taking a toll on nearly half of all endangered species. And they think that alien species have played a role in 24 of the 30 known fish extinctions that have already occurred in the United States.

In addition to these ecological costs—which affect humans and wildlife alike—alien species are creating big economic costs for people. The Atlantic comb jelly, for example, devastated stocks of anchovies and other native fish in the Black Sea, leading to an estimated $350 million loss in fisheries revenue over the past 18 years. In the United States, scientists recently estimated that alien species as a whole have cost the country $137 billion annually in economic losses. (See the box titled "Counting Costs" for a tally of the economic costs of some marine invaders.) Unfortunately, the problems with exotic species are likely to worsen in coming years. Among recent troublesome discoveries has been the arrival of the rapa whelk, an Asian snail with a huge appetite, in the waters of the Chesapeake Bay. In the bay, which is one of the nation's largest seafood nurseries, the rapa whelk could cause significant ecological destruction and could devastate the region's clam industry. The whelk could also cause further damage to the Chesapeake Bay oyster population, which has been devastated from historic overfishing and a parasitic disease.

Counting Costs

Ecologist David Pimentel and his colleagues recently released a list of 25 "Unwelcome Visitors" to the United States, along with their estimated associated costs. The list includes the following marine species:

Zebra mussels: $5 billion over 10 years

Asian clams: $1 billion annually

Green crabs: $44 million annually

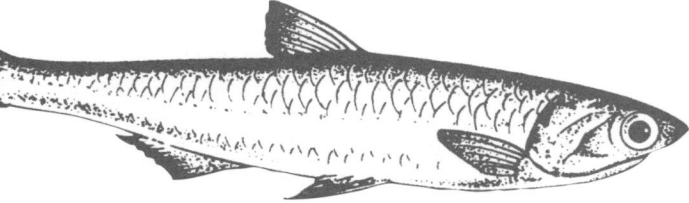

anchovy

Aliens on the Move

The Niña, the Pinta, and the Rat?

Alien invasions are hardly a new phenomenon. In fact, seeds and insects have been catching a lift to new homes for as long as the wind has blown and birds have flown. The collision of continents and the freezing of connecting water bodies have long given even large, land-bound organisms a natural bridge to new lands.

But once humans came along, the opportunities for species relocation picked up dramatically. People purposely carted cats, pigs, goats, cows, and other prized domesticated animals to new lands. And they unintentionally brought along a slew of stowaways, such as disease microbes, burrs, lice, and rats. Both the livestock and the stowaways were aliens in their new homes, capable of affecting the species that already lived there. For example, invasive pigs, rats, dogs, and cats, which arrived by boat along with early human settlers, decimated many Pacific Island bird species that had never developed (or that had lost) defenses, such as flight. By the fourteenth century, alien stowaways were routinely hitching free rides from ocean to ocean. Christopher Columbus and other European explorers took along more hitchhiking species than they intended on their expeditions to America, and they undoubtedly took some hitchhikers home.

THE FAR SIDE® BY GARY LARSON

© 1991 FarWorks, Inc. All Rights Reserved/Dist. by Creators Syndicate

The Far Side® by Gary Larson © 1991 FarWorks, Inc. All Rights Reserved. Used with permission.

How poodles first came to North America

Ever since, species have continued to crisscross the globe as steadily as humans have traveled it. What's different today is that the rate of travel and international trade (much of it by way of the world's oceans) has increased greatly. That, and the fact that our aquatic systems are also being compromised by other human activities, means that the pace of alien invasions has picked up as well. The problem is compounded because alien species seldom disappear from a place where they have successfully taken hold. That means that their numbers—and their effects—tend to accumulate over time. What's more, species that have been introduced to a new region often spread relentlessly. For example, the European green crab first arrived in the United States in the nineteenth century, landing somewhere between New Jersey and Cape Cod. By the 1960s, it had spread north through Nova Scotia, damaging New England's once lucrative soft-shell clam fishery.

Norway rat

403

Aliens à la Mode (of Transport)

Today, alien species reach their new environments through all sorts of channels—by sticking to the soles of travelers' shoes, escaping from pet stores, and slipping into airplane cargo.

For alien marine species, though, the primary mode of transport is the ballast water of ships. Ballast water is the water ships take on at the beginning of voyages to help them maintain balance on the sea. Unfortunately, that water can also contain thousands of aquatic species— from crabs to clams to sea squirts—all of which are subsequently discharged with the ballast water on arrival in the next port. Ballast water has transported the Atlantic comb jelly from the East Coast of the United States to the Black Sea, carried the Chinese mitten crab and hundreds of other exotic species into San Francisco Bay, and brought the ruffe and round goby to the Great Lakes.

Another common way that new species are introduced into ecosystems is through *aquaculture*—the farming of aquatic plants and animals, such as fish, shellfish, and seaweed. This industry has been responsible for both intentional and unintentional introductions. The Pacific oyster was introduced intentionally to the Washington coast early in the twentieth century to launch oyster farming in the area. But several unwanted guests were packed in the same crates as the oysters—including the Japanese oyster drill and *Spartina*, or cordgrass—both of which are causing huge problems. Also, large numbers of Atlantic salmon are now being raised in netlike pens on the Pacific coast of North America. Unfortunately, these alien salmon regularly escape and pose a threat to already endangered species of native salmon in the region.

Comb jellies, which are native to the Atlantic Ocean, are unwelcome guests in several marine ecosystems around the world. Comb jellies can produce up to 3,000 eggs per day, causing biologists to include them on a list of the world's worst invaders.

Those of us who maintain home aquariums may also be responsible for releasing alien aquatic species to the wild, especially freshwater plants and animals. The swamp eel is one example of an aquarium escapee. Another is the common goldfish, which has become a nuisance species in parts of eastern Washington.

Sometimes nonnative species are intentionally introduced in an effort to control the growth and spread of other introduced species. This practice is referred to as biological control. But, unfortunately, it doesn't always work—and may make matters worse. Grass carp, introduced to control unwanted aquatic plants in inland lakes, had a bigger appetite than expected and decimated native plant species. Rosy wolf snails, introduced to control the spread of the giant African snail in the waters off the Hawaiian island of Oahu, ended up compounding the problem by feeding on native snails too. As a result, many of the snails unique to the Hawaiian Islands are gone.

Restaurants and seafood markets sometimes play a role in introducing nuisance species. Alien species can hitch a ride in seaweed, seawater, and other materials used to transport live seafood. They can then reach open water again if the materials are disposed of improperly.

Another way alien species are spread is by ship-bottom fouling. Fouling occurs when species cling to the hulls of boats, which then—unintentionally—transport the species to new places. In Hawaii, most of the introduced invertebrates have arrived this way.

Channels, canals, and locks also bring nonnative species into new waters. During the 1920s, the Atlantic sea lamprey followed the Welland Canal around Niagara Falls to reach its new home in Lake Erie. By 1946, it had made its way to Lake Superior.

The ABCs of Successful Invasion: Learn from the Best!

The green crab is a great (and terrible!) example of a highly adaptable invasive species. Consider these qualities:

- *High reproductive rate: One female can lay up to 200,000 eggs per year.*

- *High dispersal potential: Green crab larvae can remain ocean-bound in algae and plankton for up to 80 days, giving them the potential to travel up to 1,200 miles during the larval stage.*

- *Rapid growth rate: Green crabs reach maturity in two to three years.*

- *Extremely broad habitat tolerances: Green crabs can live in a variety of habitats, including those with a great range of salinity and temperatures from 32° F to 91° F.*

- *Extremely broad diet: The green crab is omnivorous and opportunistic, able to survive on a variety of foods that includes plants, insects, clams, oysters, worms, snails, mussels, urchins, sea stars, fish, and other crabs.*

Some species are purposely stocked to enhance recreational fisheries: European brown and coastal rainbow trout are now reducing native cutthroat trout in many U.S. rivers where the trout were introduced for the benefit of sport fishers. In the 1950s, three fish species were intentionally introduced to Hawaii's coasts to supplement fishery stocks. These species are now taking their toll on native fish and corals.

Last but not least, people sometimes introduce alien species by dumping their live bait overboard at the end of the day of fishing.

Friend or Foe?

Needless to say, alien species reach new territory through about as many pathways as one can imagine. But what happens once they get there? Does every species run rampant in its new environment, causing loss and destruction? Fortunately, the answer is no. Many alien species simply can't survive in their new territory. If you introduced a humpback whale to a freshwater pond, it would clearly die for lack of food and space. (While the introduction of alien species can cause a great deal of damage, relatively few have proven to be harmful—at least so far. The problem is that while some species may seem to be harmless, it may take several years for the extent of the damage to become apparent.)

Furthermore, some introduced species can even be seen as beneficial. The intentional introduction to the United States of wheat, soybeans, poultry, and other agricultural products has helped to feed generations of people here and in other countries.

The presence of harmless or beneficial alien species doesn't diminish the serious and lasting damage that other aliens cause to their new environments. To counter some of this damage, scientists are trying to understand more about what makes certain alien species spread so easily. As mentioned, scientists think that, in some cases, species whose numbers were kept in check by predators in their native habitat grow relatively unchecked at least in part because of few predators in their new home.

Another major factor that allows nonnative species to flourish is their rapid ability to adapt to changing environments. Adaptability is the product of a variety of factors, including high reproductive rates, rapid growth rates, high dispersal potential, and flexibility in terms of food and habitat. For example, the swamp eel is able to breathe air, it can live easily in only a few inches of water, and it eats a wide variety of prey. All of these factors help the eel to adapt to and thrive in a variety of settings. The ailanthus tree, a common invasive plant, can grow in very small spaces and requires minimal nutrients—in fact, scientists have even found these plants growing under subway grates. (For more on the adaptability of invasive species, see "The ABCs of Successful Invasion" on page 256.)

Aliens Solutions

Fighting Back

With all the problems associated with species invasions, it should come as no surprise that people are working hard to reduce the number of further invasions and to control existing problems. Unfortunately, we have yet to come up with any surefire control mechanisms. Invasive species are by definition quick spreading, and it's hard to guess where and when they'll show up next.

Still, many countries (including the United States) are trying to slow the influx of alien species by conducting regular border inspections. In the United States, several federal agencies enforce a battery of national and international regulations to help ensure that troublemakers don't make it past the border. In Australia and New Zealand, public officials have taken an even more demanding approach, allowing only carefully reviewed and tested "safe species" to be imported, unless the person bringing in the exotic species can prove that it will not be harmful. Records show that this approach has slowed invasions by about 30 percent. In the past, these kinds of stringent rules have often met with great resistance in the United States by nursery and pet store owners who don't want to see their product lines limited, but even these groups have become more supportive of controls in recent years.

Another preventive measure that many ecologists support is reducing the introduction of marine species through ships' ballast water. Already, many places require that ships exchange their ballast water while out at sea in hopes that most of the imported organisms will die in the nutrient-poor, highly saline waters. But since this approach is hard to monitor, and some people think it's risky, some scientists are now looking into the possibility of sterilizing or filtering ballast water before it's released.

Until recently, conventional wisdom held that preventive measures were the only way to stop species invasions. In the past, it was difficult to imagine any more aggressive method of controlling pest species than physically removing invasive plants from woodlands, wetlands, and other habitats, which proved to be marginally effective as well as extremely expensive and time-consuming. But recent efforts show that it is not impossible to stop an invasive species—even in aquatic environments—once it has taken hold.

In the case of the striped mussel that invaded the Australian marina, authorities took radical action by poisoning the waters with copper and chlorine, killing off all of the organisms in the marina. Already, native species have begun to reclaim the bay, while the exotics have not been seen again.

In a similar move, authorities in southern California decided that they had only one option if they hoped to stop the spread of a South African worm that was infecting the cultured abalones: They killed the infected abalones, as well as more than a million black turban snails that were known to host the worm. To date, this remedy has prevented the spread of this invasive species.

abalone

Unfortunately, efforts to restore habitats once exotics have invaded are extremely expensive, especially considering that these habitats are being degraded by human activities at the same time. The Australian marina remedy cost about $1.5 million, and the need for constant monitoring by divers continues to add to that price. So, experts agree that the cheapest and easiest way to control exotics is still to prevent them from coming in at all.

Future Stocks

It is difficult to anticipate how alien species will affect wild and human communities in coming years. Some scientists predict a gradual reduction in biodiversity, leading to the dominance of only the most adaptable species, among them rats, pigeons, starlings, cockroaches, house sparrows, raccoons, and humans. Others are hoping that increased efforts to prevent and control invasions will pay off. In 1999, U.S. President Bill Clinton signed an executive order instructing all federal agencies to stop activities that might be helping exotic species to spread. The order also called for the formation of a federal council to devise a management plan for invasive species.

For some, the best way to reduce the negative impacts of exotic species is to take a diversified approach: Prevent every invasion possible and aggressively combat those that accidentally occur. Because these efforts will require the support and involvement of millions of people, public education is also an important part of combating invasive species. A good model, say some, is Australia, where invasive aliens have devastated native species. As one person puts it, "the average taxi driver" in Australia is already well informed about the power and destruction of exotic species. Perhaps with good education, public support, and enough financial investment, alien species can be kept well enough at bay that our own coasts, lakes, and rivers continue to abound with rich and varied life.

kudzu vines

Kerry Britton, USDA Forest Service, Image 0002156, www.invasive.org, February 6, 2003.

BIOFACT

Once established in a new, nonnative habitat, kudzu vines grow incredibly fast—extending as much as 60 feet per season, which amounts to about 1 foot per day of growth.

1 America's "Most Wanted"

SUBJECTS
science

SKILLS
gathering (reading comprehension, researching, collecting, identifying main ideas), interpreting (summarizing, relating, inferring, drawing conclusions, identifying cause and effect), presenting (reporting, explaining, clarifying)

FRAMEWORK LINKS
3, 23, 44, 48, 49, 65

VOCABULARY
alien species, biodiversity, ecosystem, exotic species, habitat, introduced species, invasive species, nonnative species, native species

TIME
two to three sessions

MATERIALS
copies of "America's 'Most Wanted'" (page 264), "The Plaintiffs" (page 265), and "Case Profile" (page 266)

CONNECTIONS
Start off with "Sea for Yourself" (pages 72-81) so students have the chance to explore local aquatic habitats and experience firsthand any invasive species that may be affecting your area. Follow up with "Decisions! Decisions!" (pages 284-289) to help students weigh pros and cons of programs designed to rid areas of harmful alien species.

 **AT A GLANCE**

Track down information on one of several "most wanted" alien species, and find out how it is causing problems in the United States.

 **OBJECTIVES**

Define alien species. Describe some environmental problems associated with alien species. Give examples of several alien species that are causing problems in the United States.

For most people, the word "aliens" conjures up images of green-skinned creatures with antennas that invade Planet Earth from outer space. In the field of ecology, though, aliens are something much less extraordinary. They're the purple loosestrife growing in our marshes and lakes, the starlings pecking at our lawns, and the zebra mussels clinging to piers in the Great Lakes and the Mississippi River. Generally speaking, alien species are organisms that have been transplanted to a new environment, intentionally or unintentionally.

Your students may not know much about alien species before you start this unit, and they may not realize what a menace some of them pose to natural communities around the country and the world. This activity will introduce your students to some of the most meddlesome invasive species found on land and in water in the United States. The activity will also help your students explore some of the native species that are being the most severely affected by alien invasions.

Before You Begin

Make one copy for each student (or each team of students) of the "America's Most Wanted," "The Plaintiffs," and "Case Profile" handouts. If you prefer to have students research local alien species, include several invasive species affecting your area on the list of "Most Wanted" species. Be sure to include a "plaintiff" species affected by each of these species. Arrange for student to have Internet access.

What to Do

1. Hand out copies of "America's Most Wanted."

Tell your students that this handout describes nine of the "most wanted" plant and animal species. What are these species? Why are they wanted? These are questions that your students, as bio-detectives, will have to figure out.

2. Distribute copies of "Case Profile" and "The Plaintiffs."

Explain to the students that their job is to do the following detective work: Gather information (individually or in teams) to complete a "case profile" for one of the "most wanted" species. Based on this information, students should be able to pick out which species on the "plaintiffs" list is being harmed by their "most wanted" species. Assign one "most wanted" species to each student or team of students. The students should gather information from books, magazine articles, and the Internet. Although it will probably be easiest for them to search the Web using the name of their species as a search term, you may also want to direct them to the more general alien species Web sites listed at the end of this activity.

3. Report back with results.

Have your students share the results of their investigations. Go through one "plaintiff" at a time and ask the students whose "most wanted" species are contributing to its decline. What can the group tell the rest of the class about the "most wanted" species that they researched? At some point, it should become clear to the class that all of the "most wanted" species have something in common. What is it? *(They have all come—or have been brought—to the United States [or a particular part of the United States] from another country or region. All of them are harming species in their new environment.)*

4. Discuss alien species.

Explain to the group that these introduced species are called *alien species* (see list of other names on page 252). Review the definition of an alien species. *(It's a species that has been transplanted into a new environment, either intentionally or unintentionally. For example, dandelions, ginkgo trees, and oxeye daisies are all alien species: They originated on other continents and reached North America with the help of people.)* Can your students think of any alien species other than those on the list you have given them? Explain that they may know this group of organisms by some of the other names used to describe them; for example, exotic species, introduced species, or nonnative species.

BIOFACT

The European starling was first introduced to the United States in 1890. Wishing to better the new world by populating it with all of the bird species featured in William Shakespeare's plays, Eugene Scheffland let loose 100 starlings in New York's Central Park. It is now found throughout the United States, Canada, and Caribbean.

U.S. Invasive Species

The following is a partial list of troublesome alien species found in different regions of the United States.

Northeast

Green crab, gypsy moth, purple loosestrife, Dutch elm disease, hydrilla, mute swan, West Nile virus, European starling, hemlock woolly adelgid, Asian long-horned beetle

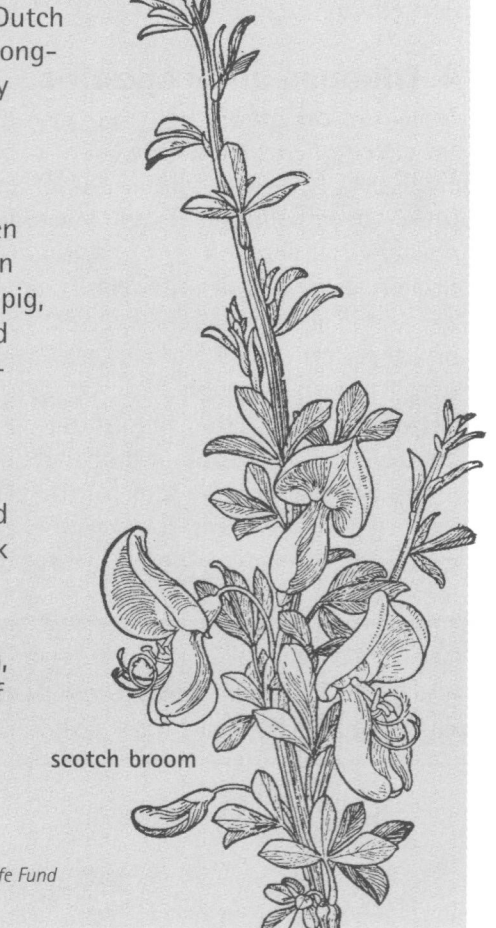

water hyacinth

Southeast

Green crab, melaleuca, water hyacinth, Chinese tallow, hydrilla, English ivy, kudzu, fire ant, Brazilian pepper, nutria

Midwest

Zebra mussel, rusty crayfish, sea lamprey, gypsy moth, purple loosestrife, Dutch elm disease, Asian long-horned beetle, leafy spurge

West

Scotch broom, green crab, Chinese mitten crab, goldfish, wild pig, hydrilla, Africanized bee, fire ant, cheatgrass, bullfrog

Southwest

Fire ant, Africanized honey bee, tamarisk

Hawaii

Mongoose, wild pig, wild goat, rosy wolf snail, invasive marine algae, giant African snail

scotch broom

(**Note:** When using the term "alien" in this discussion, you should be careful to be sensitive to students who are legal or illegal aliens of the United States.)

Now explain that those alien species that are harmful to native species are called *invasive* aliens. Did the students realize that invasive alien species are causing so many problems for native species and native habitats? Explain to your group that invasive aliens are among the top threats to biodiversity—the variety of life on Earth. (You might use this opportunity to review the five greatest threats to biodiversity: **H**abitat loss, **I**ntroduced species, **P**opulation growth, **P**ollution, and **O**ver-consumption of natural resources. These threats can be remembered by the acronym HIPPO.) Some people think alien species are actually the biggest threat to Earth's biodiversity. While not all alien species are destructive, the ones highlighted in this activity are among the most troublesome in our country.

Did the students find any information about the ways alien species affect humans? *(Answers will vary.)* Your students will see that the biodiversity loss associated with alien species can have economic, social, and even political consequences.

In the rest of this case study, students will be learning more about invasive alien species—what they are, what they do, and how they can be controlled—with a focus on those alien species that affect aquatic ecosystems.

The HIPPO Dilemma

To understand the role people are playing in biodiversity loss, it helps, to think of something called the HIPPO dilemma. This term doesn't refer to hippopotamuses; rather, it's an acronym for the main threats to biodiversity:

Habitat loss
Introduced species
Population growth
Pollution
Over-consumption

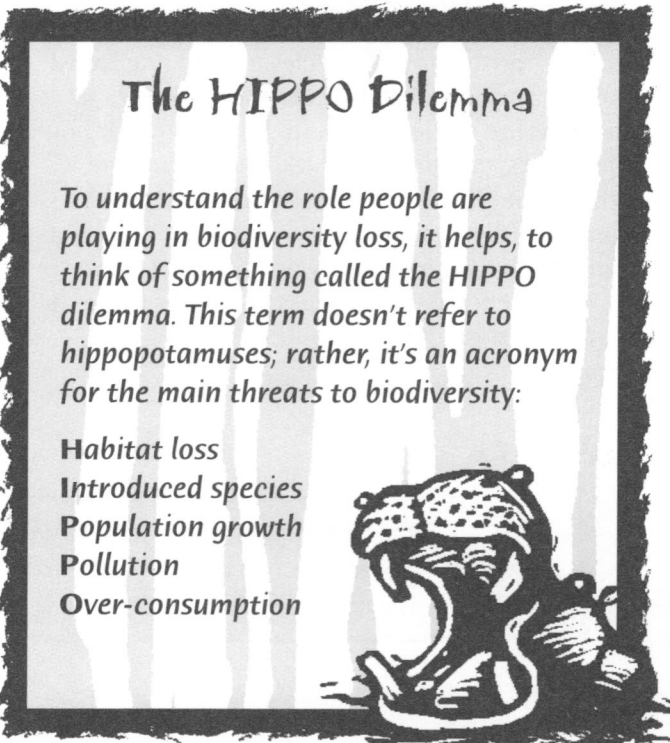

WRAPPING IT UP

Assessment

Have each student create a "most wanted" poster for a problem alien species that is not featured in the activity. (Refer to pages 290-291 for helpful resources and the list on page 262.) The poster should include a picture, the name of the species, and basic information such as the background of the species and what kind of trouble it is causing. (The posters could also be placed in a gallery of "most wanteds" when completed.)

> **Unsatisfactory—**The species is not pictured or the information is only partially presented.
>
> **Satisfactory—**The species is pictured and basic information is presented regarding the origin of the species and some of the problems it has created.
>
> **Excellent—**The species is pictured with information about its origin (including the ecosystem), the regions of impact, and the types of problems it is creating in these new environments.

Portfolio

Have students write a short piece on their "Most Wanted" species, describing how that species is affecting one of the plaintiffs highlighted in the activity. Include this writing in their portfolios.

Writing Idea

Have students interview local resource managers, such as park rangers or scientists working in city, county, or state parks, to find out about troublesome invasive species. Each student should write a short piece on a local invader, which can be included in the school newspaper as well as in a local park or garden club newsletter.

Extension

Have your students gather newspaper, magazine, and Internet articles on alien species. Create an "invasive alien species" bulletin board to keep your group focused on the topic for the duration of the unit. (See below for alien species Web sites.)

European starling

Invasive Alien Species Web Sites

www.invasivespecies.gov

www.issg.org

www.nps.gov/plants/alien

http://tncweeds.ucdavis.edu

www.usgs.gov/invasive_species/plw

www.invasive.org

www.aphis.usda.gov

www.great-lakes.net/envt/flora-fauna/invasive/invasive.html

www.hear.org/AlienSpeciesInHawaii/index.html

www.protectyourwaters.net

www.anstaskforce.gov

Purple loosestrife
(Lythrum salicaria)

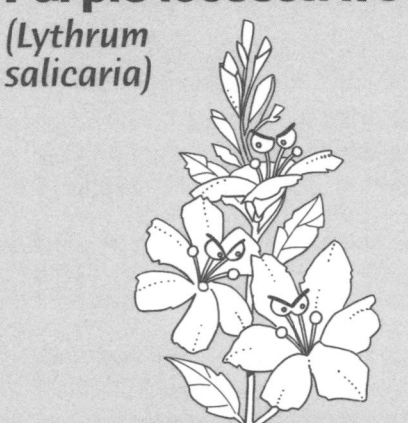

Asian long-horned beetle
(Anoplophora glabripennis)

Bullfrog
(Rana catesbeiana)

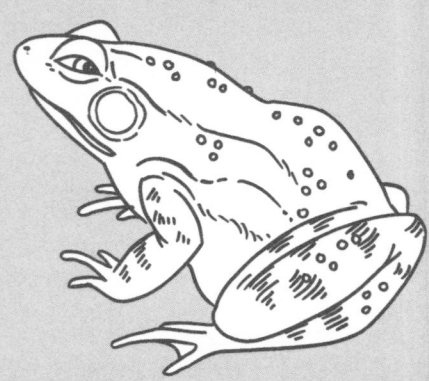

Asian swamp eel
(Monopterus albus)

Indian mongoose
(Herpestes auropunctatus)

Fire ant
(Solenopsis invicta or Solenopsis wagneri)

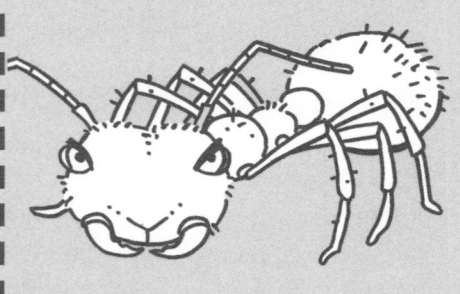

Zebra mussel
(Dreissena polymorpha)

European starling
(Sturnus vulgaris)

Chinese mitten crab
(Eriocheir sinensis)

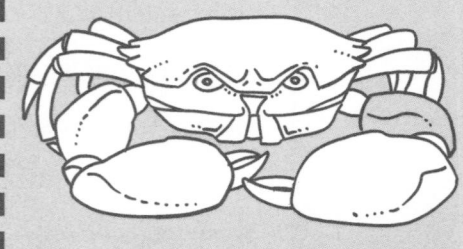

Least bittern
(Ixobrychus exilis)

Leopard frog
(Rana pipiens)

Largemouth bass
(Micropterus salmoides)

Horned lizard
(Phrynosoma platyrhinos)

Red-headed woodpecker
(Melanerpes erythrocephalus)

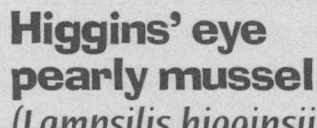

Higgins' eye pearly mussel
(Lampsilis higginsii)

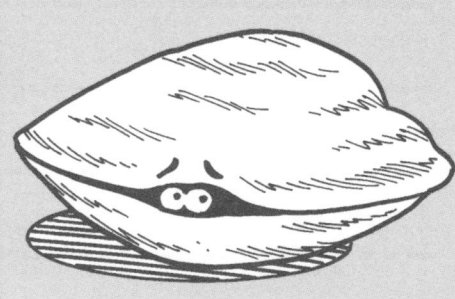

Coho salmon
(Oncorhynchus kisutch)

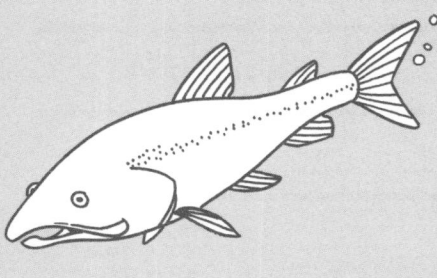

Nene
(Branta sandvicensis)

Maple tree
(Acer spp.)

Aliases:

_____

_____

Origin:

_____

_____

U.S. states or regions where it has been seen:

_____

_____

_____

Distinguishing features:

_____

_____

_____

Preferred habitat:

_____

_____

_____

Which plaintiff is suffering declines because of this species?

_____

_____

_____

_____

ANSWERS TO CASE PROFILES

1. **Asian swamp eel** (also known as the rice eel, ricefield eel, belut, and rice paddy eel). Originally from subtropical and tropical climates in East Asia, now found in Georgia, Florida, and Hawaii. Extremely hardy eel, reaching lengths of three feet or more. Breathes under water but can travel across dry land, sometimes in packs of up to 50 eels, breathing air through its snout. Prefers tropical freshwater habitat but can tolerate high salt and freezing temperatures. Probably released by aquarium owners. A threat to frogs, aquatic invertebrates, and native fish such as largemouth bass, on which it feeds. Arrived in Hawaii around 1900, arrived in Georgia and Florida around 1996. Major concern is the threat the swamp eels now pose to native species in the highly diverse ecosystem of the Florida Everglades. (Plaintiff: largemouth bass)

2. **Zebra mussel** (also known as the moule zebra). Native to Caspian and Black Seas, now found in the Great Lakes and all major river systems east of the Rockies. Black- or brown-and-white-striped bivalve, reaching lengths of about one inch. Found in freshwater lakes and rivers. Arrived 1980s, probably in ballast water of ships. Introduced to smaller lakes by traveling on boat hulls and trailers. Suffocates and starves native mussels and has pushed the Higgins' eye pearly mussel, a native of the Mississippi River, to the brink of extinction. Negatively affects fish, including walleye and trout, by extracting plant life from the water, which disrupts food chains by depriving prey species of an important food source. (Plaintiff: Higgins' eye pearly mussel)

3. **Asian long-horned beetle** Originally from China, this beetle is now breeding in New York City and Chicago. It is also found in other places in the United States, but so far it has not developed sustainable populations in other areas. About one to one-and-one-half inches long,

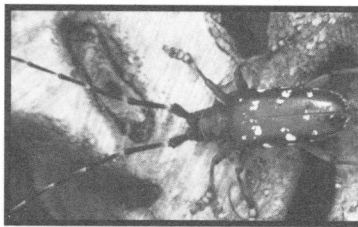
Kenneth R. Law, USDA APHIS PPQ, Image 4798040, www.invasive.org, April 2, 2003

black and shiny, with white spots. Adults make a hole in the bark of hardwood trees and lay their eggs. Immature beetles hatch and feed on the interior tissue of trees, then move to the exterior of trees to feed during adulthood. First found in New York City in 1996 and in Chicago in 1998. (Probably arrived in wooden crates or packing materials.) These beetles can destroy many hardwoods—maples, birches, horse chestnuts, poplars, willows, elms, ashes, and black locusts—so they have the potential to be devastating to city streets where these trees grow. Asian long-horned beetles could also threaten maple sugaring and other businesses dependent upon hardwood trees. (Plaintiff: maple tree)

4. **Purple loosestrife** Native to Europe and Asia and now established in 40 U.S. states. Tall plant with purple flower. Grows in wetland habitats. Arrived on the East Coast of North America in 1800s. Frequently brought to the United States, even now, for gardening and landscaping. Replaces native grasses with impenetrable stands that are not suitable as food, cover, or nesting sites for many native animals, including ducks, geese, rails, bitterns, muskrats, frogs, toads, and turtles. (Plaintiff: least bittern)

5. Indian mongoose A native of India, the mongoose was introduced to the West Indies and from there to Hawaii in 1883. It can now be found on the islands of Hawaii, Maui, Molokai, and Oahu, but has been successfully kept out of Kauai. The mongoose is a small, slender carnivore with short legs. It inhabits tropical forests. Introduced intentionally to Hawaii to help control snakes and rodents, the mongoose has ended up doing great damage. In particular, it threatens native bird populations, including the nene, which is the state bird of Hawaii. (Plaintiff: nene)

6. European starling (also known as the English starling and the blackbird). A native of western and central Europe, the European starling is now common throughout the United States, especially around farmlands and in urban and suburban areas. A mid-sized perching bird, the starling has black feathers with a green and purple iridescence. Eugene Scheiffelin, who was determined to bring to the United States every bird mentioned by Shakespeare, intentionally introduced starlings to New York City's Central Park in 1890. From an initial population of 100, starlings expanded and now number around 200 million. Starlings are aggressive birds: They compete with native birds for nesting holes, drive species such as bluebirds, flickers, and crested flycatchers away from areas near their nests, and even attack existing nests of other birds, destroying the eggs. Starlings have been known to drive red-headed woodpeckers and other native birds from their nesting cavities. (Plaintiff: red-headed woodpecker)

7. Bullfrog A native of the eastern and central United States, the bullfrog was intentionally introduced to western states to help control agricultural pests. It continues to spread by way of the aquarium trade and by its own overland travel. Bullfrogs are the largest of the true frogs in North American, weighing up to one pound. They live near water and are usually found around lakes, ponds, rivers, or bogs. Bullfrogs prey on snakes, worms, insects, tadpoles, crustaceans, and other bullfrogs. There even have been reports of bullfrogs eating bats. In areas where they've been introduced, bullfrogs pose a threat to native reptiles and amphibians, including snakes, turtles, and other frogs. In some parts of the United States, bullfrogs are believed to be responsible for declines in native leopard frog populations. (Plaintiff: leopard frog)

bullfrog

8. Fire ants Although there are some native fire ants in the United States, the red imported fire ant is a native of South America. It is now found throughout the southwestern and southeastern United States. These imported fire ants live and forage for food in underground tunnels and are found everywhere from roadsides to backyards. The worker ants are dark, small, and very aggres-

Imported Fire Ant Station Archives, Image 1148020, www.invasive.org, February 6, 2003

sive, and they sting relentlessly. It is believed that fire ants first came to the United States around 1920, either in the soil of potted plants or on ships that arrived in Mobile, Alabama. When Edward O. Wilson, the well-known ant biologist, was just a high school student in Alabama, he was the first person to study and document these imported fire ants. Fire ants are known to attack young livestock, young horned lizards (horny toads), native fire ants, quail chicks, and other species. They also compete with small reptiles and birds for the same food sources. (Plaintiff: horned lizard)

9. Chinese mitten crab First discovered in San Francisco Bay in the early 1990s. These crabs are believed to have been introduced either accidentally in ship ballast water or intentionally by someone interested in starting a mitten crab fishery. (They are a delicacy in Asia.) Mitten crabs are now spreading quickly and pose many threats. They burrow into banks and levees, causing erosion. They also cause hardships for local fishers when they get caught in trawls and fishing nets. The crabs eat a wide variety of plants and animals, including the eggs and young of local salmon, which are already suffering population declines. (Plaintiff: coho salmon)

BIOFACT

Introduced to south Florida, the Asian swamp eel can survive without food or water for more than a month, live in extreme temperatures, and change from female to male. The eel has no known natural predators in the south Florida ecosystem, and has developed a voracious appetite for native animals including frogs, shrimp, and crayfish.

2 Drawing Conclusions

SUBJECTS
art, science

SKILLS
gathering (reading comprehension, researching), analyzing (identifying components and relationships among components, reasoning, confirming), presenting (illustrating, articulating, explaining)

FRAMEWORK LINKS
12, 22, 23, 44, 48, 49, 65

VOCABULARY
ballast water, *ginkgo, London plane tree, wetland*

TIME
three sessions

MATERIALS
copies of "Aliens from the Deep" (page 275) and "Alien Impacts" (pages 276–277), paper and drawing materials for making comic strips

CONNECTIONS
Before conducting this activity, use "America's Most Wanted" (pages 260–269) to help students learn about the factors that may make alien species invasive. Afterward, encourage students to use "Aliens Among Us" (pages 280–283) to find out more about local nonnative species that may be causing trouble.

AT A GLANCE
Create a comic strip that depicts an alien species arriving in a new habitat. Then draw a conclusion (literally!) of what impacts this alien might have had.

OBJECTIVES
Name some ways that introduced species behave in a new environment. Describe some characteristics of an alien species that are likely to cause problems in its new environment. Explain why scientists can't always predict how an alien species will affect its environment.

Because some alien species have the ability to create serious problems for their new environments, we tend to think of them all as dangerous. But in fact, many alien species don't last long at all in their new environment. Others prove to be harmless in their newfound habitats. And some even provide significant benefits. Although scientists can never be sure how an introduced species will do in its new home, certain characteristics raise red flags (see "Friend or Foe" on page 257). In this activity, your students will put on their analytical thinking caps to see if they can anticipate how each of the species described ended up affecting its new environment. Then your students will use the answers to see if they can figure out what some of the "red flag" characteristics might be.

English ivy

Before You Begin

Make enough copies of "Aliens from the Deep" (page 275) and "Alien Impacts" (pages 276-277) so that each student or each team of three students has one.

What to Do

1. Hand out copies of the "Aliens from the Deep" comic strip.

Have the students read the "Aliens from the Deep" comic strip. Then have them answer the following questions: What are the aliens in the comic strip called? *(They're a kind of snail called rapa whelks.)* How did they reach the Virginia port? *(They traveled by ship from the Black Sea.)* You can discuss how the aliens were discharged along with the ship's ballast water—water that is taken on by a ship in one port to balance the ship for traveling across the sea. The water is discharged when the ship reaches the new port.

> **Can your students define the word population?** *(A population is a group of individual organisms of the same species living in the same area. For example, all the zebra mussels in a lake are a population.)*

2. Discuss alien impacts.

Ask the students if they have any idea what will happen to the alien species in the species' new home. Explain that scientists are never positive about how well a species will do in its surroundings and how much of an effect it will have on its new habitat. For example, some species will simply die off because they don't have the conditions they need to survive. Others (such as oxeye daisies, ginkgo trees, and London plane trees) will establish

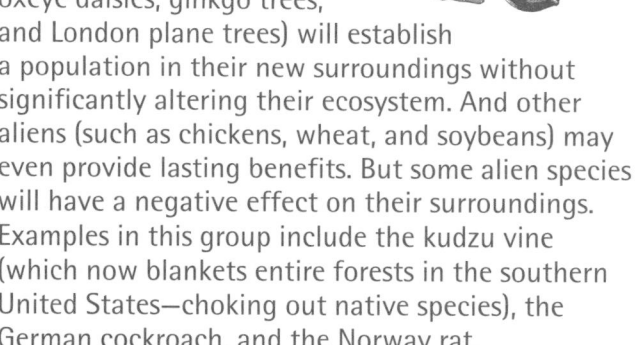

ginkgo

a population in their new surroundings without significantly altering their ecosystem. And other aliens (such as chickens, wheat, and soybeans) may even provide lasting benefits. But some alien species will have a negative effect on their surroundings. Examples in this group include the kudzu vine (which now blankets entire forests in the southern United States—choking out native species), the German cockroach, and the Norway rat.

Make sure that your students know that the rapa whelk featured in the comic strip is a real species. Scientists don't yet know exactly what effect the rapa whelk will have on the coastal waters of the Atlantic, but they are concerned that it will be a big problem. That's because it has characteristics that scientists think make a species more likely to be a menace. In this activity, your students will be learning more about different alien species and seeing if they can figure out what some of those menacing characteristics are.

3. Hand out "Alien Impacts" worksheet.

Give a worksheet to each student or, if you prefer, to each group of three students. Tell the students that they will be making a two-part comic strip of one of the scenarios on the handout. (You can assign the scenarios or allow the students to select the one they prefer.) In the first part of the comic strip, they should illustrate the information provided. Tell them to use the picture to create a good cartoon likeness of the species described. They may want to find other illustrations or photos of their species to complete their cartoons. Encourage the students to have fun with the comic strips, exaggerating characteristics for effect, making the scenes very active and exciting, and so on.

4. Share comic strips.

Have at least one student share his or her comic strip for each species on the "Alien Impacts" sheet. Have the students speculate what might have happened next. Ask for a rationale for each idea and what impact they predict the alien introduction would have. Share the information about "what really happened" (pages 278-279) after each species has been discussed.

For the second part of the comic strip, have the students illustrate what happened following the alien introduction in the first part of their comic strips.

5. Review characteristics of a problem-causing alien species.

As a group, ask the students to think about some of the characteristics that led certain species to become a nuisance. Ask for a volunteer to record the answers on the board. The answers might include, but are not limited to, the following:

■ *reproduces rapidly* (e.g., hydrilla, green crab, nutria)

■ *easily adapts to a variety of habitats* (e.g., hydrilla grows in many kinds of water bodies and does especially well in lakeside habitats; nutria can live in freshwater and saltwater habitats)

■ *eats a lot or eats many different kinds of organisms* (e.g., Nile perch, green crab, nutria)

■ *has the ability to move quickly and easily* (e.g., green crab juveniles are very mobile for 80 days; on the other hand, the black-tailed jackrabbits' movements were constrained by roads or water on all sides)

■ *has very few predators* (e.g., Nile perch, green crab)

■ *is living in an environment that has already been significantly changed by humans* (Some scientists speculate that disturbed habitats are more likely to suffer from the introduction of alien species such as purple loosestrife.)

You should remind your students that scientists still cannot really predict whether an alien species will flourish or fizzle out in its new habitat. But the characteristics listed above are red flags—indications that a species is more likely to be a problem than not!

BIOFACT

Hydrilla, an invasive aquatic plant infesting nearly half of Florida's waterways, forms green carpets of foliage thick and strong enough for ducks to walk on!

> *"Biological invasion by alien species is now recognized as one of the major threats to native species and ecosystems, yet awareness of the problem is alarmingly low. The effects on biodiversity are immense and often irreversible."*
>
> **—IUCN Species Survival Commission**

6. Discuss alien species around the world.

Ask your students if they were surprised by the impacts that some of these alien species had on their environment. Then explain that some scientists think that alien species are one of the most significant causes of species decline worldwide. Some scientists think that alien species might even be more of a problem for threatened and endangered species than habitat loss or pollution. Aliens are contributing to species loss and ecosystem changes in both terrestrial and aquatic environments in every part of the world.

In addition, alien species are causing billions of dollars of damage each year to people by wreaking havoc on fisheries, damaging equipment, requiring expensive cleanups, and so on.

Have your students heard much about alien species in their area? In preparation for "Aliens Among Us" (Activity #3) encourage your students to collect information or articles about alien species in your community or region.

7. Describe the final assignment.

As an in-class exercise or as a homework assignment, have the students imagine they are asked to assess the likely threat of the alien species in the "Aliens from the Deep" comic strip. What things would they need to know about the species to complete their assessment? *(Answers will vary but may include: what it normally eats, whether those foods are available in the Atlantic port, which other organisms it is now consuming, how rapidly it reproduces, how it moves, whether it has any local predators, whether others of its species have ever been seen in these or nearby waters, and so on.)*

WRAPPING IT UP

Assessment

Using the student-created comic strip, consider the concepts of (1) how the species was introduced, (2) the hypothesized behavior of the species in the new environment, and (3) the reasoning used by the student.

Unsatisfactory—Addresses only one or two of the criteria or is not adequate in addressing any one of the three.

Satisfactory—Adequately addresses all three criteria.

Excellent—Reveals scientific method in thinking through the introduction, the behavior, and the impact of the species on the environment.

Portfolio

Students should include copies of their comic strips in their portfolios.

Writing Idea

Have students write a sensationalized newspaper article describing the alien "invasion" depicted in their comic strips. (For examples, see "The Natural Inquirer" in *WOW—A Biodiversity Primer* by WWF.)

Extension

Students can conduct further research on one of the species described briefly in the "Alien Impacts" worksheet. Have students use the Internet, books, newspapers, and other sources to round out their research. They may also want to contact resource managers working at government agencies, such as the U.S. Fish and Wildlife Service or U.S. Department of Agriculture, to gather information on these alien species.

Alien species reach new homes through many different pathways and with very different results. Some die off quickly, some persist but never cause any trouble, some bring people benefits, and some cause significant environmental disruption. Can you guess how each of the following ended up affecting its habitat?

1.

In the 1960s, scientists in Florida discovered a nonnative aquatic plant called hydrilla growing in local waterways. Once it is rooted in the bottom of lakes and ponds, hydrilla grows upward to form a dense mat on the water's surface. It can grow in waters with low or high nutrient levels, and it doesn't require much sunlight. It has four modes of reproduction, including resprouting from a very small plant fragment. It can grow as much as an inch a day. **What do you think happened?**

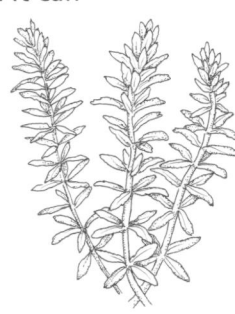

2.

In the 1950s, a crate broke open at what is now the John F. Kennedy International Airport in New York City, allowing dozens of black-tailed jackrabbits to leap to freedom. These large hares normally live in prairies, meadows, and pastures of the western United States. However, after this incident, the jackrabbits found themselves on a wide, grassy area bordered on three sides by water and on the fourth side by a network of highways. Jackrabbits in their native environment eat greens in the summer and woody vegetation in the winter. They mate year-round, producing one to four litters a year with one to eight young per litter. Jackrabbits are eaten by hawks, coyotes, owls, and badgers. **What do you think happened?**

3.

Around 1990, the Massachusetts Division of Fisheries and Wildlife reported that a Nile monitor lizard had escaped from a home where it had been kept as a pet. Nile monitor lizards are the largest lizards in Africa, and they can reach lengths of between five and six feet. They must live near water, but they can survive in a wide range of habitats. They are carnivores and eat all sorts of animals, including insects, crustaceans, and reptiles. These lizards have even been known to rummage through human trash heaps looking for food. **What do you think happened?**

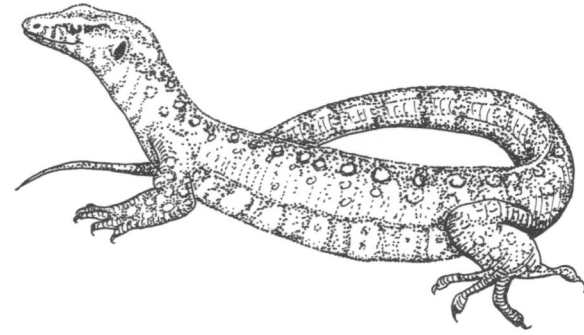

4.

In 1962, Africa's Lake Victoria contained a rich diversity of native fish, including hundreds of species of small fish called cichlids (SIK-lidz). British businessmen introduced Nile perch to Lake Victoria to create commercial and sport fishing opportunities for the local people. Nile perch are large fish that consume a diversity of other fish. **What do you think happened?**

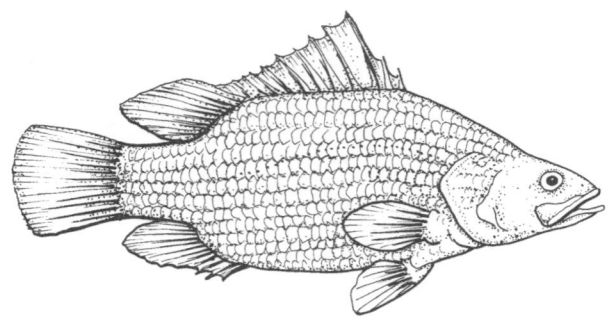

5.

The European green crab reached the Atlantic coast of the United States in the eighteenth century, landing somewhere between New Jersey and Cape Cod. The green crab is a voracious eater and can open or crush oysters, mussels, and clamshells, and it even consumes other crabs. Females produce 200,000 eggs each year, and their young are very mobile for the first 80 days of life. **What do you think happened?**

6.

In the 1890s, native oysters off the U.S. Pacific Northwest coast were being overharvested. In the 1920s, in an effort to boost the oyster industry, people imported a species of large Japanese oyster to the region. They are still being imported today. This oyster matures in about one and one-half years in warm waters. In cold waters, it can take four to five years to mature. An oyster produces millions of eggs, but only a small portion of these survive. **What do you think happened?**

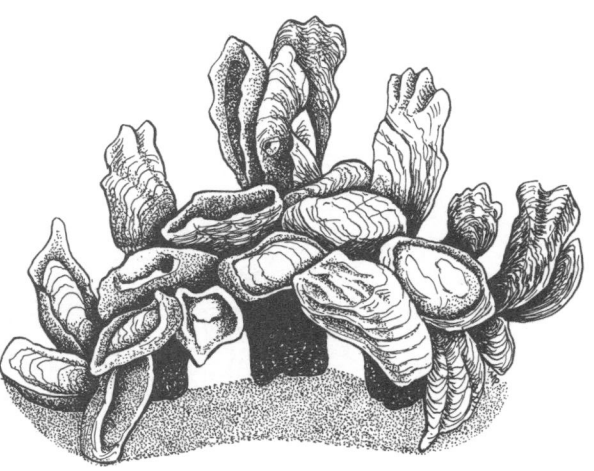

7.

In the 1930s, imported South American rodents called nutria were released into the U.S. Gulf Coast region in hopes that they would someday supply the fur industry with an alternative to mink. Nutria are aquatic animals that are able to survive in freshwater and saltwater habitats. They reach sexual maturity as early as four months of age, and they breed year-round, producing four to six young per litter. They feed on almost any green plant found on land and in the water, including some grains. They may eat an amount of food each day that's equal to one-fourth of their body weight. Their primary predators are trappers and alligators. **What do you think happened?**

8.

Hunters brought ring-necked pheasants from Asia to the United States from the 1800s to early 1900s. Wildlife management agencies and individuals alike hoped that these ground-dwelling birds would reproduce and provide a good target for hunters. Ring-necked pheasants feed on a wide variety of plants, including berries, grasses, and cultivated crops. They spend a lot of time in the open, but they require hedges or wetland areas for shelter. Females lay about 10 to 12 eggs each year, many of which are eaten by predators. **What do you think happened?**

ANSWERS TO ALIEN IMPACTS: WHAT REALLY HAPPENED

1. Hydrilla has been called "the perfect aquatic weed" because it is capable of spreading so fast and so far. (In its native Asian habitat, hydrilla is kept in check by the periodic dry seasons, which cause it to die off because of lack of water.) Although ill suited to deep water, it is perfectly suited to shorelines, shallow ponds, and the many shallow lakes that cover the state of Florida. It has also spread rapidly along the coastal areas of the United States. Where hydrilla has become established, it forms dense mats on the water's surface. It displaces native plants, impedes water flow, suppresses some native fish populations, and interferes with aquatic recreation activities such as fishing, scuba diving, and boating. While some people favor hydrilla because it provides food for waterfowl, it poses significant problems for the majority of wildlife in the ecosystems it has invaded.

2. The black-tailed jackrabbits have been able to get enough food to survive and reproduce at Kennedy Airport, and have established a moderate-sized but harmless population. The hares are prevented from spreading into other regions by the road and water barriers on each side of the airport.

3. A single Nile monitor lizard may have survived for a while in Massachusetts, but it would not have presented a real threat. First of all, unless it was a pregnant female, it could not have reproduced without a nearby mate. More importantly, it would not have made it through a cold winter without shelter.

4.
Nile perch were part of the Lake Victoria ecosystem for about 20 years before they began to feed heavily on native fish. Since then, more than 80 native species have become extinct. But are the Nile perch to blame? Some scientists think that pollution, overfishing, and other environmental problems are responsible for most of the decline in fish populations. In fact, things have gotten so bad in Lake Victoria that adult Nile perch are now eating juveniles of their own species, and some people think the perch may also be headed toward extinction. Another environmental problem came about after the Nile perch were introduced. When the local people began eating the perch, they found that these oily nonnative fish had to be dried over an open fire, so the people decimated the forests all around the lake in the process of gathering firewood. Until then, local people had eaten native fish, which dried easily in the sun.

5. The green crab has proven to be an extremely damaging invasive species. Spreading slowly up the Atlantic Coast, the crab has destroyed Maine's soft-shell crab industry and now threatens scallops from Maryland to Massachusetts. The green crab is a popular seafood in parts of Europe, but not in the United States.

6. The Pacific, or Japanese, oyster is in certain ways an example of a successfully introduced species. Even today, these oysters are cultured in Willapa Bay and Puget Sound and produce a significant portion of Washington's $40-million-a-year oyster revenues. Unfortunately, importers didn't just bring Pacific oysters: The oysters' packing crates also contained the Japanese oyster drill. Oyster drills bore into oyster shells and then eat the oyster, so the drills have to be handpicked out of oyster beds. What's more, some of the earliest shipments of oysters brought *Spartina*, or cordgrass, to the region—an invasive species that displaces oyster beds.

7. Nutria have spread steadily throughout the Gulf Coast region since their early introductions in the 1930s and 1940s. Although populations are somewhat controlled when nutria are trapped for their meat, these nonnative mammals still cause a great deal of damage to their newfound habitat. Nutria reduce native vegetation, destroy sugar cane and rice crops, and are believed to have contributed to declines in native muskrats and water fowl.

8. Ring-necked pheasants are said to be among the most successful of introduced game animals. Despite annual restocking by wildlife agencies and hunting organizations, they have never become a serious nuisance. In fact, the pheasants can create significant economic benefits because of the value they provide as a prized upland game bird. Their populations are kept in control by predation, and in some places their populations are declining because of bad weather, pesticide use, and the reduction of suitable habitat. (Pheasants thrive where crops are interspersed with hedges and wetlands, but they suffer when wetlands are drained and hedges are removed.)

3 | Aliens Among Us

SUBJECTS
science, social studies

SKILLS
(Skills will vary depending on which activity idea is chosen.)

FRAMEWORK LINKS
23, 48, 49, 63

VOCABULARY
invasive, native species, nonnative species

TIME
two to three sessions

MATERIALS
copies of information about local invasive species from Web sites or a field guide that tells whether plants are native or nonnative to your area

CONNECTIONS
"America's Most Wanted" (pages 260–269) will provide students with clues of characteristics they should be looking for in nonnative plants and animals that will help assess whether those plants and animals might become invasive problem species. "Sea for Yourself" (pages 72–81) also complements this activity: It will allow students to further explore the local area and talk to resource-management professionals about their concerns with invasive species.

AT A GLANCE
Learn more about alien species in your area and their effects on local habitats.

OBJECTIVES
Name several nonnative species in your area. Describe one problem caused by alien species that affects people in your community. Describe some of the different ways that alien species can affect people in other communities.

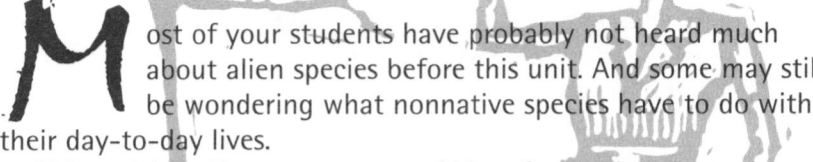

Most of your students have probably not heard much about alien species before this unit. And some may still be wondering what nonnative species have to do with their day-to-day lives.

This activity offers you a range of ideas for making the topic more relevant for your group. One option is to take your students into the schoolyard or a nearby natural area to have them document the kinds of species (native and nonnative) they find there. Another idea is to take a guided field trip with a local expert who can introduce your students to some alien species in your area.

We encourage you to adapt and expand on these activity ideas to make them useful for and pertinent to your group of students. Or, if you prefer, you can come up with something entirely different to get your students focused on how "alien invasions" are relevant to their lives.

> *"The world's biodiversity is being homogenized at an ascending rate as more and more alien species invade even relatively natural environments."*
>
> –E.O. Wilson, biologist

Before You Begin

Encourage your students to collect information and articles about alien species in your area. If you are unsure of your ability to identify local invasive plants, you may want to invite someone who knows them to help with Option #1.

What to Do

1. Select one of the activity ideas below.

Choose one or more of the following options to introduce your group to local alien species.

Option One: Field Investigation

Spend one class period preparing your students for the outdoor work. See "Taking Your Class Outside" (pages 78-79) for some practical tips on preparing for outdoor trips with your students. Also, introduce them to Web or field-guide information about alien species in your area. In the next session, take your students outside to investigate the relative abundance of native and nonnative plants in your schoolyard. Try to identify as many different plant species as possible. (Have students sketch any plants they cannot identify.) You may want to have them keep track of whether each plant identified was "abundant," "occasional," or "rare" at their field site.

On the third day, ask your students to use Web information and field guides to determine which species are native and nonnative. Have them compile their results to see how many different plants they identified. Based on these identifications, the students should figure out the relative proportion of native to nonnative plants. (This presents many opportunities for students to work on their graphing and presentation skills.) **Note:** Students should not conclude that abundance or scarcity of a species will help determine whether it is native or nonnative.

Afterward, discuss the results with the students. Were they surprised by how many species were nonnative? Do they think nonnative species are more common in developed or undeveloped areas? *(Some alien species seem to do especially well in disturbed environments, and people are more likely to introduce them [intentionally or unintentionally] to human-dominated environments. So, developed areas do generally have a higher proportion of alien species than wild or undeveloped environments do. Make the point that nonnative species can be harmful because they use resources that native species need. When native plants and animals die, other animals that depend on them for food may also die, and the ecology of the entire area can be changed by the invasive species.)*

The European green crab is quicker and more flexible than other crabs—one crab can eat 40 half-inch clams per day and devour crabs as large as itself!

Option Two: Field Trip

Contact a local nature center, natural history museum, conservation group, county extension service, or wildlife agency to find an area naturalist who can take you and your students on an alien species field trip. The trip can be through a nearby natural area (such as a river, pond, lake, bay, field, or forest) or through city streets in your neighborhood. The naturalist may want to focus on all the nonnative species you encounter or may want to highlight a particular invasive species problem and how it's affecting the community. You could also visit an area where people have removed invasive species and a restoration program is underway.

Option Three: Volunteer Day

Have your students volunteer for a day with a local group working to combat alien invasions. Students can spend the day clearing invasive plant species and seeing firsthand the effects these invaders have had on the local ecological community. Be sure to find out from the project organizer what the students need to wear to protect themselves while helping.

2. Discuss the following questions, which are appropriate for use with all options.

- What kinds of alien species live in your area?

- Do those species seem to have a positive or negative effect on the environment? How can you tell?

- Do the alien species have an impact on people who live in your area? In what way(s)?

- What can be done to help native species when aliens invade their area?

WRAPPING IT UP

Assessment

Have students write a news article focusing on a nonnative species in your area and describing how it is affecting other species as well as the people in your community. Encourage the students to make the story compelling yet factual. Ask them to create a story that encourages the reader to think about the issues.

Unsatisfactory—Incomplete article; includes one or more components but does not tie them together.

Satisfactory—Identifies the species, where it originated, how it came to its present location, and its basic effects.

Excellent—Includes critical thinking regarding the species' introduction, effects and management.

Portfolio

Students should include the news articles from the assessment in their portfolios.

Writing Idea

Based on the students' findings during their local explorations, have them write a plan for a program aimed at reducing the number of invasive species in their area. Be sure that students include an educational component in their plan, helping other community members understand the impact that alien species may be having on the local environment.

Extension

Have students conduct an investigation of their own yard or another area close to home, looking specifically for invasive alien species. What actions would they propose to take to help control those species?

BIOFACT

Africanized honey bees are related to southern African bees, which were imported to South America in 1956. Scientists in Brazil were trying to breed a bee species that would be better adapted to the tropics. But, by accident, some of the bees escaped and bred with local Brazilian bees. Since 1957, both pure African bees and their hybrid offspring, the Africanized honey bee, have traveled throughout South, Central, and North America, moving more than 200 miles per year.

SUBJECTS
social studies, science, art (drama), language arts

SKILLS
interpreting (reasoning, identifying problems, drawing conclusions), evaluating (critiquing, assessing), presenting (acting), citizenship skills (working in a group, evaluating a position, evaluating the need for citizen action)

FRAMEWORK LINKS
37, 47, 48, 55, 59, 60, 62, 63, 69, 70, 71

VOCABULARY
ballast water, *ecological restoration*

TIME
two to three sessions

MATERIALS
copies of the cards on pages 288–289

CONNECTIONS
To learn about the problems associated with certain alien species, start with "America's Most Wanted" (pages 260–269). For other activities that deal with conservation-related conflict resolution, try "What's a Zoo to Do?" and "Thinking About Tomorrow," both in Biodiversity Basics.

AT A GLANCE
Play the part of community decision-makers trying to take action on alien species.

OBJECTIVES
Identify several methods for controlling alien species. Explain controversies and considerations surrounding the economic, ecological, ethical, social, and aesthetic issues involved in controlling alien invasions and restoring native species.

In recent years, scientists have shown that invasive alien species are one of the most significant causes of species extinctions around the country and the world. Not surprisingly, then, a great deal of effort has gone into trying to prevent new invasions and control those that have already occurred.

In general, preventing an alien invasion is less expensive than combating one that's already well underway. Preventive approaches include customs inspections at airports and seaports, treatment of the ballast water in ships, bans on dumping bait or aquarium fish into waterways, and so on. (For more on prevention techniques, see "Alien Solutions" on pages 258-259.)

Prevention techniques are sometimes controversial. In the United States, nurseries and pet shops will often oppose restrictions on certain plant and animal imports because these restrictions limit the stores' opportunities for sales. Likewise, many people have challenged policies requiring ships to exchange their ballast water out at sea, pointing out that it can be hazardous to the people on board because it makes the ship unstable.

But the most intense controversies—and the most significant costs—generally arise over efforts to control alien species invasions once they occur. For one thing, control techniques are far more radical than prevention: They include tactics such as trapping invasive wildlife, introducing new alien species to control existing ones, and poisoning waters where aquatic aliens have invaded.

These techniques are often very expensive. They can also be objectionable to local people who have come to appreciate the alien invaders for their aesthetic, recreational, or nutritional value. And these methods are often filled with scientific uncertainty. For example, biologists concerned about purple loosestrife invasions in U.S. wetlands have begun introducing an alien beetle that is known to feed on loosestrife in its native habitat. Although scientists have conducted rigorous laboratory tests, many people worry that the beetle itself may end up becoming a nuisance in its new environment.

Despite all these challenges, most conservationists agree that controlling alien invasions and restoring native species is essential if we hope to sustain the nation's—and the world's—biodiversity. With that in mind, this activity gives your students a chance to explore some of the many approaches that people are taking to control invasions and restore native ecosystems. It also encourages them to grapple with some of the many complexities involved in these efforts.

Before You Begin

Make enough copies of the cards on pages 288-289 so that each scenario goes to one-fourth of the students.

What to Do

1. Discuss alien control efforts.

Ask the students if they've thought about what can be done to reduce the impact of alien invasions. Can they think of ways that people can prevent alien invasions from happening in the first place? Have they heard of any methods for stopping alien species invasions once they've already occurred? Explain that they'll be taking a closer look at ways to control alien species in this activity.

2. Organize the class into four teams.

Explain to the group that they'll be acting out four true stories of conflicts involving alien species management. Divide the students into four teams and have them gather in different parts of the room. (Note: "Swan Song" is written at an easier reading level and describes a less complex issue than the other three.) Give each member of Team One a copy of "Marina Sauce," Team Two a copy of "Ballast-ing Acts," and so on.

Have the team members read through their scenarios. Then have them come up with a skit, as directed on the cards, and assign roles to different members of their group. Ideally, all students should be involved in the role-play. Tell the students to try to pick a situation that will be dramatically interesting while getting to the core of the conflict described. (Note: As an alternative, you can have the students read and respond to their scenarios without having them perform. But role-plays often help get students involved in the material and provide an opportunity for interactivity with the entire class.) You will probably want to give the students at least 20 minutes to prepare their skits.

3. Conduct role-plays.

Have the four groups take turns presenting their role-plays. After each performance, have the rest of the class jot down their answers to the following questions:

a) What is the nonnative species described in the scenario, and what problems is it causing?

b) What actions have been proposed to control the invasive species?

c) Why do some people disagree with these actions?

When the students have completed their skits, have them return to their seats for a group discussion.

4. Discuss complexities of ecological restoration.

Remind your students that all four role-plays portrayed attempts at ecological restoration—combating invasive species in their new environments and helping restore native habitat and species. Review the four scenarios and have students share their responses to the three questions. Then ask them what they would do if they were a decision-maker in each of these instances. Solicit feedback on one scenario at a time, asking for a show of hands for different options. Ask the students to explain their responses.

This discussion is an appropriate place to encourage students to listen to and show respect for differing opinions. You might want to see if the students can identify the values that inform their decisions in each of these cases. Did they find themselves most concerned about ecological health? Aesthetics? People's welfare? Jobs? Costs? Why might it be useful to identify the underlying values of the different people involved in these conflicts before trying to reach a resolution?

5. Discuss scientific uncertainty.

Ask the students to think about the role of scientific uncertainty in these disputes. In which examples did a lack of scientific certainty about the invasive species (and its effects) seem to affect the viewpoints of people involved? In which cases did lack of scientific certainty about ways to control the alien species play a role? Encourage the students to think about the pros and cons of taking action when you lack scientific certainty about the effects of that action.

6. Wrap up the activity.

To wrap up this activity, have the students write a brief paragraph explaining whether they think alien species control is important and describing why it is often challenging. Then have them list explanations and examples of how alien species control can be a challenge (1) economically, (2) ecologically, (3) ethically, (4) socially, and (5) aesthetically. You might want to review these explanations as a group, or collect them to use as an evaluation.

BIOFACT

Round gobies, invasive fish from Eurasia, can spawn up to five times per mating season. Gobies build elaborate rock nests in lake bottoms to house their fast-growing families and aggressively guard these nesting areas, to the detriment of native fish.

WRAPPING IT UP

Assessment

Have students write a short reflective essay on a local nonnative species. The essay should explore why trying to control this plant or animal might be controversial in economic, ecological, ethical, social, and aesthetic terms.

Unsatisfactory—The essay includes some of the elements but fails to create a cohesive, reflective piece.

Satisfactory—The essay includes all of the required elements.

Excellent—The essay reveals critical thinking, logical reasoning, and creativity in how the elements are constructed.

Portfolio

Include the paragraphs students have written for the wrap-up and the assessment in their portfolios.

Writing Idea

Have students write a paragraph on one of the skits, answering the questions posed under section 3 (questions a, b, and c).

Extensions

- Have your students investigate a local project to control alien species. What are the pros and cons of this effort? If your students can find people who support it and others who oppose it, have them interview both groups.

  - See if your students can find a local restoration project that needs volunteers. Plan a weekend workday to help with this project.

BIOFACT

In 1988, a large population of the invasive and highly destructive gypsy moth was detected in Utah. Resource managers immediately jumped into action by implementing an extensive public education campaign, an integrated pest management initiative, and a program that used cutting-edge pheromone trapping technology. The program seems to have been a success: In 2001, only one moth was found in the treatment area.

Marina Sauce

When divers explored a marina off the Australian town of Darwin, they saw something they knew didn't belong: tiny mussels related to the zebra mussel, which have clogged up North America's Great Lakes. Hundreds of millions of these mussels were clinging to piers and boat hulls and lines where there had been none just six months earlier. The **divers**, who were part of an inspection team from a scientific agency, identified the mussels as coming from Central America. They believed the mussels had arrived on the hull of a yacht. Within a week, **government officials** took drastic action. First, they refused to let any boats in the area leave the port, despite the objections of **boat owners**. Then they poisoned the waters of the marina with chlorine and copper. The poison killed all of the mussels, but it also killed everything else that lived in the marina. Now, though, the native species are coming back and the Central American mussels haven't returned. But was it worth the $1.5 million price tag? What happens if the mussels show up again? And is it ecologically acceptable to poison natural areas and kill off so much life? These are issues that have yet to be resolved.

What to Do

The main characters in your scenario are shown in bold type above. Use them, along with any others characters you would like, to portray some of the conflicts and complexities involved in getting the Central American mussels out of the Darwin marina.

Ballast-ing Acts

In certain parts of San Francisco Bay, you can find as many as 3,000 Asian clams per square foot! The Asian clam is just one of the 250 alien species threatening the habitats and fisheries of the bay. Many of these aliens originally traveled to San Francisco Bay with ships delivering goods from across the ocean. These huge ships take water, called ballast, into large tanks to help balance their load. When the ships arrive in a new port, they discharge the water before reloading. But as this water is discharged, so are all the creatures that were taken in at the last port and that survived the trip. As many as 50 species have been found in the ballast water of a single ship. There may not be anything that eats these alien species in their new home, and they may start to take over and crowd out the species that belong there. For this reason, concerned people in the San Francisco Bay area, as well as in other ports, are trying to figure out a way to get the aliens out of the ballast water. Some **conservationists** advocate having ships pause before they get to port, dump out their ballast water (since most coastal species cannot survive that far out at sea), and replace it with deep-sea water. But some **ship captains** say that it's a dangerous practice, leaving ships temporarily unstable. They think it would be foolish to comply with the regulations. **Biologists** are researching the possibility of using filters, UV radiation, or other processes that could strain out or kill organisms in ballast water. But more research (and funding) is needed to see if these methods will work.

What to Do

The main characters in your scenario are shown in bold type above. Use them, along with any others characters you would like, to act out a situation that shows the conflict over ballast water treatment in the San Francisco Bay area.

Discord over Cordgrass

The Willapa Bay region of Washington State is one of the most productive coastal areas in the world. Oyster, clam, and salmon fisheries are a huge part of the economy. Thousands of birds come here to feed and nest. Unfortunately, the health of this region is threatened by several exotic species, including *Spartina*, or cordgrass. *Spartina* was brought to the region in the 1800s as packing material for eastern oysters, which were introduced to revive the oyster industry after native oysters were harvested to near extinction. Unfortunately, *Spartina* has grown out of control in its new home, reducing the habitat for native crabs, snails, salmon, shorebirds, and other organisms.

Not surprisingly, people are working hard to get rid of *Spartina* to protect the ecosystem and the economy. They've tried mowing and hand-plucking the grass, and some people are advocating the introduction of a *Spartina*-eating insect to the area. But so far the most cost-efficient control method seems to be an herbicide that, when dumped into the water, kills off much of the grass. Many **people involved in the oyster industry** favor this control method. But members of the **Shoal-water Bay Indian tribe**, which has a reservation on the edge of the bay, are concerned. They don't think it makes sense to put a poison in the water, since it can be absorbed by other living things in the bay. They're especially concerned because members of their tribe have been having serious health problems. Instead, the tribal members and others insist, people should investigate other ways of controlling *Spartina*. They suggest, for example, that **people in the paper industry** could mow and hand-cut the *Spartina* and make it into paper and other products. This wouldn't get rid of the grass as cheaply or quickly as the herbicides do, but it could be much better for people and the ecosystem over the long run.

What to Do

The main characters in your scenario are shown in bold type above. Use them, along with any others characters you would like, to portray the conflict over controlling *Spartina* grass in Willapa Bay.

Swan Song

Almost everyone agrees that mute swans are majestic, beautiful birds. What they don't agree on is whether mute swans belong on lakes and bays in the Northeast and Mid-Atlantic states. Mute swans were introduced to this country from Europe in the 1800s, and they are continuing to spread. Now biologists say that these invasive swans are damaging wetlands. They say the birds force out native water birds, making it impossible for the native birds to nest and reproduce.

In Vermont, **wildlife managers** concerned about the effects of mute swans recently began shooting and killing the swans on their lakes after other control methods didn't work. Some **local residents** were furious, especially because the shootings were unannounced and because they happened right in front of their eyes.

What to Do

The main characters in your scenario are shown in bold type above. Use them, along with any others characters you would like, to act out a situation depicting the conflict over mute swans on Vermont's lakes.

mute swan

Alien Species Resources

Here are just a few resources to help you design and enhance your Alien Species Case Study. Because this is an area of great ecological concern, there are many other resources available on alien species. For a list of general marine biodiversity resources, see the Resources section on pages 360-369.

Organizations

Aquatic Plant Management Society, Inc. (APMS) is an international organization of scientists, educators, students, commercial pesticide applicators, administrators, and concerned individuals interested in the management and study of aquatic plants. The society published an activity booklet for fifth graders called *Understanding Invasive Aquatic Weeds*, which is available on line. Click on "Publications" and then on "Activity Book." **www.apms.org**

National Invasive Species Council (NISC), formed by an Executive Order of the President in 1999, ensures that federal agency activities dealing with invasive species are properly coordinated. A comprehensive online information system can be found at **www.invasivespecies.gov**

National Sea Grant College Program has an Exotic Species Resource Center that produces educational materials. Sea Grant is a program of the National Oceanic and Atmospheric Administration of the U.S. Department of Commerce. The New York Sea Grant program operates the Aquatic Nuisance Species

Clearinghouse, an international library of research, public policy, and educational publications that pertain to invasive marine and freshwater aquatic nuisance species in North America. To access the clearinghouse visit **www.cce.cornell.edu/aquaticin vaders/nan_ld.cfm** or access the general Sea Grant College Program site at **www.seagrantnews.org/ education**.

U.S. Environmental Protection Agency and the **Ocean Conservancy** highlight ways for volunteers to monitor invasive species and collect samples in their jointly produced *Volunteer Estuary Monitoring: A Methods Manual*. **www.epa.gov/owow/estuaries/ monitor/chptr19.html#nonindigenous**

Western Society of Weed Science (WSWS) is made up of weed science professionals working in the western United States and offers educational resources such as *A Kid's Journey to Understanding Weeds*. **www.wsweedscience.org/publications/ wyo_ed_desc.html**

Curriculum Resources, Books, and Web Sites

Alien Invaders—A Rivers Curriculum Project investigates issues surrounding the zebra mussel, which has invaded the waters of the United States and Canada. Project materials and zebra mussel monitoring devices are available from the Rivers Project. **www.siue.edu/OSME/river/Ordering%20 Materials/Order.html**

America's Least Wanted: Alien Species Invasions of U.S. Ecosystems, a program of NatureServe and the Nature Conservancy, provides a list of their "dirty dozen" alien species and describes the ecosystems these aliens affect. **www.natureserve. org/publications/americasleastwanted.jsp**

Alien Species Resources (Cont'd.)

Black Sea Battle is a story produced by the *Why Files in Education*. The *Why Files* is an NSF-funded project that uses news and current events as springboards to explore science, health, environment, and technology with activities that are correlated to national standards. *Black Sea Battle* tells the story of the invasion of jellyfish into the Black Sea and the resulting effects on the native fish population.
www.whyfiles.org/055oddball/fish.html

The Bridge—Ocean Sciences Education Teacher Resource Center is a growing collection of the best marine education resources available on line. It provides educators with a convenient source of accurate and useful information on global, national, and regional marine science topics and gives researchers a contact point for educational outreach. Under the "Ocean Science Topics," select "Biology," then "Exotics." Produced by Sea Grant Marine Advisory Services, Virginia Institute of Marine Science, College of William and Mary, Gloucester Point, VA 23062. **www.vims.edu/bridge**

Center for Invasive Plant Management, in cooperation with Michigan State University, provides links to K-12 teaching resources on invasive species, including books and program ideas. The site is also a resource for invasive species grants, management ideas, identification information, and control methods. **www.weedcenter.org/ education/educationhome.html**

Exotic Aquatics and Zebra Mussel Mania Traveling Trunks are part of the "Traveling Trunks" education program produced by the Illinois-Indiana and Minnesota Sea Grant Programs. Trunks are filled with educational materials that can be used by students to gain hands-on experience with aquatic exotic plants and animals. **www.siue.edu/ OSME/river/Ordering%20Materials/order_ Zebra.html**

Exotic Aquatics on the Move: Lesson Plans (for grades 6 through 12) is a collection of activities focusing on nonnative species in the marine environment. The resource is available on CD-ROM and in print from Washington Sea Grant Program Publications. (206) 543-0555. **www.seagrant.umn.edu/pubs/mailorder.html**

The Problem refers to the introduction of nonnative species through ballast water, which has been called one of the four greatest threats to the world's oceans. A concise explanation of the problem of exotic species introduction to new areas via ballast water can be found at **globallast.imo.org/problem.htm**.

The Sea Grant Nonindigenous Species Site (SGNIS), produced by the Great Lakes Sea Grant Network, is a national clearinghouse for information on zebra mussels and other aquatic nuisance species. **www.sgnis.org**

Wild Things 2001: Investigating Invasive Species is a middle-school curriculum guide created by the U.S. Fish and Wildlife Service. The guide emphasizes field trips and outdoor activities as a way to learn about alien species in natural habitats. **www.wildthingsfws.org/WT2001/guide_ instructions.htm**

You Ought to Tell Somebody! Dealing with Aquatic Invasive Species video gives an overview of the invasive species problem and provides identification and natural history information about one significant northwestern threat, the Chinese mitten crab. Available from Oregon Sea Grant, 322 Kerr Admin. Bldg., Oregon State University, Corvallis, OR 97331-4501. $18.95 **www.seagrant.oregonstate.edu/sgpubs/multi media.html**

Case Study
Salmon

When Europeans first arrived in the
Pacific Northwest in the mid-1700s, they
described rivers so crowded with salmon
that a person could almost walk across on
their backs. Today, though, salmon are
scarce or even extinct over much of their
range. In this case study, your students
will chart the life cycle of salmon and
evaluate the impacts of development,
pollution, and other threats. At the same
time, they'll study salmon-based cultures
and see how native peoples and other
groups are working to sustain salmon
populations and the vital traditions
that surround them.

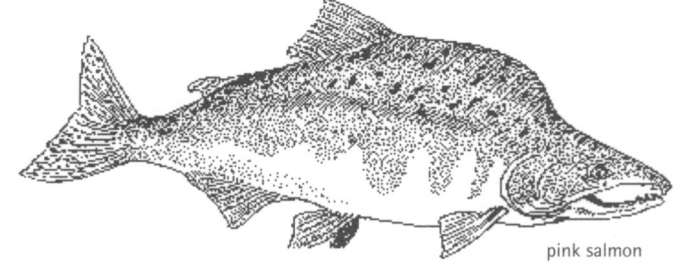

pink salmon

Robert W. Hines, US Fish and Wildlife Service

"We learn a lot of lessons from watching animals. The salmon are one of our best teachers. We learn from them that we have to do certain things by the seasons. We watch the salmon as smolts going to the ocean and observe them returning home. We see them fulfill the circle of life, just as we must do. If the salmon aren't here, the circle becomes broken and we all suffer."

—Leroy Seth, Nez Perce tribe

Salmon on the Web

When you take a bite of salmon, you're connecting with a fish that has powerful links to the history, cultures, ecology, and economies of North America and other parts of the world.

Did you know, for example, that many native groups of the Pacific Northwest Coast have developed their traditions, art, and stories around salmon? Did you know that salmon provide essential food for everything from seabirds to seals to grizzlies? Did you know that U.S. consumption of salmon doubled between 1987 and 1995? And did you know that salmon are seriously threatened in many parts of their range because of a host of human activities?

For all of these reasons, salmon make a terrific topic of study—one that will open your students' eyes to the fascinating ways that human lives intersect with marine biodiversity.

Earth is our home page, hit "save"!

To save paper and make our materials more widely available, we've put the Salmon Case Study on the Web at www.worldwildlife.org/windows/marine in a PDF format.

www.worldwildlife.org/windows/marine

Here's what you'll find when you visit the Salmon Case Study site:

● BACKGROUND INFORMATION

Several fact-filled pages provide an overview of salmon and related issues, including a description of salmon species, the life cycle of a salmon, habitat requirements of salmon, and the major threats to salmon survival.

● ACTIVITIES

1) A Fish Tale

The remarkable life cycle of salmon offers a vivid portrait of the connections among wild species and spaces, and of the wonders of the natural world. In this activity, your students will listen to a story that portrays the salmon life cycle—from birth to death. As your students listen to the story, they will arrange a series of salmon pictures in the proper sequence and learn some fascinating salmon facts.

2) Salmon Scavenger Hunt

What is there to know about salmon? A lot! Salmon predators, threats, and restoration efforts are just some of the topics covered in this Salmon Scavenger Hunt. Students will form teams and compete to amass points and knowledge as they search for salmon information on the Web and in other resources.

3) Much Ado About O$_2$

Salmon can only survive in water if the water has a high level of dissolved oxygen. But the dissolved oxygen content of water can vary substantially, especially in response to certain human activities. In "Much Ado About O$_2$," two hands-on experiments will help your group understand the connections between dissolved oxygen and temperature change, as well as dissolved oxygen and aquatic plants.

4) Salmon People

Salmon have been central to the cultures and economies of indigenous groups in the Pacific Northwest for hundreds of years. In this activity, your students will read and discuss two native myths: "Salmon People" comes from the Haida tribe of the Queen Charlotte Islands in British Columbia and Prince of Wales Island in southeastern Alaska. "The Girl Who Swam with the Fish" is a myth from the Athabaskan peoples of western Alaska. Students will compare the attitudes and beliefs expressed in these two myths, and they will then read a recent article about efforts by native groups to restore salmon, learning more about the role salmon continue to play in these communities.

5) Make Way for Salmon

Salmon face countless threats as they make their way from freshwater streams out to the ocean and then back again to their natal streams to spawn. Your students will apply their accumulated knowledge of salmon ecology and current threats by creating their own board game depicting the life and times of salmon.

● RESOURCES

We've provided a list of curriculum resources, books, Web sites, and contact information for organizations working on salmon-related issues to help you learn more about salmon natural history and conservation.

Mini Case Studies

Inspired by what they've learned about marine biodiversity, your students might want to tackle other research projects to find out more about the amazing variety of creatures, habitats, and issues that make our oceans of life so interesting. In this section, you'll find short descriptions of marine topics for your students to explore. We hope that these thought-provoking ideas, as well as suggestions for how to best guide your students' research efforts, will be useful in extending your marine unit. Whether you use them as a supplement to activities you've already completed or as an introduction to future activities, we hope these mini case studies will pique your students' curiosity and help them to learn more about marine biodiversity.

blue crab

© WWF-Canon/Martin Harvey

"Never before has a wake-up call from nature been so clear, never again will there be better opportunities to protect what remains of the ocean's living wealth."

–Sylvia Earle, marine biologist

#1 Giant Squid
A Creature of Mythic Proportions

What is 60 feet long from the tip of its long arms to its tail, has eyes the size of volleyballs, and has never been seen alive in its deep-sea habitat? The incredible giant squid! Most of us know giant squids only through novels and science fiction movies. Scientists know them only from carcasses washed up on shore, floating in the sea, or hauled up in fishing nets, and remains found in the stomachs of their main predator, the sperm whale. Today, the search for a live giant squid goes on, making this a perfect topic for introducing your students to the wonders and ongoing mysteries of the deep blue sea.

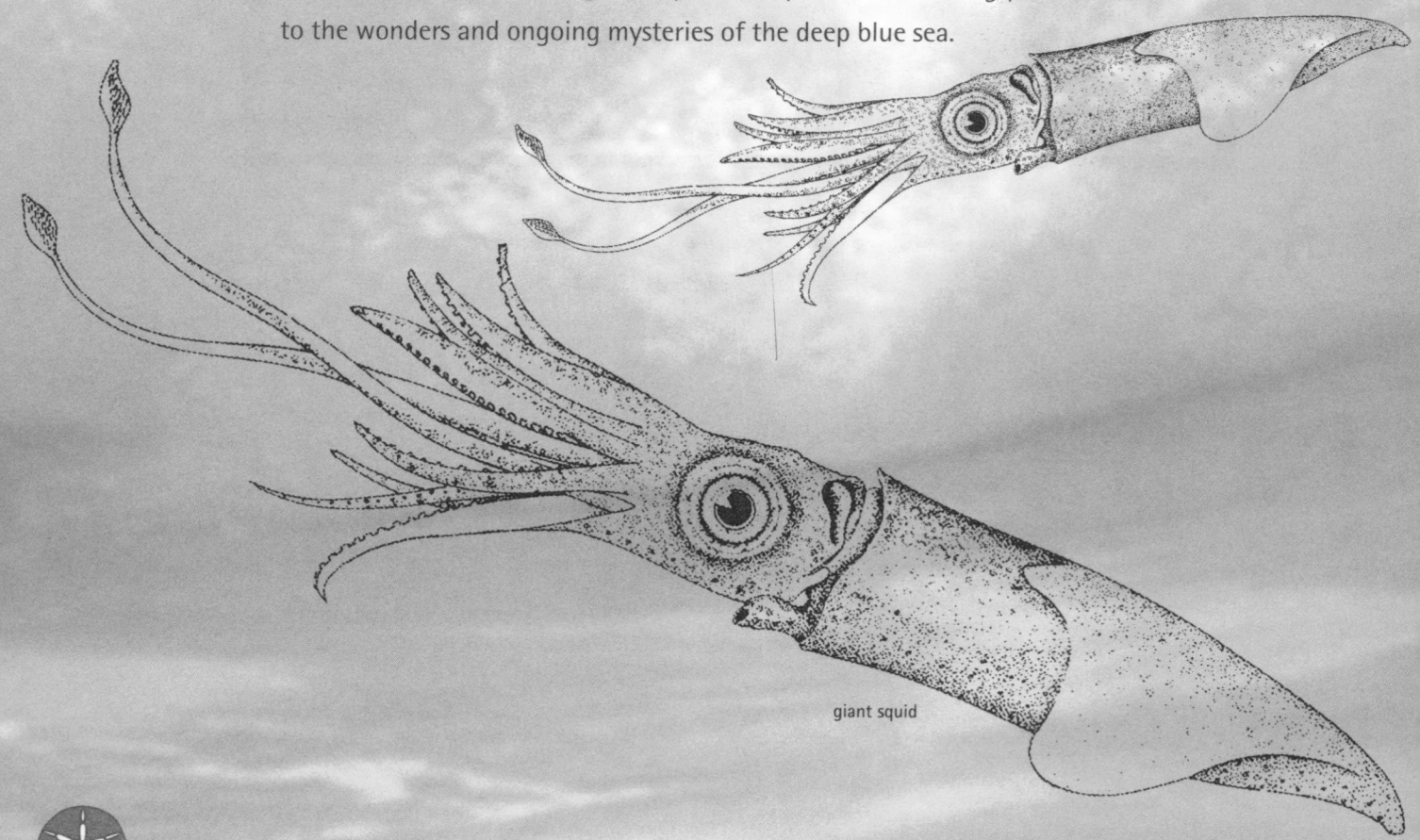

giant squid

Activity Ideas

■ **Legends of "The Beast" (language arts, science)**
Read and analyze ancient myths, contemporary fiction, and movies about giant squids. How do those sources characterize this rare creature? Do you think the accounts are accurate? (You can find references to squid myths on both the Smithsonian and UnMuseum Web pages, listed in the resource section below.)

■ **The Squid Life (science)**
Gather information on the amazing natural history—and diversity—of squids around the world, which include hundreds of species ranging in length from three-fourths of an inch to the 60-foot giant squid.

■ **In Search of . . . (geography, science, technology, language arts)**
Use the Web and other sources to track the adventures of marine scientists trying to learn more about giant squids. Then create a fictional journal about the activities, technologies, and adventures common to squid research. Be sure to include a page imagining the day on which scientists finally discover a live giant squid.

■ **Life Preservers (science, technology)**
Studying the nerve fibers, or axons, of squids has led to advances in treating multiple sclerosis and other diseases. These axons are easy to study because, unlike the microscopic axons of humans, they're so large they're visible to the naked eye. What other benefits do squids provide? Create an ad campaign for squids based on the information you find.

squid

RESOURCES

- *Octopus and Squid* by James C. Hunt. Monterey Bay Aquarium, 1997.
- *The Search for the Giant Squid* by Richard Ellis. Penguin, 1999.
- The Giant Squid (The UnMuseum). **www.unmuseum.org/squid.htm**
- In Search of the Giant Squid (Smithsonian Institution). **seawifs.gsfc.nasa.gov/squid.html**
- Monterey Bay Aquarium Research Institute (Monterey Bay Aquarium). **www.mbari.org**

BIOFACT

Giant squids have eyes the size of a volleyball—the largest of any animal on Earth.

#2 The Deep Sea

Exploring the Great Unknown

Even as scientists map everything from human genomes to distant planets, the deep sea remains largely a mystery. This zone—which begins roughly 660 feet below the surface—covers more than half of the planet, and yet we've explored only about one percent of it.

As your students dive into this deep-sea unit, they'll encounter strange geologic formations: deep trenches, vast canyons, newly discovered thermal vents, and more. They'll meet creatures with fascinating adaptations to their dark environment: glowing bodies, hairs that help them "feel" prey in the dark, and flexible jaws for gobbling up almost anything that bumps into them. Your students will also learn more about the scientists who are studying this region and, in the process, have come to appreciate just how much we have yet to learn about our planet's biodiversity.

Activity Ideas

■ **Charted Territory (language arts, science)**

Listen to an account of deep-sea exploration (read aloud by the teacher or a student), and draw pictures of what you think the deep sea looks like. (For ideas of pieces to read aloud, see the Resources section below.) Then, based on information you find on the Web and in other sources, draw a more accurate diagram of the ocean and its deep-sea regions. Chart depths and use words or pictures to indicate significant features (for example, where photosynthesis ends, where people can no longer swim without pressurized equipment, and so on).

■ **Design a Fish (science, art)**

List the environmental conditions in the deep sea. Then create a fish that is specially adapted to survive in this setting. Compare it to some real-life organisms to see how well you anticipated the characteristics of deep-sea creatures.

■ **Wonders of the Deep Sea World (art, science)**

Make 3-D models or posters of the unusual topographic features found in the deep sea, from thermal vents to cold seeps. Or make models showing the movement of tectonic plates on the ocean floor over time.

■ **Time Capsules (science, language arts)**

Create a time capsule with sample images and information about what scientists now know about the deep sea. Then list your predictions of what we'll discover about the region in the next 10 years. (Plan to open your time capsule at that time!)

hatchet fish

RESOURCES

- "The Blue Planet" by the BBC. (Discovery Channel series.) Available at **www.bbcamericashop.com**
- *The Deep Sea* by Bruce Robinson and Judith Connor. Monterey Bay Aquarium Press, 1999.
- *Down to the Sunless Sea* by Kate Madin. Turnstone Publishing Group, 2002.
- Black Smokers (American Museum of Natural History). **www.amnh.org/nationalcenter/expeditions/blacksmokers/black_smokers.html** (virtual tours of deep-sea vents)
- The Deep Sea (Smithsonian Institution) **seawifs.gsfc.nasa.gov/OCEAN_PLANET/HTML/oceanography_deep_sea.html**
- Extreme 2000: Voyage to the Deep (University of Delaware). **www.ocean.udel.edu/deepsea/home/home.html** (deep sea expedition aboard ALVIN; includes QuickTime videos, phone recordings, and daily logs)
- Into the Abyss (NOVA). **www.pbs.org/wgbh/nova/abyss**
- Mysteries of the Deep (Monterey Bay Aquarium). **www.mbayaq.org/efc/efc_se/se_mod.asp**

BIOFACT

Deep in the sea, large chimney-like structures spew out dark, ultra-heated water that looks like smoke. One such "black smoker" off the Oregon coast towers more than 13 stories and goes by the nickname "Godzilla!"

#3 Whales
Giants of the Sea

If you have been on a whale-watching trip and seen humpbacks leap and swim, or if you have seen whales on TV or video, chances are good you've been dazzled by their size and beauty. But whales aren't just fascinating biological specimens; they're also the subject of myths and legends, the source of a host of products, and the cause of international controversies.

If you organize a unit on whales, your students are sure to be excited by it. As you work through the unit, you can tap that excitement to get them engaged in critical thinking about ethics, international cooperation, and endangered species management.

right whale

blue whale

Activity Ideas

■ **Whoppers (mathematics, science)**

Compile a list of measurements of whale body parts. Then create a fun poster comparing those to commonly known objects. How big is a whale heart? How heavy is a whale brain? How many people could fit end-to-end on a blue whale tongue?

■ **Whale of a Catalogue (social studies, language arts)**

One reason whales have been hunted for so long is that their baleen, blubber, and other body parts have been used by people around the globe. Create a catalogue of whale products and indicate when and where they were used and by whom.

■ **Ups and Downs (science, social studies, mathematics)**

Divide into teams and research the history of an assigned whale species. Is its population dwindling? Rebounding? Graph the population change as best you can, and create a brief summary explaining the cause of your species' decline or rise.

■ **All Together Now (social studies)**

Gather information on some of the international controversies surrounding whale hunting. Then organize a mock international meeting. You might debate the ethics of Japan and Norway continuing to hunt whales in spite of the ban other countries recognize, or you might discuss the controversy over whether native peoples in the United States and Russia should still be allowed to hunt. What values inform each player's position? What compromises might be reached?

RESOURCES

- *DK Handbooks: Whales, Dolphins, and Porpoises* edited by Mark Carwadine et al. DK Publishing, 1995.
- *Gray Whales* by Alan Baldridge and David Gordon. Monterey Bay Aquarium, 1991.
- *Men and Whales* by Richard Ellis. The Lyons Press, 1999.
- *Whales and Dolphins in Question: The Smithsonian Answer Book* by James G. Mead et al. Smithsonian Institution Press, 2002.
- Bridge: Ocean Sciences Teacher Resource Center (Virginia Institute of Marine Science). **www.vims.edu/bridge** (marine education resources, including information and materials on whales)
- WhaleNet. **whale.wheelock.edu**
- Whales: A Thematic Web Unit (University of Virginia). **curry.edschool.virginia.edu/go/whales** (award-winning Web site with numerous links to whale-related sites)

BIOFACT

The tongue of a blue whale weighs as much as an entire Asian elephant.

#4 Marine Pharmacology

Finding Cures in Ocean Organisms

Do you know that we use medicines derived from corals to treat arthritis, from marine sponges to treat bacterial infections, and from sea squirts to battle cancer? Like the tropical rain forest, the ocean is a storehouse of potential treatments for many diseases.

Using ocean products for medical purposes is nothing new; for example, the Inupiat Indians of Alaska have long used whale blubber to cure common ailments. Only recently, though, have modern scientists begun trolling the seas in search of sources of new medicines. Some scientists are even consulting native healers to help target promising species.

Organizing a unit on marine medicines will help highlight some of the connections between biodiversity and cultural diversity. And it will showcase yet another way that marine biodiversity benefits us all.

sea squirt

Activity Ideas

■ **Marine Medicine Cabinet Scavenger Hunt (science, social studies)**
Create a report on a medicine derived from marine species such as sponges, sea squirts, corals, sharks, or squids. How did scientists locate this compound? How effective is it? How widely is it used? How can its use be sustainable?

■ **Casting for Cures (science, language arts)**
Choose an ocean organism with interesting physical characteristics. Then write a creative "fantasy" report explaining what medicinal properties you think it could have and how it can be used for healing.

■ **Native Know-How (social studies, language arts, science)**
Read articles about native people's knowledge of the medicinal uses of plants and animals (from both marine and terrestrial habitats). What plants and animals are used in healing? What are different people's perspectives on native medicines? How is native knowledge influencing the search for new marine medicines, and vice versa?

■ **A Career for What Ails You (social studies)**
Finding possible cures from marine organisms takes the efforts of many people—from chemists to marine biologists to investors. Pick a job you think you could do that would help in the search for marine medicines. Write a job description for it, including the skills and education you'd need and what you'd work to accomplish.

RESOURCES

- "Drugs from Slugs?" by Mary Cay Carson. *Ranger Rick*, February 1999, pages 32-35.
- *Kuuvaymiut Subsistence: Traditional Eskimo Life in the Latter Twentieth Century* by D. B. Anderson et al. Shared Beringian Heritage Program, U.S. National Park Service in cooperation with the Northwest Arctic Borough School District, 1998. Available through the Northwest Arctic Borough School District, P.O. Box 51, Kotzebue, AK 99752. (907) 442-3472. www.nwarctic.org
- Bridge: Ocean Sciences Teacher Resource Center (Virginia Institute of Marine Science). www.vims.edu/bridge (marine education resources, including articles and activities from National Sea Grant Research, American Chemical Society, and other organizations)
- The Healing Qualities of the Sea (Green Reef: Reef Briefs). www.greenreefbelize.com/reefbriefs/briefs34.html
- Life on Sea Life Yields Drug Leads (Nature). www.nature.com/nsu/010614/010614-1.html
- U.S. Scientists Lead Way in Marine Pharmacology (Sea Grant). www.seagrantnews.org/news/cal.html

BIOFACT

Once scientists have created a medicine from marine organisms, they can often copy the chemicals so that no additional plants or animals have to be disturbed or killed to make more of the medicine.

#5 The Bering Sea

Life at the Top of the World

I f you've eaten ocean fish lately, chances are good it came from the Bering Sea. Today, more than 50 percent of all fishery products consumed in the United States come from this region.

The Bering Sea is a fascinating place, and not just because of its phenomenal fisheries. It's also home to rare sea mammals and seabirds, as well as native peoples still trying to subsist on local resources. At the same time, many environmental problems threaten this region, most notably overfishing, toxic pollution, and global climate change. By looking at this region as a whole, your students will be able to examine the many intricate connections between social, ecological, and economic health. And they'll see how supporting all three of these aspects contributes to the region's sustainability.

Stellar sea lion

Activity Ideas

■ **Exploring Fiction (language arts, social studies)**

Research one of the explorers who came to the Bering Sea region (such as Captain Cook or the Russian explorers) and write a piece of historical fiction describing what the exploration team might have seen.

■ **Cultures in Change (drama, social studies)**

Use the Web to locate information about the native peoples of the Bering Sea. How did different groups traditionally use the resources of the region? How have their lives changed? How do these changes affect the sustainability of the region? Create skits that depict some of the past and current events that have been particularly significant to the region's native peoples, being sure to represent differences among the groups.

■ **Fish Markets (social studies, science, economics)**

Create a seafood menu showing some of the major fish found in the Bering Sea. Instead of describing a seafood dish, though, provide background about each fish. Sketch out its food web, being sure to include humans. Make a diagram of the fishing method used to catch it. What are the potential effects of this gear on the fish's habitat? Find out if the fish's populations are stable or dwindling. Is this fishery sustainable?

■ **Predict Your Own Future
(science, social studies, language arts)**

Gather information on a current problem in the Bering Sea, such as climate change, overfishing, or Steller sea lion decline. What is the source of the problem? How is it affecting the region's ecological, social, and economic health? Create a story about this problem with two different endings, one positive and one negative. Explain what actions would lead to each of these endings.

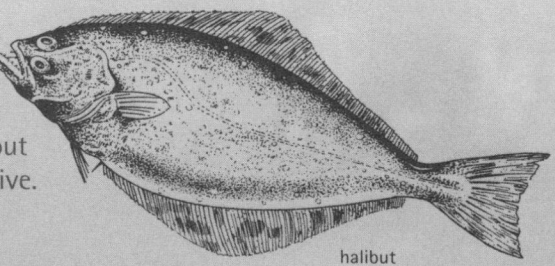

halibut

RESOURCES

- *Arctic Environment Atlas* by Kathleen Crane and Jennifer Lee Galasso. Office of Naval Research, Naval Research Lab, Hunter College, 1999.
- *Science Under Sail, Russia's Great Voyages to America 1728-1867* by Donna Matthews and Barbara Sweetland Smith. Anchorage Museum of History and Art, 2000.
- Alaska Sea Grant College Program. **www.uaf.edu/seagrant**
- Arctic Circle (University of Connecticut). **borealis.lib.uconn.edu** (comprehensive cultural, natural, and social information on and for arctic peoples)
- Arctic Information Web Site (ARCUS). **www.arcus.org**
- Natives Gain Voice on Sea (NOAA). **www.pmel.noaa.gov/bering/pages/trad_article.html**
- Stellar Sea Lion (Alaska Department of Fish and Game). **www.state.ak.us/local/akpages/FISH.GAME/wildlife/geninfo/game/sealion.htm** (information on sea lion decline)
- Tales from the Bering Sea (World Wildlife Fund). **www.worldwildlife.org/beringsea**

BIOFACT

The Bering Sea supports 70 percent of the world's northern fur seal population and the highest concentration of Pacific walruses.

#6 Oil Spills
Trouble on the Water

Oil spills are a bleak and familiar sight to anyone who watches world news. Most of us have seen pictures of oil-covered beaches and oil-soaked animals, as well as the rescuers who work to clean everything up. Oil spills result in an incredible 37 million gallons of oil polluting the oceans every year.

It's hard to know who's most to blame for oil spills, since just about all of us depend upon oil to meet our basic needs. Whose responsibility is it to make sure oil spills don't happen? Whose responsibility is it to clean them up? In a unit on oil spills, your students can pursue those questions while also exploring the benefits and limits of technological solutions, such as new tanker designs and skimmers. Studying oil spills also provides a great opportunity for your students to think about the complex consequences of natural resource extraction and our consumption of oil.

Heidi Snell/Charles Darwin Foundation

Activity Ideas

■ **Oil in a Day's News (social studies, geography, language arts)**
Oil spills have been occurring all around the world for decades. Gather news articles on oil spills and create a map-based poster showing where and when the spills have occurred and the damage they caused to wildlife and local people.

■ **Spill Simulations (science, technology)**
Learn about the methods used to clean oil from water and wildlife. How effective are these efforts? Create demonstrations of one or more of these methods using containers of water, motor oil, sponges, feathers, dishwashing fluid, and other materials. Can you come up with a better cleanup approach?

■ **Think Tankers (science, technology)**
What does an oil tanker look like? What design ideas have or could be implemented to make them safer? Use these ideas to design a safer tanker. Is it indestructible? What special features does it have? What are the benefits and limits of tanker technologies?

cormorant

■ **Act Now! (social studies, science)**
More oil reaches waterways each year through runoff and improper disposal than through all the year's commercial oil spills combined. Find out more about the people who benefit from oil extraction and the people who contribute to oil pollution. What can businesses do? What can you do at home? Decide what the responsibilities are of individuals, communities, and corporations to reduce oil pollution. Write a letter or design a brochure to promote one aspect of this responsibility.

RESOURCES

- "A Better Boat: Industry Puts Money, Effort into Preventing Another Exxon Valdez." *Anchorage Daily News*, May 27, 2001, page G-1.
- *Oil Spill!* by Russell G. Wright. Event-Based Science Series, Innovative Learning Press/Addison Wesley, 1995. (Oil spill curriculum)
- *Oil Spills* by M.K. Anderson. Franklin Watts, 1991.
- *Out of the Channel: The Exxon Valdez Spill in Prince William Sound* by John Keeple. HarperCollins, 1991.
- *The Wasted Ocean* by David K. Bulloch. Lyons and Burford, 1989.
- Bridge: Ocean Sciences Teacher Resource Center (Virginia Institute of Marine Science). www.vims.edu/bridge (marine education resources, including oil spill activities from the Gulf of Maine Aquarium, Ocean Planet, Oil Spill Awareness through Geoscience Education [OSAGE], and others)

 **BIOFACT**
Genetic engineers have developed oil-eating bacteria that aid in oil spill cleanups.

#7 Penguins

Keepers of the Southern Oceans

mperor. Rockhopper. Chinstrap. Jackass. Macaroni. Even the roll call of
penguin species reflects the charisma and striking appearance of
these birds.

Creating a case study around penguins is a great way
to introduce your students to species and habitat
diversity. Despite what many people think,
penguins live only in the southern hemisphere; you won't find any in the Arctic. How
do they survive antarctic temperatures?
What adaptations enable them to spend up
to three-quarters of their lives at sea? In
addition to answering these questions, your
students can take a look at the problems threatening
penguins and get a glimpse of the major threats
facing all biodiversity worldwide.

emperor penguin

Activity Ideas

■ **Meet the Penguins! (science, language arts)**

Get to know the diversity of penguins by making a penguin field guide—or even a penguin family album. Divide into teams and have each team cover one species. Include natural history information and illustrations or photographs. Or devise a key to identifying a dozen or so species of penguins.

■ **Habitat Maps (geography, science)**

In teams from the first activity, locate the summer and winter habitat of your assigned penguin species. Create colored labels (one color per species) and pin the labels on a map, connected by string to the summer and winter habitats of your chosen species. Then explain the adaptations that enable your species to live in its different habitats.

■ **Black and White and Read All Over (language arts, science)**

Collect articles about problems that threaten penguins. Group them according to the five major threats to wildlife worldwide: habitat loss, introduced species, pollution and climate change, population growth, and over-consumption. Which of these caused the most penguin losses in the past? Which do now? What role do penguins play in their ecosystems? How might their demise affect other species?

■ **Penguin Pals (social studies, science, language arts)**

Find out what scientist Dee Boersma is doing to learn about and help penguins (see www.artsci.washington.edu/newsletter/Autumn99/Boersma.htm and other sites). What does her work involve? What has she learned so far? Is she hopeful that penguins will fare well in the future? Conduct a Web search to see if you can find other people working to help penguins around the world.

rock hopper

■ **Big as Life (art, science)**

To visualize how big some species get, create life-sized paper models of penguins using large sheets of paper.

RESOURCES

- *Penguin Planet: Their World, Our World* by Kevin Schafer. Creative Publications International, 2000.
- *Penguins of the World* by Wayne Lynch. Firefly, 1997.
- Be a Penguin! (New England Aquarium). www.neaq.org/scilearn/kids/beapeng.html/index.html
- Focus on Penguins. (Monterey Bay Aquarium). www.mbayaq.org/efc/efc_fo/fo_penguins.asp
- Penguins (Henson Robinson Zoo). www.hensonrobinsonzoo.org/home_n.html

BIOFACT

When they're cold, penguins on land lie on their bellies so their black backs absorb the sun's warm rays. When they're hot, they lie with their white tummies facing up, reflecting the heat away.

#8 Islands
Oases in the Ocean

If you've ever read through a list of priority ecoregions or biological hotspots compiled by groups such as World Wildlife Fund or Conservation International, you're bound to notice that many of the places on the list are islands. That's because islands tend to be inhabited by unique species that have had thousands of years to evolve apart from their relatives on nearby continents. In addition, many islands are facing severe environmental problems that threaten their biodiversity. Alien species, coastal development, pollution, and sea level rise are among the many threats to island life.

You and your students can take a virtual trip to diverse, exotic places by spending a few weeks studying islands. In the process, you'll explore some of the most important topics in biodiversity conservation.

Activity Ideas

■ **Form an Isle (art, science)**
What are some of the different ways that islands form? Make a model showing a particular kind of island (such as volcanic, coral atoll, or barrier) and explaining how it formed.

■ **Isolated Incidents (science, language arts)**
Pick an island or island chain, such as Hawaii, Madagascar, Galápagos, Fiji, or the Florida Keys, and then use the Web to gather information on species found on or near the island or chain. Are any of these species unique to the island or chain? Why do land plants and animals on islands tend to be unique? How did island plants and animals get to the island in the first place?

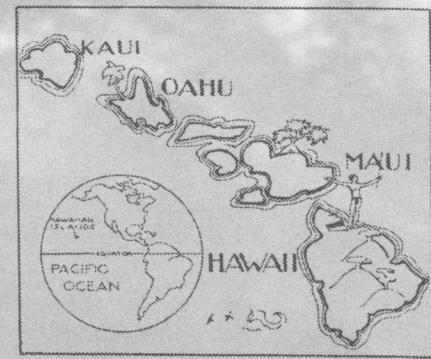

■ **Rethinking Islands (science)**
Choose an environmental problem, such as global warming, alien species, habitat loss, or pollution, and look at how it affects islands. Are the impacts of these problems more severe on islands than on other land areas? Why? In what ways are wild habitats such as tropical rain forests, mountain tops, and prairies similar to islands? In what ways is Earth an island?

■ **Island Hoppers (social studies)**
Plan an ecotourist trip to an island—such as a photo safari, field study, or service project. You can find a list of such trips through the Sierra Club, School for Field Studies, World Wildlife Fund, Earthwatch Expeditions, and other groups. Create a travel itinerary and figure out how you would get to the island, what it would cost, and what you would need. Write what you would expect to learn and accomplish on this visit.

RESOURCES

- *Ecology and Evolution: Islands of Change* by Richard Benz. NSTA Press, 2000. (activity book)
- *Eyewitness—Island* published by Eyewitness, 1997. (35-minute video on islands)
- *Geography Detective: Islands* by Philip Steele. Zoe Books, 1996. (activity book)
- *Song of the Dodo: Island Biogeography in an Age of Extinction* by David Quammen. Touchstone Books, 1997.
- Channel Islands National Park (National Park Service). **www.nps.gov/chis**
- Irreplacable Islands: A Vision for the Future of the Galapagos (World Wildlife Fund). **www.worldwildlife.org/galapagos**
- Madagascar: A World Apart (PBS). **www.pbs.org/edens/madagascar**

BIOFACT

Ninety-seven percent of the reptiles and mammals found on the Galápagos Islands are endemic, which means that they do not exist anywhere else on Earth.

#9 Sea Turtles
Rare and Remarkable Reptiles

Sea turtles are among the most irresistible, and most endangered, of marine creatures. Every one of the eight species of sea turtles is threatened or endangered. Among the greatest threats to sea turtle survival are habitat loss or degradation on the beaches where females lay their eggs, egg predation by humans and other animals, trade in turtle products, ocean pollution, and drowning in fishing gear.

Your students are sure to enjoy an introduction to the fascinating life history of sea turtles. They can also explore the variety of sea turtle threats and some of the things people are doing to help. In the process, they'll learn more about the work of marine scientists by following along with sea turtle tracking programs.

leatherback sea turtle

World Wildlife Fund

Activity Ideas

■ Spotlight on Sea Turtles (science, language arts)

Create a scrapbook or Web site with photos and information on each of the eight different sea turtle species. Include interesting details on the natural history of the species, as well as information on its status and threats.

■ Hatchling Happenings (science, art, mathematics)

Gather information on sea turtle hatchlings. What special adaptations enable them to find their way from their nest to the sea? What hazards do they face? Create a board game in which players are baby sea turtles in their first days of life. Or try Project Aquatic WILD's "Turtle Hurdles" activity (see page 362 for more information on this publication).

green sea turtle

■ Sea Turtle Sizes (science, art)

To get a better idea of the range of sizes, create life-sized paper models of various species of turtle, both hatchlings and adults.

■ Turtle Trackers (science, social studies)

Track the movements of a radio-tagged sea turtle (see Resources below). How far does the turtle travel? Is there any pattern to its movements? What threats does this turtle face at different parts of its journey? Why are scientists interested in tracking sea turtles?

■ Adopt a Sea Turtle (social studies)

Find a way to lend sea turtles a hand. You might want to "adopt" a sea turtle, write a letter to the editor of your local newspaper, raise money to protect sea turtle nesting beaches, or create an educational brochure introducing your friends and family members to the wonders of these marine reptiles.

RESOURCES

- "Sea Turtles: In a Race for Survival" by Anne and Jack Rudloe. *National Geographic* 185 (2), February 1994.
- *Turtle Watch* by George Ancona. Macmillan Publishing Co., 1987.
- *Turtles of the World* by Carl H. Ernst and Roger W. Barbour. Smithsonian Institution Press, 1989.
- *Zoobooks: Turtles* by Timothy Levi Biel. World Color, 1993.
- Bridge: Ocean Sciences Teacher Resource Center (Virginia Institute of Marine Science). **www.vims.edu/bridge** (marine education resources with many links to sea turtle topics; see *Spotlight on a Scientist: Kate Mansfield,* which includes Kate's spreadsheet of satellite data on a sea turtle's movements. Students can create maps based on this information.)
- Mote Marine Laboratory. **www.mote.org** (includes information on the sea turtle rehabilitation hospital)
- Sea Turtle Migration: Tracking and Coastal Habitat Educational Guide (Caribbean Conservation Corporation). **www.cccturtle.org/eduform.htm** (downloadable sea turtle tracking activity for teachers)
- Sea Turtles (SeaWorld). **www.seaworld.org/infobooks/SeaTurtle/home.html**
- Turtle Trax. **www.turtles.org** (provides links to numerous sea turtle sites; includes a "Kidz Korner")

BIOFACT

Leatherbacks are the largest of all sea turtles, growing up to 7 feet long and weighing as much as 1,200 pounds.

#10 Shipwrecks
From Refuse to Reef

Shipwrecks have long been an object of fascination for divers, history buffs, and treasure seekers. But did you know they also create micro-habitats for many marine species? Barnacles, corals, sponges, and myriad other marine creatures attach themselves to sunken ships, and, in turn, support a food web. For this reason, sunken ships and even out-of-use oil platforms are not considered ocean trash—in fact, they're called artificial reefs! Not only do they create habitat for marine species, but they also attract snorkelers and divers (and, with them, ample tourist dollars). No wonder, then, that creating artificial reefs is part of many regions' plans for restoring marine life.

Your students will get a good picture of creative environmental restoration by studying shipwrecks and other artificial reefs.

Activity Ideas

■ **Wrecked! (social studies, language arts, geography)**
Play the part of a reporter gathering information on ships that have sunk in different parts of the world. Are they causing any problems? Providing any benefits? Is anyone paying attention to them? Why? Identify the location of these ships on a world map.

■ **Underwater Archaeology (social studies, language arts)**
Visit a Web site on underwater archaeology to learn more about the historical treasures found under the sea. Write a brief article on one of these discoveries.

■ **The Shipwreck Community (art, science)**
Create a model of a sunken ship covered with different marine organisms. In what ways do sunken ships make good habitats for these species? Describe the larger food web that is created around this habitat.

■ **Restoring Reefs (science, language arts)**
Investigate the ways that people are working to create more artificial reefs. Then make a travel brochure that entices people to visit artificial reefs and describes their ecological function.

RESOURCES

- Artificial Reefs (Texas Parks and Wildlife). **www.tpwd.state.tx.us/fish/reef/artreef.htm** (article on "Rigs to Reefs" artificial reef program)
- Marine Artificial Reefs (Sea Science). water.dnr.state.sc.us/marine/pub/seascience/artreef.html (article on artificial reefs in South Carolina)
- Ocean Explorer: Explorations (NOAA). oceanexplorer.noaa.gov/explorations/monitor01/monitor01.html (highlights recent underwater archaeological efforts)
- Organization of Artificial Reefs. www.oar-reefs.org
- Rig Diving! (U.S. Department of Interior Minerals Management Service). **www.gomr.mms.gov/homepg/lagniapp/lagphoto.html** (describes the exploration of an artificial reef)
- Rivers to the Ocean Webcast (Texas Parks and Wildlife). **www.tpwd.state.tx.us/expltx/eft/waterways/webcast.htm** (Webcam of an artificial reef)

BIOFACT

The city of New York plans to sink old subway cars as artificial reefs in the mid-Atlantic.

#11 Marine Protected Areas

Saving Space for Sea Species

Your students are probably familiar with the idea of protected areas on land, such as wildlife refuges, national parks, city parks, and national forests. But they may be less familiar with the dazzling variety of marine protected areas (MPAs) around the world. In the United States alone, these marine refuges range from the tiny 14-acre Farnsworth Bank Ecological Reserve in Los Angeles County to the 131,800-square-mile MPA off the coast of Hawaii. Like their terrestrial counterparts, marine protected areas are designed for a variety of purposes. Some protect coral reefs, while others protect sea turtle nesting sites or the habitat of a variety of marine species. Others are designed to keep fish stocks thriving. And still others serve primarily as recreation areas.

Not surprisingly, managing a protected area is at least as difficult as convincing people to establish one. In this unit on marine protected areas, your students will evaluate some of the challenges facing MPA managers. They'll also learn why such challenges don't discourage the many conservationists who believe that marine protected areas are an essential component of any effort to protect marine biodiversity into the future.

hooded seal

Activity Ideas

- **Underwater Gems Gallery (social studies, art)**
  Create a gallery of images and information about marine protected areas around the world. Each student or team should tackle a MPA and include information about its size, location, biological resources, and management goals.

- **By Land and by Sea (science, social studies)**
  Create a group list of all the ways that marine protected areas are similar to or different from protected areas on land. Think about issues such as access, boundaries, development, pollution, and alien species.

- **Mapping MPAs (social studies, science)**
  Use World Wildlife Fund's Global 200 ecoregion information or another source to locate some of the most ecologically important and threatened marine areas around the world. Select one of those regions to study. What are the major threats to this region? Are any marine protected areas located in the region? How might MPAs help with the environmental problems in this region? What problems can't MPAs address?

- **By Design (science, social studies, language arts, art)**
  Using the information from the previous activity, design a MPA. Describe its borders, its purposes, and, most importantly, your management plan. Who is likely to support your MPA? Who might fight it? On what basis would you argue for the establishment of this MPA? Write a report that summarizes your idea.

RESOURCES

- Endangered Seas (World Wildlife Fund).
  www.panda.org/endangeredseas
- Endangered Spaces: The Global 200 (World Wildlife Fund).
  www.worldwildlife.org/global200/spaces.cfm
- Marine Protected Areas of the United States. **mpa.gov**
  (Web site on marine protected areas, including regional maps of MPAs around the United States)
- The MPA Library (Marine Protected Areas of the United States). **mpa.gov/mpaservices/library/websites.html**
  (a list of Web sites that offer information and activities on marine protected areas)
- Sanctuaries and Reserves Division (NOAA). **whale.wheelock.edu/whalenet-stuff/NOAA_Sanctuaries.html**
  (a list of NOAA marine sanctuaries)

triggerfish

BIOFACT

Less than one-half of one percent of Earth's seas lie within a marine protected area.

Appendices

"Too much civilization, accompanied by too little education, is creating havoc with these beautiful underwater habitats. It is high time that federal and public efforts converge to protect these fragile environments."

—Bruce Babbitt, Former Secretary, U.S. Department of the Interior

Appendix A

Putting the Pieces Together

This section provides ideas about how to link activities to build effective units focused on marine biodiversity and other biodiversity-related issues. Whether you're a nonformal educator or an art, science, language arts, or social studies teacher, we encourage you to adapt the activities, combine them with other resources, and devise organizing themes that will best meet your particular educational objectives. Although each of the case studies (Coral Reefs, Shrimp, Sharks, Alien Species, and Salmon) make excellent units in themselves, we have created some additional units that organize activities from a variety of the case studies around a central theme. (See pages 326-337.) But we know that no one can create better lessons for your students than you can—whether you teach in an aquarium, zoo, school, museum, or community center. We encourage you to tailor the sample units we've provided to suit your particular needs, or—better yet—to develop your own marine and biodiversity units that draw on local issues as well as content that is relevant to and engaging for your students. And check out our Web site at **www.world wildlife.org/windows/corr.html** for correlations between the activities in *Oceans of Life* activities and the national science and social studies standards.

Below is an overview of the educational strategies we used to develop the sample unit plans. For more in-depth information on these topics, see the background section (pages 20-57) and "Resources" (pages 360-369).

The Thinking Behind the Sample Units

1. Framing the Unit

For the *Windows on the Wild* modules, we've developed a content framework organized around four major themes: *What is biodiversity? Why is biodiversity important? What is the status of biodiversity? How can we protect biodiversity?*

Specific learning objectives accompany each of these themes, and the activities in this and other *WOW* modules tie to concepts in the biodiversity framework (on the Web at **www.worldwildlife.org/windows**) as well as to the key marine concepts and ideas (page 14).

In addition, we've developed a skills framework that is adapted from the North American Association for Environmental Education's (NAAEE) *Environmental Education Guidelines for Excellence: What School-Age Learners Should Know and Be Able to Do*, and the thinking and process skills developed by the Association for Supervision and Curriculum Development and the American Association for the Advancement of Science. We've also incorporated skills from the Middle School Association and the national standards work in civics, science, mathematics, reading, language arts, and geography. Our skills framework focuses on thinking, process, and citizen action skills that we believe lead to a more informed and engaged citizenry.

Together, the content and skills frameworks have guided the development of the activities in all *WOW* modules as well as in the marine biodiversity units that follow.

2. Starting with a Storyline

We've built each of our sample units around a storyline. Storylines are organizing questions or attention-grabbing issues that help focus a course of inquiry. For example, the "Value of Marine Biodiversity" unit on pages 334-335 uses an ecosystem-services approach as the storyline. The questions "What are the many different reasons that we value biodiversity?" and "What is the range of services that marine diversity provides?" become unifying threads for understanding the importance of conserving marine species and ecosystems.

You can use storylines at the lesson level to help students understand individual concepts, or at the unit level to give sets of activities an added dimension of relevance or interest. Either way, storylines can function as effectively in the classroom as they do in a good mystery: enhancing curiosity and making students eager to find out more.

3. Building on Concepts

In the following sample units, we've organized the content in a way that builds on previous material. That may sound simple, but it's not always easy to do with a topic as multidimensional as marine biodiversity. In developing your own units, you'll need to create a sequence of activities. To help in that process, the "Connections" section at the beginning of each activity lists other activities that are good to do before or after that activity. But in some cases, you will need to insert activities from other sources to fill in some of the "holes." And in other cases, your students will need prior knowledge to get the most out of the activity.

4. A Focus on Learning Styles and Multiple Intelligences

One of our goals in developing this module was to make marine biodiversity issues come alive for all students—regardless of how they learn best. To do that, we've tried to cater to a variety of learning styles and intelligences. You'll find a mix of activities and teaching strategies—from hands-on discovery to small group discussions—and you'll see these learning styles highlighted in different ways throughout the sample units.

In addition to recognizing different learning styles, we have also tried to emphasize strategies that promote the variety of strengths your students already have and help them develop new strengths.

5. Thinking, Questioning, and Creating

We developed *Oceans of Life* to get students to think, question, and be creative. Throughout this module you will see a variety of activities that encourage questioning, problem solving, issue investigation, creative writing, and other critical and creative thinking skills. We encourage you to build on what we've developed to help your students learn to be critical thinkers and use their creativity to come up with new ways of seeing the world. If you're new to teaching, there are many resources that can give you ideas about how to develop more insightful questions, inspire creativity, encourage effective writing, and support your students to think at a higher level. For

example, writing down questions before you start a unit allows you to think more carefully about the types of questions you pose and can help you get your students to think at a higher level.

Encouraging students to brainstorm, without passing judgment on other people's ideas, can help them feel more uninhibited in coming up with creative ideas. We also encourage you to use the writing suggestions at the end of each activity as a way to emphasize different types of writing and let kids know that you think writing is important— regardless of what subject you teach.

6. Curriculum Connections

The sample units on the following pages provide a variety of ideas for integrating marine biodiversity issues across the curriculum. Although we know it's difficult in some schools to link science, mathematics, language arts, and social studies, we also know that many educators around the country are doing just that.

No matter what or where you teach, there are creative ways to bring biodiversity and marine-related issues into your teaching. For example, you might already teach a unit on cultural connections to nature in your social studies class. Or perhaps you conduct programs on the water cycle at a nature center. In these and other cases, the activities in this module (and in other *WOW* modules such as *Biodiversity Basics* and *Wildlife for Sale*) can help you add a new twist to your teaching by tying your activities to biodiversity and marine issues. You may

> *"Our oceans are our national trust and are calling out for our help. This generation has a duty as good stewards to respond to this challenge for the sake of our children."*
>
> **–Leon Panetta,
> Chair of Pew Oceans Commission**

want to incorporate some activities from *Oceans of Life* into a unit you already do. You can also use the activities in this module to teach specific thinking and process skills (which are highlighted at the beginning of each activity) or improve writing skills (ideas are highlighted at the end of each activity).

7. Language Learners and Biodiversity

We know that typical middle school classrooms are made up of students with a variety of language backgrounds. The same holds true of young people taking part in nature clubs, zoo and aquarium programs, and museum courses. In developing this module, we incorporated suggestions from linguistic experts to help address the needs of students who speak English as a second language and students reading below grade level. And on pages 346-347, we've listed some of the tips from linguists and educators about how best to think about language issues as you plan your units. We've grouped the tips into listening and speaking skills, vocabulary, reading, and writing.

8. An Experiential Design

We relied on several experiential design models to develop the activities and units in this module. We know that students and adults learn better when they are actively engaged in their learning and have a chance to take part in varied experiences.

In many cases, one activity does not include all the steps of the experiential learning cycle. So in our sample units we have combined and sequenced activities in a way that allows students to experience an exciting introduction (which can take the shape of a question, a quiz, or some other activity that gets them engaged); a chance to take part in some experience (an activity, experiment, or investigation); time to process what they did (often a small- or large-group discussion); an opportunity to reflect and generalize (what did they learn, what insights did they gain that they can apply in other situations, what did others think and feel?); and finally, a chance to apply and practice (an opportunity to take action based on what they learned, preferably through a community-based project or activity). We encourage you to explore the many models that exist for designing effective

experiential units. The diagrams below show two experiential learning models.

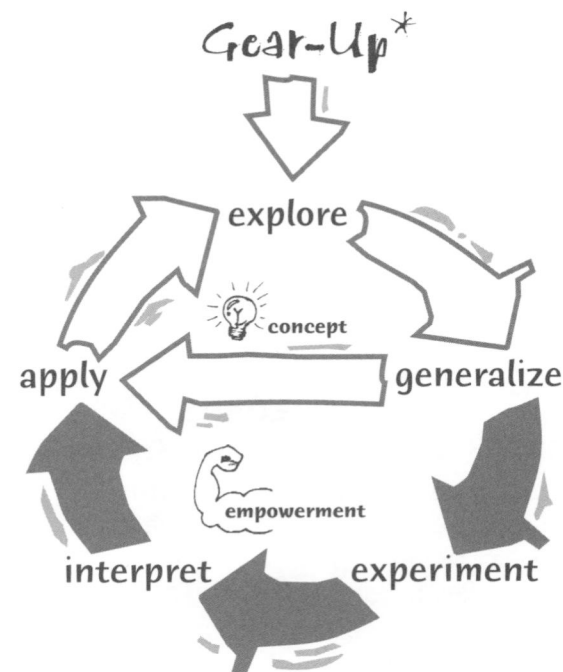

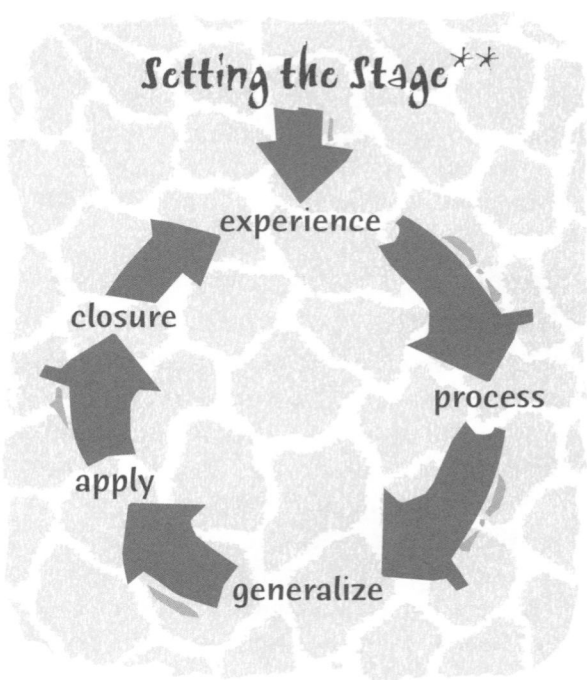

*Adapted from Alaska Science Consortium
**Adapted from Training Resources Group (TRG)

9. Working Together

Solving the world's complex problems will take people with diverse ideas coming together to forge creative solutions. We believe that one of the most important skills we can teach young people is how to work effectively in groups. Cooperative learning helps students practice active listening, discussing, questioning, consensus building, defining personal positions, and more. Working together also helps students learn to respect other's opinions, beliefs, and values. At the same time, they learn how to agree to disagree and work together to understand controversial situations. Each of our sample units includes group experiences to help students practice the teamwork skills that they will use for a lifetime.

10. Wrapping It Up

You'll see that the *Oceans of Life* activities have suggestions for how to assess student learning. In our sample unit plans, we often use one of the final activities as an assessment to pull the unit together and provide closure for your students. We believe that the best wrap-ups (to an activity, a unit, or a course) help students see how much they have learned, how all this knowledge fits together, and how well they can apply it to a final product or project. We also believe that the best assessments go beyond multiple choice questions and show how students have mastered knowledge or skills by having them apply their learning to a real-world situation.

11. Action Strategies

The first introductory unit (pages 326-328) includes ideas of marine-related community action projects for your students to undertake. We believe that it's important for your students to have a grounding in biodiversity and marine issues before tackling local, national, or global action projects. However, you can add an action component to any of the units, including the ones you develop on your own.

12. Going Beyond Basics

We used a variety of resources to develop *Oceans of Life*, and we list many of them in each activity. As you plan your units, we encourage you to bring in other resources—from books to speakers to all sorts of videos, CD-ROMs, and multimedia programs. See pages 360-369 for a list of additional resources you may want to use. We also encourage you to use newspapers, magazines, and the Internet to bring current events into your teaching.

 **BIOFACT** Tiny one-celled plants cause the sparkles that can sometimes be seen in the ocean (or along the dampened shoreline) at night. These dinoflagellates flash a light when the water is disturbed.

Introducing Marine Biodiversity
version 1 (about 20 sessions)

This interdisciplinary unit uses activities that involve science skills and content, and highlight language arts, social studies, and art. The unit also includes a field trip to a local seashore or educational institution. The first two activities introduce students to the concept of marine biodiversity. In the next activities, students find out why marine biodiversity is important. Students then learn about the threats to marine biodiversity. Finally, students explore steps people are taking to protect marine biodiversity, and they create materials to educate others. **Note:** Use the Activities Index, page 375, to locate each *Oceans of Life* activity.

Sessions 1, 2, and 3: "Biodiversity Break-Down"

Use this activity to introduce the unit. Emphasize the three different levels of biodiversity: genetic, species, and ecosystem (step 1). Hand out the "Describing Diversity" sheet (step 2) before the end of the first session, and have the students finish it for homework the first night. In the second session, discuss ecoregions (step 3) and have students select a marine ecoregion to research (step 4). Give them time to begin their research at the end of the second session, and have them continue their research and work on their posters during the third session. Having access to the Internet would aid their research at those times. Students should finish their posters for homework.

Session 4: "Biodiversity Break-Down" and "Sea for Yourself"

Wrap up "Biodiversity Break-Down" by having each group give a brief presentation about its poster (step 5). Then allow five minutes for the assessment. Use the rest of the session to prepare students for a field trip or virtual tour (options are discussed in "Sea for Yourself").

Session 5: "Sea for Yourself"

There are three options for this activity. In the first option, you can take advantage of a nearby and accessible seashore for a field trip to pique your students' interest in marine biodiversity. In option two, your students conduct a marine biodiversity scavenger hunt at a local educational institution, such as a nature center, aquarium, or natural history museum. If neither of these two options is possible,

option three is for your students to take a virtual tour of marine biodiversity, using the Internet.

Session 6: "Coral's Web"

Now that your students have been introduced to marine biodiversity, the next several sessions will be spent exploring why biodiversity is important. Through "Sea for Yourself," students will have become familiar with some marine species. In "Coral's Web," students will revisit some of those species, while also learning about others. After going through the outside forces in step 5, assign the extension as homework. Have students include a color picture of their species.

Session 7: "Services on Stage"

Use this activity to further examine why marine biodiversity is important. Finish after step 5. At the end of the unit, your students will be developing educational materials similar to those discussed in step 6.

Session 8: "Coral's Web" (cont'd.)

Using the short reports and pictures that your students compiled, create the food web bulletin board outlined in step 7. Then, facilitate the wrap-up discussion in step 6. Have students complete the assessment.

Session 9: "Sizing Up Shrimp"

Use this activity to further examine some of the different ways that people value marine biodiversity. Assign the assessment as homework.

Sessions 10 and 11: "Postcards from the Reef"

In these sessions, students will return once again to coral reefs. In session 10, students discover where coral reefs are located and the conditions they need to survive (steps 1-4). In session 11, students learn about the benefits of coral reefs as well as the threats they face (steps 5-6). Have students complete the assessment. Use session 11 to transition from looking at the topic of why marine biodiversity is important, to examining the threats it faces.

Sessions 12 and 13: "America's 'Most Wanted'"

Use this activity to introduce the threat that alien species pose to marine biodiversity. Having access to the Internet will aid student research in session 12. At the end of session 13, or for homework, have students complete the assessment using an alien species different from the one that they researched in class.

Session 14: "Where the Wild Sharks Are"

Use this activity to help set the stage for session 15 by familiarizing your students with shark diversity. After discussing results, assign the "Shark Mystery Challenge" for homework over the next several nights.

Sessions 15 and 16: "Sharks in Decline"

This activity highlights the threats that sharks face, mostly from overfishing. After discussing the simulations (step 6) at the end of session 15, assign "Fishing Worksheet B" for homework (step 7). Discuss the students' answers at the beginning of session 16 (steps 7-8). Also go over answers to the "Shark Mystery Challenge" assigned in session 14. If time permits, introduce the culminating project.

Sessions 17, 18, 19, and 20: "Reef.Net"

Using "Reef.Net" as a template, introduce the next assignment. You may not want to limit student projects to only coral reefs. Instead, have them focus on a topic that has something to do with any aspect of marine biodiversity. Have the students use the first three sessions as work sessions for their projects. On session 20, decide on an appropriate way to have students share what they have learned with the rest of the community.

Unit at a Glance — Introducing Marine Biodiversity—version 1

Session 1	Session 2	Session 3	Session 4	Session 5
Biodiversity Break-Down Step 1	Steps 3-4	Work on posters	Step 5, Assessment **Sea for Yourself** Step 1 →	→
HW: Step 2		HW: Finish posters		
Session 6	Session 7	Session 8	Session 9	Session 10
Coral's Web Steps 1-5	**Services on Stage** Steps 1-5	**Coral's Web** Step 7, Step 6 Assessment	**Sizing Up Shrimp**	**Postcards from the Reef** Steps 1-4 →
HW: Extension			HW: Assessment	
Session 11	Session 12	Session 13	Session 14	Session 15
→ Steps 5-6, Assessment	**America's "Most Wanted"** Steps 1-2 →	Steps 3-4, Assessment →	**Where the Wild Sharks Are**	**Sharks in Decline** Steps 1-7 →
	HW: Finish research		HW: Shark Mystery Challenge →	HW: Fishing Worksheet B →
Session 16	Session 17	Session 18	Session 19	Session 20
→ Steps 7-8 Discuss Shark Mystery Challenge	**Reef.Net** Steps 1-2 →	Step 3	Step 3	Step 4 →

Introducing Marine Biodiversity
version 2 (about 12 sessions)

This unit is centered on activities from the Shark Case Study and is supplemented with activities from the other case studies to provide a broad-based introduction to marine biodiversity. It stresses the value of marine biodiversity and the many benefits humans derive from it. This unit is interdisciplinary in scope, focusing on content and skills important in social studies, science, drama, and language arts.

Session 1: "What Do You Think About Sharks?"

Everyone has an opinion about sharks. Use this activity to pique your students' interest in marine biodiversity. You can use the results of the "surveys" as an initial indication of your students' feelings toward and knowledge of sharks.

Sessions 2 and 3: "Services on Stage"

There are many things to consider when trying to sum up the value of something like a shark. Use this activity to highlight some of the general services marine biodiversity provides, some of the benefits we get from marine biodiversity, and some of the ways we value it. Allow plenty of time for students to work on their skits. Some students may have to wait until session 3 to present their skits. Skip steps 6 and 7 (unless you want to spend the extra time to have the students develop ad campaigns). In your discussion of the importance of marine biodiversity, challenge the students to relate the "services" connected to food, medicine, and culture with sharks. You can interject background information from the Sharks Case Study to keep things rolling. Save time at the end of session 3 for the assessment.

Sessions 4 and 5: "Postcards from the Reef"

Use this activity to further highlight the value of marine biodiversity. If you're short on time, you could assign step 5 for homework at the end of session 5, and start "Salmon People" after the wrap-up discussion (step 6).

*Activity in the Salmon Case Study found at www.worldwildlife.org/windows/marine.

Sessions 6, 7, and 8: "Salmon People"*

Use this activity to highlight the cultural value a marine species can have for a group of people. Read and discuss stories in session 6 (steps 1-5). Assign question sheet "When a Fish is More than a Fish" for homework. Finish that session by breaking students into groups and giving them the presentation assignment (step 6). Have students use session 7 and the beginning of session 8 to develop and practice their presentations. Students will then use the bulk of session 8 to make presentations to each other. Have them start on the assessment at the end of the session and finish it for homework.

Sessions 9 and 10: "Sharks in Decline"

Now that your students have looked at the value of marine biodiversity, they will be able to more fully appreciate some of the threats that marine species face. Run through all of the fishing simulations (steps 1-5), and discuss the results (step 6). Assign "Fishing Worksheet B" for homework (step 7). In session 10, discuss the students' answers as well as the worldwide status of sharks (step 8). If time permits, introduce the final project in "Rethinking Sharks."

Sessions 11 and 12: "Rethinking Sharks"

Discuss the reading ("Sharks in Culture"). Then announce the final assignment about portraying their views toward sharks. You can narrow the scope or focus of the assignment to best address the skills you want to stress. Give students the balance of class time in session 11 and all of session 12 to finish their

project, or allow more time if necessary and available. You can use this project and its assessment as an assessment for the entire unit. You could also revisit "What Do You Think About Sharks?" (from session 1) to see if students' attitudes and knowledge toward sharks have changed.

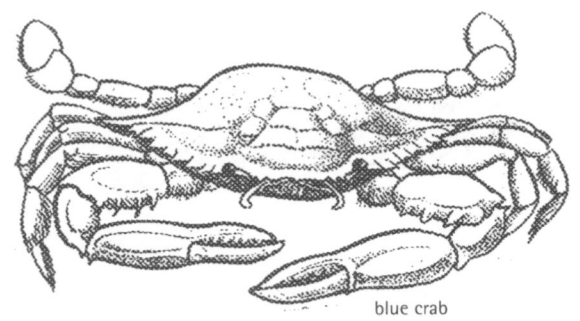
blue crab

Unit at a Glance — Introducing Marine Biodiversity—version 2

Session 1	Session 2	Session 3	Session 4	Session 5
What Do You Think About Sharks?	**Services on Stage** → Steps 1-4	Steps 4-5, Assessment	**Postcards from the Reef** → Steps 1-4	Steps 5-6, Assessment
Session 6	Session 7	Session 8	Session 9	Session 10
Salmon People* → Steps 1-5 **HW:** Complete question sheet	Step 6	Step 6 **HW:** Assessment	**Sharks in Decline** → Steps 1-6 **HW:** Fishing Worksheet B	Steps 7-8 **HW:** Read "Sharks in Culture"
Session 11	Session 12			
Rethinking Sharks → Steps 1-2	Step 3, Assessment			

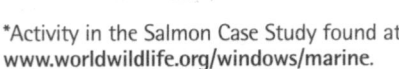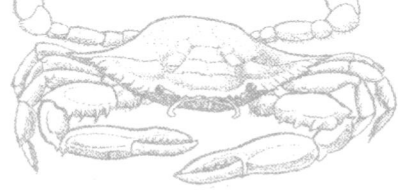

*Activity in the Salmon Case Study found at
www.worldwildlife.org/windows/marine.

Marine Biodiversity: Emphasis on Science
Linked with Biodiversity Basics (about 20 sessions)

This unit uses marine biodiversity as a context to teach science skills and concepts. The unit emphasizes research, experimentation, and analysis skills.

Session 1: "Endangered Species Gallery Walk"**

Assign this poster project as a way to get students interested in the unit. Have students select marine species or island species. Additional marine species not on the list in *Biodiversity Basics* could include the vaquita, West Indian manatee, olive ridley sea turtle, Steller's sea lion, North Island Hector's dolphin, marine otter, giant sea bass, sand tiger shark, smalltooth sawfish, Atlantic halibut, Amsterdam albatross, and Madagascar fish eagle. (Some of these species are illustrated on the "Joy to the Fishes and the Deep Blue Sea" poster.)

Sessions 2 and 3: "Much Ado About O_2"*

Students will be conducting experiments during these two sessions. The second experiment will run throughout the unit, with students recording observations and measurements twice a week. Let students know that the results of the experiments will be discussed later in the unit.

Session 4: "Ocean Explorers"

Use this outdoor simulation to help your students learn about conditions of underwater exploration and why it's so difficult to discover and research new species in the ocean.

Sessions 5-6: "Sizing Up Species"** and "Endangered Species Gallery Walk"**

"Sizing Up Species," from *Biodiversity Basics*, will give your students a better understanding of taxonomy.

*Activity in the Salmon Case Study found at
www.worldwildlife.org/windows/marine.
**Biodiversity Basics* activity

It also helps set a context for their endangered species project. Complete Parts I and II in session 5, and Part III in session 6.

Also in session 6, collect and evaluate the endangered species posters from session 1 and create a scavenger hunt. Be sure to come up with different categories for the scavenger hunt. For example, you might use "list three marine mammals and the reasons they're endangered" and "list three species that are endangered because of pollution." Try to design the scavenger hunt to help students see examples that fully represent the HIPPO dilemma outlined in the activity.

In session 5, make sure students check their O_2 experiments and record their observations and measurements.

Session 7: "Endangered Species Gallery Walk"** (cont'd.)

Display the posters in an accessible area. Encourage students to look at as many posters as possible when filling out their scavenger hunt sheets. At the end of session 7 make sure students check their O_2 experiments and record their observations and measurements.

Session 8: "Coral's Web"

Use the first six steps of this activity to emphasize the important role that species diversity plays in maintaining healthy ecosystems.

Sessions 9 and 10: "Secret Services"**

Now that students have been introduced to a variety of marine animals and their interrelationships, this activity will help students discover the benefits that healthy ecosystems provide. Most of the experiments simulate functions of wetlands, which are critical to maintaining healthy marine biodiversity. In session 10, remind the students to record their observations and measurements from their O_2 experiments.

Sessions 11 and 12: "America's 'Most Wanted'"

Your students will have been introduced to alien species in the "Endangered Species Gallery Walk." "America's Most Wanted" explores some native species that are being severely affected by alien invasions. Students should begin their research in session 11 and finish it for homework (steps 1-2). In session 12, complete steps 3-4, and have students record their observations and measurements from their O_2 experiments.

Session 13: "Decisions! Decisions!"

In this activity, your students will apply the information they learned in "America's Most Wanted" as they participate in several role-plays about what actions we should take to control certain alien species.

Session 14: "Catch of the Day"

Students will have learned from the "Endangered Species Gallery Walk" that overharvesting is another threat to marine biodiversity. The first part of "Catch of the Day" (steps 1-5) will help them understand that how we harvest our food from the ocean has an impact on biodiversity. In the second part (step 6), students will have the chance to design a turtle excluder device, or TED.

Session 15: "Sharks in Decline"

These simulations will illustrate how other fishing techniques affect bycatch. Since students will have already done the trawling simulation in the previous session, you should have enough time to finish and discuss "Fishing Worksheet B." Remind students to record their observations and measurements from their O_2 experiments.

Session 16: "A Fish Tale"*

This unit ends with a mini case study on salmon. This first activity will familiarize students with the salmon's life cycle through a guided imagery exercise. Make arrangements for your students to use the Internet for research during session 17.

Sessions 17 and 18: "Salmon Scavenger Hunt"* and "Much Ado About O_2"*

In session 17, have the students do their research and complete their scavenger hunts (steps 1-2). In session 18, review findings and threats to salmon (steps 3-4). Also have students record their final observations and measurements from their O_2 experiments, and discuss the results (see "Much Ado About O_2" steps 4-7). Encourage students to discuss the connections between the experiments, salmon, and marine life in general.

Sessions 19 and 20: "Make Way for Salmon"*

In this activity, the students will create board games to help them apply the information they learned in the last part of the unit.

*Activity in the Salmon Case Study found at **www.worldwildlife.org/windows/marine.**

Marine Biodiversity: Emphasis on Science

Unit at a Glance

Linked to "Biodiversity Basics"

Session 1 **Endangered Species Gallery Walk** **HW:** Endangered species project	**Session 2** **Much Ado About O₂** ✓ Experiment 1	**Session 3** ✓ Experiment 2	**Session 4** **Ocean Explorers**	**Session 5** **Sizing Up Species** Parts I and II ✓ Experiment 2
Session 6 Part III **Endangered Species Gallery Walk** Step 3	**Session 7** Step 4 ✓ Experiment 2	**Session 8** **Coral's Web** Steps 1–6	**Session 9** **Secret Services** Step 1	**Session 10** Step 2: Presentations ✓ Experiment 2
Session 11 **America's "Most Wanted"** Steps 1–2 **HW:** Step 2: Research	**Session 12** Steps 3–4 ✓ Experiment 2	**Session 13** **Decisions! Decisions!**	**Session 14** **Catch of the Day** Steps 1–6	**Session 15** **Sharks in Decline** Steps 1–4, 6–8 ✓ Experiment 2
Session 16 **A Fish Tale**	**Session 17** **Salmon Scavenger Hunt** Steps 1–2 **HW:** Finish Scavenger Hunt	**Session 18** Steps 3–4 **Much Ado About O₂ (Cont'd.)** Steps 4–7 Final observations Wrap-up discussion	**Session 19** **Make Way for Salmon**	**Session 20** Work on games

*Activity in the Salmon Case Study found at
www.worldwildlife.org/windows/marine.
**Biodiversity Basics* activity

The Value of Marine Biodiversity

Linked with Biodiversity Basics (about 12 sessions)

This unit explores various cultural perspectives and values about marine biodiversity. It is an interdisciplinary unit that stresses social studies, language arts, and science content and skills.

Session 1: "Going Under"

Use this activity to introduce the unit by helping students explore their own feelings about the marine environment. Have them finish their story for homework.

Session 2: "Biodiversity Break-Down"

Complete the first three steps of this activity to introduce your students to the three levels of marine biodiversity.

Session 3: "Services on Stage"

Use the first five steps of this activity to introduce some of the services that marine biodiversity provides and how we benefit from them.

Sessions 4 and 5: "The Culture/Nature Connection"**

Use this activity to help students examine the value of biodiversity from different cultural perspectives, including their own. Assign the family heritage research (Part II) as homework between the two sessions. Assign the assessment as homework at the end of session 5.

Session 6: "Ten-Minute Mysteries"**

Begin by asking your students if they thought the family heritage/biodiversity connections in the previous activity seemed somewhat outdated or old-fashioned. If so, this activity will show some modern-day biodiversity connections. (If you don't have the time to complete all of the mysteries, make sure you do mysteries 2, 7, and 9, as they relate specifically to marine species.) Have them finish the question sheet for homework.

Sessions 7 and 8: "Salmon People"*

Use this activity to continue your examination of cultural connections to biodiversity. If time permits, you may want to give the students an extra session to work on their presentations.

Session 9: "Sizing Up Shrimp"

This activity will help students examine how different values placed on certain species (such as shrimp) can lead to conflict. Assign the assessment as homework.

Session 10: "Rethinking Sharks"

This activity will give your students yet another opportunity to look at traditional views about biodiversity. It also will give them an opportunity to clarify their own feelings and attitudes about sharks. If students don't finish their projects in class, have them complete the assignment for homework.

Session 11: "The Spice of Life"**

This activity will help students determine or clarify why biodiversity is important to them. The assessment for this activity can be completed as a homework assignment.

Session 12: "Imagining Oceans"

Use the general idea of this activity to have your students write stories about their own connections with biodiversity. In addition to the story starters given in this activity, other suggested topics can be found in the writing ideas section of "Rethinking Sharks," "The Culture/Nature Connection,"** and "Salmon People."* This writing assignment and the assessment for "Spice of Life"** can serve as an assessment for the unit.

*Activity in the Salmon Case Study found at
www.worldwildlife.org/windows/marine.
**Biodiversity Basics activity

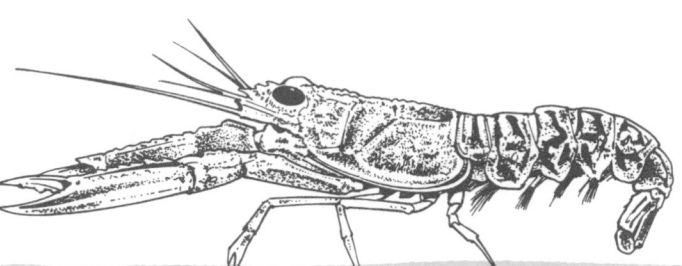

Unit at a Glance

The Value of Marine Biodiversity
Linked to "Biodiversity Basics"

Session 1	Session 2	Session 3	Session 4	Session 5
Going Under **HW:** Finish story	**Biodiversity Break-Down** Steps 1-3	**Services on Stage** Steps 1-5	**The Culture/Nature Connection**** Part 1 Part II (Steps 1-2) **HW:** Conduct family heritage research	⟶ Part II (Steps 3-4) Part III **HW:** Assessment
Session 6	Session 7	Session 8	Session 9	Session 10
Ten-Minute Mysteries** (Particularly mysteries 2, 7, and 9) **HW:** Finish question sheet	**Salmon People*** ⟶ Steps 1-5	Step 6: Presentations	**Sizing Up Shrimp** **HW:** Assessment	**Rethinking Sharks** **HW:** Complete assignment
Session 11	Session 12			
The Spice of Life** **HW:** Assessment	**Imagining Oceans** **HW:** Complete writing assignment			

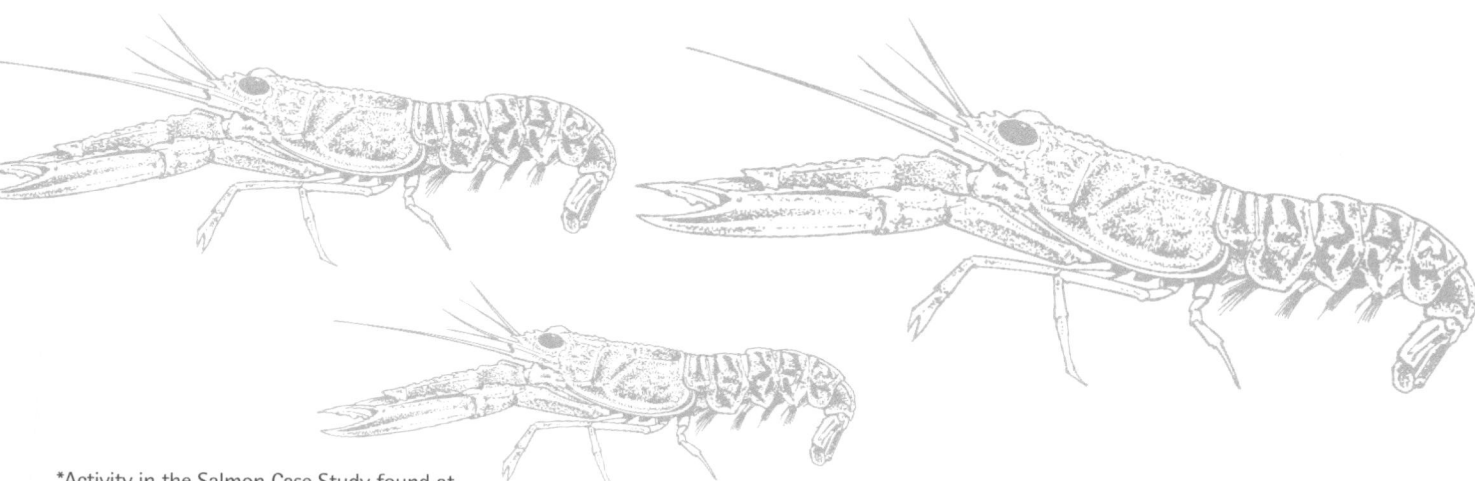

*Activity in the Salmon Case Study found at
www.worldwildlife.org/windows/marine.
***Biodiversity Basics* activity

Marine Biodiversity for Nonformal Settings

(14 sessions)

This unit, designed for use at nonformal institutions such as zoos, aquariums, and nature centers, uses activities and simulations that can be completed in one or two sessions and require no research or homework. The first two sessions introduce students to the idea of marine biodiversity. After that, the activities focus first on coral reefs, then on sharks, alien species, and shrimp.

Session 1: "Sea for Yourself"

Use option 1 or option 2 (field trips) to get students excited about marine biodiversity.

Session 2: "Services on Stage"

Use the first five steps of this activity to give students a chance to act out some of the services that marine ecosystems provide.

Session 3: "Coral's Web"

Use this activity (steps 1-6) to explain some of the many connections that exist among marine species.

Session 4: " Postcards from the Reef"

This activity builds on the previous one by introducing students to reefs around the globe, the conditions reefs need to thrive, and some of the threats they face.

Session 5: "Ocean Explorers"

Use this activity to prepare for the next one. By performing these simulations, students will gain an appreciation of the difficulties inherent in deep-sea research.

Session 6: "Where the Wild Sharks Are"

This activity shows how sharks have evolved to be able to live in all kinds of ocean environments, about which students learned in the previous activity.

Session 7: "Rethinking Sharks"

This activity will challenge students to think about sharks in a different way and give them an opportunity to express their views of this richly diverse group of fish using a medium of their choice.

Sessions 8 and 9: "Drawing Conclusions"

In this activity students will learn about the threats that alien species pose to marine biodiversity and will be challenged to think about characteristics that make introduced species likely to become problematic. Students will create comic strips (steps 1-3) depicting what they think happened when different aliens were introduced in particular areas, then they'll learn about what really happened (steps 4-5).

Session 10: "Aliens Among Us"

Now that students know about some of the problems associated with alien species, use this activity to have them go out and look for evidence of alien species at your institution or in your community.

Session 11: "Sizing Up Shrimp"

In this activity, students consider the value of shrimp from a number of perspectives. This activity also builds a context for the next activity.

Sessions 12 and 13: "Catch of the Day"

In this activity, students simulate shrimp trawling to gain a better understanding of the problem of bycatch (steps 1-6). Then they take a look at the pros and cons of shrimp farming (steps 7-10).

Session 14 and 15: "Career Choices"

Use the first six steps of this role-playing activity as a wrap-up for the shrimp activities. By looking at situations from a number of perspectives, students will gain an appreciation for the complexity of the issues surrounding marine biodiversity.

Unit at a Glance — Marine Biodiversity Unit for Nonformal Settings

Session 1	Session 2	Session 3	Session 4	Session 5
Sea for Yourself Option 1 or 2	**Services on Stage** Steps 1-5	**Coral's Web** Steps 1-6	**Postcards from the Reef**	**Ocean Explorers**
Session 6	Session 7	Session 8	Session 9	Session 10
Where the Wild Sharks Are	**Rethinking Sharks**	**Drawing Conclusions** Steps 1-3	→ Steps 4-5	**Aliens Among Us**
Session 11	Session 12	Session 13	Session 14	Session 15
Sizing Up Shrimp	**Catch of the Day** Steps 1-6	→ Steps 7-10	**Career Choices** Steps 1-4	→ Steps 5-6

Appendix B
Glossary

algae: one-celled or multi-celled organisms, typically aquatic, that make their own food through photosynthesis. All algae were once classified as plants, but many are now listed under a separate kingdom, *Protista*, which includes organisms that are neither plants nor animals because algae lack the complex cells (such as xylem and phloem) found in plants.

alien species: a plant, animal, or other organism that has been brought from its native habitat into a new area where it doesn't naturally live. Alien species can compete with and cause problems for native species. Alien species are also called introduced, exotic, and nonnative. (See *invasive* and *native species.*)

anadromous: fish species that hatch in freshwater streams and then migrate to the sea. When they reach a reproductive age, they return to their original streams to spawn. Sockeye salmon, steelhead trout, and striped bass are examples of anadromous fish.

aquaculture: the growing and harvesting of aquatic plants and animals such as fish, shellfish, and water hyacinth in ponds, lakes, or fenced-in lagoons and estuaries.

ballast water: the water ships take on before leaving port to help them maintain balance at sea. Many alien species, such as zebra mussels, have been transported to new bodies of water when the ballast water from one area is let out in a new area.

biodiversity: the variety of life on Earth, reflected in the variety of ecosystems and species, their processes and interactions, and the genetic variation within and among species.

bycatch: in commercial fishing operations, fish and other marine organisms that are captured along with targeted species. Bycatch is typically thrown overboard, dead or dying at sea. Bycatch often includes juveniles of the targeted species, as well as seabirds, marine mammals, sharks, and turtles. It is estimated that bycatch accounts for one-quarter of the global fishery catch.

climate change: see *global warming.*

conservation: protection of natural resources.

coral reef: a colony of small animals, called coral polyps, and their calcium carbonate skeletons, which build up over hundreds or thousands of years. While coral reefs cover less than one percent of the ocean area, they are home to more than thirty percent of all marine species.

crustacean: a freshwater or saltwater invertebrate that has a segmented body, a hard exoskeleton, and limbs that are paired and jointed. Lobsters, shrimp, and barnacles are examples of crustaceans.

culture: the beliefs, values, knowledge, and traditions held within a specific group of people.

dead zone: an area where the water is stripped of dissolved oxygen from the decay of excessive amounts of algae, which

causes other marine organisms to die. Excessive growths, or "blooms," are often the result of pollution. For example, excess nitrogen, potassium, and phosphorus runoff from chicken farms has caused algal blooms in the Chesapeake Bay, creating uninhabitable conditions for fish and other marine animals.

dissolved oxygen: the oxygen contained in water (not attached to hydrogen, but free flowing as oxygen gas). Dissolved oxygen levels in water can fluctuate depending on such factors as water temperature, water movement, and water quality. Decreases in dissolved oxygen levels can affect the health of many marine and freshwater species.

ecological restoration: efforts to return a damaged ecosystem to a healthier condition by creating the biological composition that predates human activity. An example of ecological restoration is the effort to restore seagrass meadows in coastal estuaries, where the meadows provide protection from erosion, and habitat for a diversity of life.

ecoregion: a relatively large unit of land or water that is characterized by a distinctive climate, specific ecological features, and plant and animal communities. North America has more than 100 different ecoregions, including the Everglades Flooded Grasslands and the California Current.

ecosystem: a community of plants, animals, and other living things that are linked by energy and nutrient flows and that

interact with each other and the physical environment. Coral reefs, deserts, rain forests, grasslands, and a rotting log are all examples of ecosystems.

endangered species: a species that is likely to become extinct throughout all or a significant portion of its range within the foreseeable future. A manatee is an endangered marine mammal.

estuary: an ecosystem in which fresh water from a river or stream meets salty water from the ocean. (The Chesapeake Bay and San Francisco Bay are examples of estuaries.) Highly productive areas, estuaries are nurseries for juveniles of many marine animals, including commercially important fish and shellfish species.

food chain: a network of organisms from producers (plants) to consumers (other plants, animals, and fungi) in which each organism feeds on or gets nutrients from the previous organism.

gene: a piece of DNA that determines a specific trait, such as hair color, eye color, or resistance to disease. Genes are passed from a parent (or parents) to offspring through sexual reproduction—in which offspring inherit genes from both parents; or asexual reproduction—in which offspring inherit genes identical to a single parent.

genetic diversity: the genetic variation present in a population or species. Differences in size, shape, color, taste, and rate of growth of the hundreds of varieties of potatoes are examples of genetic diversity.

gill net: a net that has holes that allow a fish to fit its head but not the rest of its body through the opening. When a fish tries to remove itself from the hole, the net catches and holds the fish in place. Any fish whose head is too large to fit completely into the hole, or whose body is so small that it can slip all the way through the net, will not get caught.

global warming: the increase in Earth's temperature, caused by emissions of carbon dioxide and other gases that blanket the planet and trap heat within the atmosphere.

habitat: the area where an animal, plant, microorganism, or other life form lives and finds the nutrients, water, sunlight, shelter, living space, and other essentials it needs to survive. Habitat loss (which includes the destruction, degradation, and fragmentation of habitats) is the primary cause of biodiversity loss.

hatcheries: facilities that hatch eggs, raise the young, and then release the young back into their natural habitat.

hook and line: a method of catching fish with baited hooks or artificial lures attached to a line, used by subsistence, sport, and commercial fishers.

invasive: an alien species that negatively impacts the habitat to which it has been introduced by displacing one or more native species living there.

invertebrate: an animal that does not have a backbone. The majority of Earth's animal species are invertebrates—including everything from insects to worms.

longline: a long, thin cable or monofilament that can be up to 40 miles in length. The cable, which is stretched across the ocean, has a baited hook every few feet and a float every few hundred feet. Longline fishing is generally associated with moderate to high amounts of bycatch, depending on how many hooks are used and where and when the lines are set. For example, longlines set for tuna often unintentionally catch swordfish, sharks, turtles, and seabirds.

marine: living or growing in oceans, coastal waters, and estuaries (areas where fresh water and salt water meet); not terrestrial.

marine biodiversity: the diversity of life—genes, species, and ecosystems—found in oceans, coastal waters, and estuaries (areas where fresh water and salt water meet).

native species: a plant or animal that occurs naturally in an area or a habitat. Native species are also called indigenous species.

natural resource: any aspect of the environment that species depend on for their survival, including land, soil, air, energy, and fresh water.

over-consumption: the use of resources at a rate that exceeds the ability of natural processes to replace those resources.

photosynthesis: the process by which green plants, algae, and other organisms use sunlight to produce carbohydrates (food). Oxygen is released as a byproduct of photosynthesis.

phytoplankton: very tiny (mostly microscopic) plants and plantlike organisms that drift in large quantities throughout much of the world's oceans. (See *plankton*.)

plankton: tiny, free-floating animals, plants, and plantlike organisms that rely on water currents for distribution and transportation. Many marine species, such as cod and Dungeness crabs, are planktonic for the early part of their life cycle. Plankton are an important food source for many commercial fish species, and they create the base of ocean food webs.

pool: part of a stream that offers deep water, shade, reduced current, and the protective cover of boulders and submerged logs.

predator: an organism that survives by consuming other organisms.

prey: an organism hunted or caught for food.

riffle: a shallow area of fast-moving, oxygen-rich water, which is often a breeding area for aquatic insects.

run: swift-flowing area of a stream, which is deeper than a riffle.

silt: fine particles of soil or sand that are suspended in water and deposited as sediment.

spawn: to lay eggs and fertilize them.

species: (1) a group of organisms that have a unique set of characteristics (such as body shape and behavior) that distinguish them from other organisms. If individuals within the same species reproduce, they can produce fertile offspring; (2) the basic unit of biological classification. Scientists refer to animals, plants, or other organisms by both their genus and species names. The nurse shark, for example, is called *Ginglymostoma cirratum*, with *cirratum* being its species name.

sustainability: the ability to use the Earth's resources (such as land, forests, water, and wildlife) in a way that ensures that the needs of the present are met without diminishing the ability of people, other species, or future generations to survive.

symbiotic: an ecological relationship between two organisms. The relationship may be beneficial or detrimental to one or both organisms. For example, coral polyps have a beneficial symbiotic relationship with zooxanthellae, a type of algae that lives inside them. The coral provides a habitat for the algae and the algae produce food for the coral. A parasite and its host have a detrimental symbiotic relationship. The parasite lives inside or on its host, at the expense of the host's health.

terrestrial: living or growing on land; not aquatic.

threatened species: a species that is likely to become endangered throughout all or a significant portion of its range within the foreseeable future. For example, chinook salmon are considered to be threatened.

trawl: a sock-shaped net with a wide mouth tapering to a small, pointed end (sometimes called the cod end). Trawl nets are towed behind fishing vessels at a variety of depths, from mid-waters to the ocean floor. Trawling is considered to be more indiscriminate than other fishing methods because everything in the trawl's path is scooped up in the net.

vertebrate: an animal with a backbone.

water quality: condition of the water, which is determined by several factors including water temperature, water chemistry, and the amount of silt in the water.

zooplankton: plankton that are animals. (See *plankton*.)

Appendix C
Legislation Protecting Marine Biodiversity

For your reference, we have compiled a short list of laws, conventions, and treaties that support the protection of marine biodiversity in the United States and around the world. Federal laws apply throughout the United States, and international agreements apply to countries that have accepted the terms of the agreement.

Federal Laws

Clean Water Act (Federal Water Pollution Control Act) (1972)
The objective of this act is to restore and maintain the chemical, physical, and biological quality of the nation's waters, including its seas, lakes, rivers, streams, and wetlands. With strict permit requirements, the Clean Water Act limits the amount of pollutants released into the nation's waters. The ultimate goal of this law is to maintain water quality that will support both healthy wildlife populations and recreational use by people.

Coastal Zone Management Act (1972)
This act was passed to preserve, protect, and, where possible, restore or enhance the resources of the nation's coastal zone for current and future generations. It encourages and assists the coastal states by providing them with incentives to develop management plans that make wise use of the land and water resources in the coastal zone.

Coral Reef Protection (Executive Order 13089) (1998)
This order protects U.S. coral reef ecosystems. It states that all federal agencies must identify any actions that may affect U.S. coral reef ecosystems. Federal government agencies must implement measures in cooperation with the U.S. Coral Reef Task Force as well as various other governmental, nongovernmental, and scientific organizations to research, monitor, manage, and restore ecosystems affected by problems such as pollution, sedimentation, and fishing. The federal agencies also must ensure that any actions they authorize, fund, or carry out will not degrade the conditions of such ecosystems.

Endangered Species Act (1973)
This act aims to ensure the survival of endangered species, defined as those "in danger of extinction throughout all or a significant portion of their range," and threatened species, defined as those "likely to become endangered within the foreseeable future." Under the ESA, "species" refers to any member of the animal kingdom (mammal, fish, bird, amphibian, reptile, mollusk, crustacean, arthropod, or other invertebrate) as well as any member of the plant kingdom. Habitats that are critical to the conservation of a threatened or endangered species may be covered under this act.

Federal Aid in Wildlife Restoration Act (1937)
Better known as the Pittman-Robertson Act, this act provides states with federal aid generated from an excise tax on ammunition and sporting arms. Funds are used for wildlife conservation work, including wildlife surveys, research, land acquisition, and technical assistance. This act was followed by the Federal Aid in Sport Fish Restoration Act (the Dingell-Johnson Act) in 1950, which provides federal aid (from a tax on sport fishing equipment) for the restoration of fish species.

Federal Insecticide, Fungicide, and Rodenticide Act (1947)
Originally passed in 1947 and significantly amended during the 1970s and 1980s, this act prohibits the use of pesticides that have an unreasonably adverse effect on the environment. The Environmental Protection Agency (EPA) must evaluate all pesticides for their effects on nonpest species, particularly threatened or endangered wildlife. Under this act, EPA regulates all phases of pesticide use, including sale, use, handling, and disposal.

Lacey Act (1900)

Originally passed in 1900 and amended in 1981, this is one of the world's strongest laws designed to curb illegal wildlife trade. It prohibits the importation of animals that were illegally killed, or products that were illegally collected or exported from another country. The Lacey Act also allows the U.S. government to seize illegally imported goods.

Magnuson-Stevens Fishery Conservation and Management Act (1976)

This act is America's principal marine fisheries management law. It includes provisions to prevent overfishing, reduce bycatch, and protect essential fish habitat. The act is implemented by a system of regional fisheries management councils, with oversight from the National Marine Fisheries Service. The act was amended in 1996 to emphasize the sustainability of the nation's fisheries, establish a new standard requiring fisheries to be managed at maximum sustainable levels, and ensure that new approaches be pursued in habitat conservation.

Marine Mammal Protection Act (1972)

This act makes it illegal for any person under the jurisdiction of the United States to kill, hunt, injure, or harass any species of marine mammal, regardless of its population status. The act also makes it illegal to import into the United States marine mammals, or products made from them. Certain numbers of marine mammals can be collected for scientific and display purposes, as accidental bycatch in commercial fishing operations, and in subsistence hunting by natives of the North Pacific and Arctic coasts. The National Marine Fisheries Service is primarily responsible for enforcement of this act. Congress amended the MMPA in 1994, creating a new management system to govern the incidental taking of marine mammals in commercial fishing operations. Since then, the NMFS has implemented several regulations to enforce the MMPA and its amendments.

Marine Plastic Pollution Research and Control Act (MPPRCA) (1987)

This act prohibits the disposal of plastic, or garbage mixed with plastic, into any U.S. waters. MPPRCA works in conjunction with the Marine Pollution Treaty (MARPOL) (1973/1978), which prohibits the discharge of oil, sewage, plastics, and garbage into coastal and ocean waters.

Marine Protected Areas (Executive Order 13158) (2000)

This executive order strengthens and expands the national system of marine protected areas (MPAs). MPAs are areas reserved by law or other effective means to protect all or part of a designated marine environment, and may include national parks, national wildlife refuges, and national marine sanctuaries. Led by the National Oceanic and Atmospheric Administration, and supported by state, territorial, local, and tribal governments, this order encourages the evaluation and enhancement of existing sites and the creation of new MPAs.

National Environmental Policy Act (1969)

Called "the basic national charter for protection of the environment" by the EPA, this law requires that an environmental impact statement be conducted for federal government projects that could have a significant impact on the environment. Federal agencies must carefully consider the environmental effects of their proposed actions and must restore and enhance environmental quality as much as possible.

National Invasive Species Act (1996)

This act is designed to prevent unintentional introduction and dispersal of nonindigenous species into waters of the United States, coordinate federally funded research on invasive species, carry out environmentally sound control methods, and minimize economic and ecological impacts of nonindigenous aquatic nuisance species.

National Marine Sanctuaries Act (Marine Protection, Research, and Sanctuaries Act) (1972)

To improve the conservation, management, and wise use of marine resources, this act designates marine areas of special national significance as national marine sanctuaries (sanctuaries are one type of marine protected area—see above). Under this law, appropriate federal agencies, state and local governments, Native American tribes, and others are encouraged to develop and implement coordinated plans for the protection and management of the sanctuaries in the interest of maintaining the health and resilience of these marine areas.

Oceans Act (2000)

This act builds on the foundations of the 1966 Marine Resources and Engineering Development Act by requiring the U.S. President to develop and implement a comprehensive long-range national ocean and coastal policy. The act establishes a 16-member commission on Ocean Policy to carry out a thorough review of U.S. ocean and coastal activities. This commission assesses and recommends changes to U.S. laws to improve the management, conservation, and use of ocean resources.

International Treaties and Conventions

Antarctic Treaty (Agreed Measures for the Conservation of Antarctic Fauna and Flora) (1959)

While it guarantees freedom of scientific research and exchange of data in the Antarctic region (the region south of 60° south latitude), the Antarctic Treaty prohibits all military activity, nuclear explosions, and disposal of radioactive waste in Antarctica. The Agreed Measures for the Conservation of Antarctic Fauna and Flora is an annex to the treaty. It prohibits the killing, wounding, capturing, or molesting of native birds or mammals without a permit, and it provides for the establishment of protected areas of scientific interest or ecological uniqueness.

Convention on Biological Diversity (1992)

Opened for signature at the Earth Summit in 1992 in Rio de Janeiro, the Biodiversity Treaty has as major objectives the conservation and sustainable use of Earth's diverse biological resources; equitable sharing of the benefits arising from use of genetic resources; and appropriate transfer of relevant technologies. Among the commitments made by parties are to develop national strategies for conservation and sustainable use of biodiversity; identify and monitor components of biodiversity; establish protected areas where special measures are needed; and promote protection of ecosystems and natural habitats. In 1993, President Bill Clinton signed the Biodiversity Treaty and sent it to the Senate for advice and consent, indicating that implementing legislation would not be required since existing laws were sufficient to provide for U.S. participation. In 1994, the relevant Senate Committee reported favorably on the treaty, but it did not come to a vote before the end of the 103rd Congress. As of 2003, the treaty is still pending, and the United States is therefore not a party to it.

Convention on Conservation of Antarctic Marine Living Resources (1984)

This convention establishes international mechanisms and creates legal obligations necessary for the protection and conservation of Antarctic marine living resources. Using an innovative ecosystem management approach, the convention includes standards to ensure the health of the individual populations and species and to maintain the health of the Antarctic marine ecosystem as a whole.

Convention on International Trade in Endangered Species of Wild Fauna and Flora (CITES) (1973)

This convention was formed to help prevent the depletion of wild plant and animal populations that are frequently traded on the international market. Products that caused a need for protection of traded species include ivory, rhinoceros horn, tortoiseshell jewelry, and spotted cat skins. More than 150 nations have become parties to the treaty, which means they have agreed to develop wildlife protection laws within their territories for species listed for protection.

International Convention for the Regulation of Whaling (1946)

This convention established the International Whaling Commission (IWC). The IWC grew from an initial group of 15 nations (including the United States), all of which conducted commercial whaling operations, to the present group of 48 members. Many of the current member countries have never harvested whales. The IWC meets annually to approve harvest quotas and other guidelines for whaling; however, the IWC has no direct means for enforcing its regulations. The international convention also allows the IWC to designate whale sanctuary areas, such as the creation of the Indian Ocean Whale Sanctuary (passed in 1979). Under the international convention, any nation filing a formal objection to an IWC regulatory action within 90 days is exempt from compliance. Hence, the IWC's effectiveness depends primarily upon voluntary international cooperation and sanctions that member nations may impose unilaterally. Since its creation, the IWC's objectives have gradually shifted from regulating whale harvesting to conserving whale populations. In 1982, the IWC set harvest quotas for all whale stocks at zero, beginning with the 1985/86 whaling season.

Pan American Convention (Convention on Nature Protection and Wildlife Preservation in the Western Hemisphere) (1940)

The objective of this treaty is to preserve all species of flora and fauna that are native to the Americas, as well as all areas of extraordinary beauty, areas with striking geological formations, and areas of historic or scientific value. Parties to the treaty, which include the United States and 18 other nations of the Western Hemisphere, agree to establish national parks and wilderness reserves, conserve species listed for special protection, and cooperate in research initiatives.

Ramsar Convention (Convention on Wetlands of International Importance Especially as Waterfowl Habitats) (1971)

Adopted in Ramsar, Iran, this convention seeks to protect wetlands to maintain the ecological functions and recreational opportunities they provide. The convention maintains a list of wetlands of international importance and encourages the wise use of all wetlands. Member nations must protect listed wetlands within their territories and are expected to promote the protection of all wetlands.

United Nations Convention on the Law of the Sea (1982)

This convention was formed to establish a comprehensive and elaborate regime to deal with all matters relating to the law of the sea. More than 150 countries representing all regions of the world convened to determine methods of dealing with the spectrum of governing aspects of ocean space including establishing limits or boundaries, environmental control, scientific research, economic and commercial activities, technology, and the settlement of disputes relating to ocean matters.

World Heritage Convention (Convention Concerning the Protection of the World Cultural and Natural Heritage) (1972)

This convention maintains a list of global sites that are of particular cultural or ecological significance. By signing the convention, countries pledge to conserve listed sites within their territories. Sites may be of particular architectural or religious significance, represent the traditional way of life of a certain culture, represent ongoing ecological processes or geological phenomena, or contain important and significant natural habitats needed for the conservation of biological diversity.

For More Information

Digest of Federal Resource Laws of Interest to the U.S. Fish and Wildlife Service

Originally published by the Office of Legislative Services (now the Division of Congressional and Legislative Affairs) in 1979, the digest has been updated twice since the original publication, with the latest update being in 1992. The digest provides a comprehensive list and description of all federal laws under which the Fish and Wildlife Service functions, including administrative laws, treaties, executive orders, interstate compacts, and memoranda of agreement.
laws.fws.gov/lawsdigest/indx.html

The Evolution of National Wildlife Law

This book, by Michael J. Bean and Melanie J. Rowland, is a project of Environmental Defense and World Wildlife Fund-U.S. (Praeger Publishers, 1997). It examines the historical foundation of wildlife conservation, laws governing species conservation, wildlife habitat laws and regulation, and the laws and treaties that regulate worldwide wildlife conservation.

Federal Wildlife and Related Laws Handbook

This handbook provides an overview of U.S. wildlife law and wildlife issues, statutes, cooperative agreements, and treaties. It contains a glossary, a bibliography, and selected acronyms. It was produced by the Center for Wildlife Law at the University of New Mexico and is available from Government Institutes, 4 Research Place, Suite 200, Rockville, MD 20850. It can also be ordered online at
www.govinst.com/Merchant2/merchant.mv

State Wildlife Laws Handbook

This handbook was produced by the Center for Wildlife Law at the University of New Mexico. It contains overviews of wildlife law and poaching in the United States, summaries of state wildlife laws, and a comparison of state laws. It is available from Government Institutes, 4 Research Place, Suite 200, Rockville, MD 20850. It can also be ordered online at
www.govinst.com/Merchant2/merchant.mv

Appendix D

Biodiversity Education Frameworks

Studying marine biodiversity opens doors into a wealth of topics, issues, and ideas. Investigating declines of shark populations, for example, will help your students better understand reproductive biology. Reading about the importance of salmon to native people can help open their eyes to the connections between cultural diversity and biological diversity. Exploring shrimp fisheries can provide insights into the interplay of biodiversity, economics, and values.

   To help you identify the broader themes, concepts, and skills embedded in each activity, we've developed the biodiversity education framework. The framework covers general biodiversity concepts, as well as specific concepts related to marine biodiversity. You can use and adapt the framework for any biodiversity-related lessons you teach, and we hope you'll find it especially useful for organizing your exploration of marine biodiversity.

Here's what you'll find when you visit the Biodiversity Education Framework site:

• Part I: The Conceptual Framework

In Part I of the framework, you'll find more than 80 key concepts that address biodiversity and related issues. The concepts are organized under the following major sections headings:

What Is Biodiversity?

Defines biodiversity and explores basic ecological principles as well as key ecological definitions that help us understand biodiversity.

Why Is Biodiversity Important?

Explores how biodiversity affects our lives and supports life on Earth.

What's the Status of Biodiversity?

Describes some of the most serious threats to biodiversity.

Earth is our home page, hit "save."

How Can We Protect Biodiversity?

Suggests ways for addressing the biodiversity crisis, including learning more about biodiversity and working to conserve biodiversity. Reviews predictions about the importance of maintaining and restoring biodiversity.

• Part II: The Skills Framework

In Part II of the framework, you'll find a list of skills that we think are essential for learning about biodiversity and making responsible decisions about how best to protect it. Grouped into eight major categories, this framework includes the skills students need to gather, process, and act upon information, as well as those they need to become effective and engaged citizens.

> To save paper and make our materials more widely available, we've put the "Biodiversity Education Framework" on the Web in PDF format. **www.worldwildlife.org/windows**

Language Learning Tips

If you have been working with students who need special language assistance, the following tips are probably very familiar to you. However, if you are a new teacher or community educator working for the first time with language learners, you might want to read these tips before starting a biodiversity unit.

1. Developing Strong Listening and Speaking Skills

Students' literacy begins with developing strong listening and speaking skills. To encourage speaking fluency, linguists recommend that you first ignore the students' grammatical mistakes if they do not interfere with understanding. Instead, encourage expression. The following are specific ideas to help you develop your students' ability to express themselves:

- Have students express concepts in more than one way by using both scientific and ordinary language.

- Help students "unpack" the meaning of sentences that have abstract terms by restating them in concrete, everyday language.

- Use structured organizers such as outlines, charts, graphs, symbols, and diagrams; demonstrate what you are discussing whenever possible; and show visually the relationship among terms, processes, or steps by using concrete objects.

- Ask questions often and rephrase the students' questions and answers to summarize what they've learned.

- Use cooperative group activities to help students process difficult information before they tackle individual assignments. Students can ask one another for summaries and explanations of the material, or they can practice posing and responding to specific questions.

2. Building Vocabulary

As students acquire new vocabulary related to biodiversity, it's essential that they know more than a simple definition of words. Terms such as "introduced species" or "conservation" carry with them a set of ideas and values that students need to know before they can truly understand the material. Here are some ways to help students develop vocabulary skills:

- Preview topics and vocabulary when presenting new material. Pictures and other visuals are important aids for understanding.

- Provide vocabulary games or worksheets so students can practice working with key terms.

- Focus attention on technical terms, defining them in language that is familiar. Go beyond defining, making sure students understand the range of issues relevant to the term.

- Pay attention to "cultural" vocabulary that might be unfamiliar to students from some backgrounds. For example, names of foods, current trends, and familiar expressions ("sweet tooth," "green thumb") are often unfamiliar.

- Help students see connections between verbs and nouns (consume/consumption) or adjectives and nouns (extinct/extinction).

- Have students use new terms often. Encourage them to use technical terms when they ask questions, contribute to a discussion, or write about biodiversity topics.

- Have students develop vocabulary journals in which they list any new words they encounter.

3. Creating Readers

Many of the activities in this module contain text that could be a challenge for some of your students. Several ways to help your students get through material that's dense or that contains a lot of technical terms are described below:

- Read aloud to students, and allow them to read aloud to one another (not necessarily in front of the whole class). Hearing a text read aloud while they follow along gives students a chance to hear how words are pronounced and how the bits of information are parceled out. This method can help students make associations between spoken and written forms of language.

- Provide reading strategies for tackling difficult passages. For example, you might focus students' attention on headings, provide preview questions, or highlight key points.

- Provide outlines or other reading aids for students to complete as they read.

- Pay attention to terms that occur more often in written language than in spoken language and therefore may be unfamiliar to students. Words used to compare or make causal connections (such as produces, leads to, causes, and results in) and abstract vocabulary (such as principle, example, property, explain, and generalize) are important for academic literacy.

4. Developing Writing and Editing Skills

By middle school, students are ready to go beyond the narrative and imaginative writing they did in elementary school to more advanced writing. Many of the activities in this module help develop students' skills in classifying, analyzing cause and effect, expressing their opinions, and justifying their hypotheses. You'll want to focus on different aspects of their writing, depending on the assignment. For example, you might focus on content alone for journal writing, or you might help students with organization, grammar, and vocabulary on more formal reports. In addition, consider the following writing tips:

- Brainstorm with students before they write, helping them develop the vocabulary that they will need for the writing task.

- Assign writing tasks after listening activities and oral discussions. Give students practice in writing descriptions, comparisons, captions, definitions, reports, and other types of pieces. Provide examples to help them compare their writing with that of others.

- Present models and outlines of the kind of writing you expect from students. Point out some of the key features of the grammar in the models (for example, that the verbs are in present tense or that there are numerous causal conjunctions such as if, because, or so).

- Use a process approach to writing that allows students to draft, review, revise, and rewrite the most important assignments. This approach gives them a chance to develop their writing skills by focusing on the same piece of text in different ways.

Language Learning Tips— Web Resources

AAA EFL Links. Searchable list of language links for students and teachers. **www.aaaefl.co.uk**

ALS/ESL Cybersite. Online site with resources for the student and teacher, including free software. **www.station05.qc.ca/css/ cybersite/accueil.html**

The Linguist List: ESL, EFL, and L2 Information by Eastern Michigan University and Wayne State University. Provides links to several language-related Web sites. **www.linguistlist.org/esl.html**

Appendix F

Activities Overview # Introducing Marine Biodiversity

Activities	At a Glance	Objectives	Subjects
1. Biodiversity Break-Down (pages 64-71)	Use the "Joy to the Fishes and the Deep Blue Sea" poster to learn about the three levels of marine biodiversity and some of the richest marine ecoregions around the world. Then create your own poster.	Define marine biodiversity. Identify examples of genetic, species, and ecosystem diversity in the marine environment.	science, art
2. Sea for Yourself (pages 72-81)	Take a trip to the seashore; visit a natural history museum, science center, zoo, or aquarium; or go on a "virtual tour" of oceans to get more closely acquainted with marine life and the three levels of marine biodiversity.	Learn about the variety of marine life. Find examples of genetic, species, and ecosystem diversity in the marine environment either in your local area, at a scientific institution, or through a virtual field trip.	science, art, language arts
3. Ocean Explorers (pages 82-85)	Conduct an outdoor simulation to demonstrate why it's difficult to discover new species in the ocean.	Explain some of the major differences between ocean conditions near the shore and in the deep sea. Explain how those differences can affect ocean research.	science
4. Services on Stage (pages 86-93)	Act out four short skits that demonstrate some of the many services marine biodiversity provides, then design a print or video ad that educates people about the importance of marine biodiversity.	Explain how marine biodiversity affects people's everyday lives. Describe the role of marine biodiversity in balancing the gases in the air, providing a source of new medicines, making up a large portion of the world's food supply, and providing enjoyment for millions of people.	social studies, science, language arts, art (drama)
5. Going Under (pages 94-98)	Explore different perspectives of what it's like to be under the sea.	Articulate personal feelings about the ocean and ocean exploration.	language arts, science, social studies

Time	Framework Links	Key Skills	Connections
one to three sessions, depending on amount of time allotted for research	1. 1.1, 2, 3, 4, 21, 23.1, 25, 71	gathering (researching), organizing (matching, arranging, categorizing, classifying)	To get students thinking about various aspects of biodiversity, try "What's Your Biodiversity IQ?" in *Biodiversity Basics*. "Mapping Biodiversity," also in *Biodiversity Basics*, will provide a more thorough discussion of ecoregions—how they're defined, where they're located, and which ones are the most threatened. You can also access this activity online at **www.worldwildlife.org/windows**. (Select "Activity Archive" under "*WOW* Online." Then select, "Mapping Biodiversity.")
one to two sessions	1.1, 2, 3, 4, 23.1	gathering (collecting, recording, observing, measuring), organizing (sorting, arranging, categorizing, drawing)	Before taking students on the "Sea for Yourself" field trip (whether real or virtual), use "Biodiversity Break-Down" (pages 64-71) as an introduction to the three levels of marine diversity.
one session	3, 23.1, 53.1, 53.2	gathering (simulating), organizing (plotting data, graphing), interpreting (generalizing, relating, inferring, reasoning)	After this activity, use "Going Under" (pages 94-98) as a way to help students better understand scientists' (and nonscientists') perspectives of what it's like to be under the sea.
two to three sessions, depending on amount of time allotted for research, skits, and ad campaign projects.	30, 30.1, 33.1, 33.2, 33.3, 34.1	applying (composing, creating), presenting (writing, illustrating, acting, persuading), citizenship skills (working in a group, defending a position)	For more activities on ecological services, try "Biodiversity Performs!" and "Secret Services," both in *Biodiversity Basics*.
one session	1.1, 34.1, 53.2, 56	gathering (reading comprehension), presenting (writing, describing, articulating, explaining, making analogies and metaphors)	Encourage students to further explore their own perspectives on marine creatures by trying "What Do You Think About Sharks?" and "Rethinking Sharks." To look at various perspectives on other biodiversity-related topics, check out "A Wild Pharmacy" in *Wildlife for Sale* and "Spice of Life" in *Biodiversity Basics*.

Activities Overview — Coral Reefs

Activities	At a Glance	Objectives	Subjects
1. **Build-a-Reef** (pages 110-115)	Draw or build models of a coral colony and display them in a classroom "coral reef."	Define "reef-building corals" and name several factors that limit their growth. Describe adaptations of coral. Translate information about corals into a drawing or a three-dimensional model.	science, art
2. **Postcards from the Reef** (pages 116-127)	Gather clues provided in a series of fictional postcard messages to determine the location of coral reefs around the world. Learn more about the benefits of healthy coral reefs to people and wildlife, as well as some of the threats that coral reefs face.	Describe several locations around the world where coral reefs are found. List at least three necessary conditions for coral reef formation. Name several benefits of coral reefs, and articulate activities that threaten corals.	science, social studies (geography)
3. **Coral's Web** (pages 128-139)	Use information cards to create a coral reef food web, and then explore some of the ways that natural and human forces affect this web of life.	Describe some of the ways that species in a coral reef interact with one another. Explain three specific food web relationships in a coral reef. List some of the natural and human events that can affect coral reef species and discuss why food webs magnify the effect of any such event.	science
4. **Coral Bleaching: A Drama in Four Acts** (pages 140-143)	Perform skits that show the interdependence of corals and zooxanthellae, as well as the devastating effects of rising ocean temperatures and coral bleaching.	Describe the relationship between corals and zooxanthellae, and define symbiosis. Explain some of the causes and effects of global climate change, including coral bleaching.	language arts, science, art (drama)
5. **Reef.Net** (pages 144-147)	Create a Web site or educational materials that focus on coral reef management and conservation.	Explain why reefs are important. Name some of the major threats to coral reefs and marine biodiversity. Explain some of the ways organizations are working to protect reefs. Describe what individuals can do to help protect reefs.	science, social studies, language arts, art

Time	Framework Links	Key Skills	Connections
one week for research, two sessions (longer if papier-mâché option is chosen)	1.1, 2, 3, 21, 23.1	gathering (researching, collecting), organizing (arranging, sorting, drawing), applying (designing, building, creating)	To help students better understand the important connections that exist among all coral-reef creatures, use "Coral's Web" (pages 128-139) after completing this activity. "Reef.Net" (pages 144-147) offers opportunities to creatively design a Web site and present coral reef information in a colorful, reader-friendly format—another way to further develop students' artistic skills.
three sessions	10, 13, 17, 17.1, 30.1, 33.2, 34.1, 35, 47.1, 52.1	gathering (reading comprehension), organizing (mapping), presenting (describing, explaining)	Use "Coral Bleaching: A Drama in Four Acts" (pages 140-143) to further explore climate change, one of the greatest threats to coral reefs and marine diversity. To learn about hot beds of terrestrial biodiversity, try "Mapping Biodiversity" in *Biodiversity Basics*.
two sessions	1, 1.1, 3, 10, 10.1, 12, 13, 16, 17.1, 23, 25, 29, 30.1, 33.1, 35, 47.1, 52.1	organizing (arranging), analyzing (identifying components and relationships among components), interpreting (identifying cause and effect, relating, inferring, reasoning)	"Build-a-Reef" (pages 110-115) is a good activity to get students exploring the ocean's diversity of coral reefs and their inhabitants. For more on food webs and the cascading effects of stresses on them, see "All the World's a Web" in *Biodiversity Basics*.
one to two sessions	10, 12, 13, 26, 35, 52, 52.1	analyzing (identifying components and relationships among components, discussing), applying (restructuring, synthesizing), presenting (demonstrating, acting), citizenship skills (working in a group)	To get students thinking about the many ecological services that healthy marine habitats provide, start with "Services on Stage" (pages 86-93), which will also make use of their theatrical skills. To help empower students and encourage them to think about how they can make a difference in slowing the effects of global warming, have them create educational Web sites using the "Reef.Net" (pages 144-147).
one or two nights for research, two or three in-class sessions (stretched over the span of a week)	40, 62, 63, 67, 67.1 68, 71, 72	gathering (researching, collecting), applying (synthesizing, restructuring, composing), presenting (writing, reporting, explaining, clarifying), citizenship skills (working in a group)	Use "Coral's Web" (pages 128-139) and "Coral Bleaching: A Drama in Four Acts" (pages 140-143) to give students background on threats facing coral reefs. For suggestions on ways to engage audiences in biodiversity-related issues, see "The Biodiversity Campaign" in *Biodiversity Basics*. And if your students are interested in further exploring how wildlife is portrayed by the media, try "Animal Magnetism" in *Wildlife for Sale*.

Activities Overview — Shrimp

Activities	At a Glance	Objectives	Subjects
1. Sizing Up Shrimp (pages 160-165)	Explore different perspectives on shrimp and identify the many ways that people value these sea creatures.	Identify your own views and understanding of shrimp. Be able to explain the economic importance of shrimp. Explain how different ways of valuing shrimp can lead to conflict.	social studies, language arts, science
2. Catch of the Day (pages 166-175)	Conduct a simulation to explore the problems of bycatch in shrimp trawling, then take a closer look at the pros and cons of shrimp farming.	Discuss some of the pros and cons of trawling and aquaculture in catching shrimp, including trade-offs in terms of efficiency in changing trawling net size, trawling at different depths, and so on. Describe the benefits of turtle excluder devices. Create designs for more effective trawling nets and more sustainable shrimp farms.	science
3. Career Choices (pages 176-184)	Play the role of someone whose job relates to shrimp or shrimp issues, and think through your rights and responsibilities. Analyze shrimp-related jobs in your community and elsewhere.	Describe the problems associated with fish farms, bycatch, and dead zones. Name some of the ways that people in their work lives can have an effect on these shrimp-related issues. Present and defend a position on one of these issues from the point of view of someone with a career that interests you.	social studies, language arts,
4. Imagining Oceans (pages 186-190)	Finish a "story starter" to share your visions for the future of the world's oceans.	Use vivid descriptions to create a portrait of the oceans of the future. Demonstrate an awareness of current marine issues and the kinds of actions and innovations that could help address these issues in coming years. Develop creative ideas and images to articulate a personal vision.	language arts, science, social studies

Time	Framework Links	Key Skills	Connections
one session	5, 30.1, 33.1, 34.1, 37, 40, 41, 42, 59, 60	gathering (reading comprehension, identifying main ideas), interpreting (generalizing, relating, inferring, reasoning, elaborating), evaluating (assessing, critiquing, identifying bias)	Use "The Spice of Life" in *Biodiversity Basics* to examine more general ways in which people value biodiversity. "Career Choices" (pages 176-184) can encourage students to consider economic aspects that may affect people's perspectives toward shrimp and fishing.
two sessions	12, 29, 30.1, 33.1, 35, 47.1, 50.1, 52.1, 55, 56, 65, 71, 71.2	gathering (reading comprehension, simulating, brainstorming), interpreting (relating, making models, identifying cause and effect), applying (experimenting, hypothesizing, proposing solutions, problem solving)	To learn more about fishing practices and their impacts on the environment, conduct the hands-on simulations described in "Sharks in Decline" (pages 230-238). "The Many Sides of Cotton," in *Biodiversity Basics,* offers tips on mediating conflict and addressing complex issues.
two sessions, optional third session	13, 29, 30.1, 33.1, 34.1, 35, 37, 40, 46, 47.1, 50.1, 52.1, 59, 60, 62, 63, 67, 68, 72	gathering (researching, listening), interpreting (inferring, drawing conclusions, defining problems, reasoning), applying (predicting, proposing solutions, problem solving, developing and implementing investigations), citizenship skills (working in a group, evaluating a position, evaluating the need for citizen action)	"Decisions! Decisions!" (pages 284-289) makes a good follow-up to this activity, as it asks students to consider various perspectives on the controversial topic of alien species. To focus on career development, have students try "Calculate Your Wildlife Career Profile" in *Wildlife for Sale* and "Career Moves" in *Biodiversity Basics.*
one session	40, 68, 69, 71.1	applying (proposing solutions), presenting (writing, articulating, explaining, clarifying), citizenship (evaluating the need for citizen action, evaluating results of action)	To help students better envision a positive and sustainable future, try "Thinking About Tomorrow" and "Future Worlds," both in *Biodiversity Basics.* For other creative writing exercises, use "Going Under" (pages 94-98), and "Drawing Conclusions" (pages 270-279), as well as "The Nature of Poetry" in *Biodiversity Basics.*

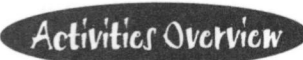 **Sharks**

Activities	At a Glance	Objectives	Subjects
1. What Do You Think About Sharks? (pages 206-214)	Explore your knowledge of and attitudes toward sharks by reading a short story.	Identify statements that are facts versus those that are attitudes. Explore personal attitudes toward and knowledge of sharks. Describe some of the ways that knowledge and attitudes are related. Name several ways that attitudes about sharks can influence our actions toward them.	language arts, social studies, science
2. Where the Wild Sharks Are (pages 216-228)	Identify shark species and determine where in the ocean each species lives.	Describe some of the different ocean zones. Describe several species of sharks and the parts of the ocean in which they live.	science, language arts
3. Sharks in Decline (pages 230-238)	Carry out group simulations of common fishing methods and assess why these methods and sharks' reproductive biology are together contributing to a rapid decline in shark populations.	Describe several methods by which sharks are captured. Discuss some of the advantages and disadvantages of each method.	mathematics, science
4. Rethinking Sharks (pages 240-246)	Read a traditional Hawaiian story about sharks and then write a poem, make a poster, draw a comic strip, or create a piece of art that portrays your views toward sharks.	Describe different cultural views of sharks. Articulate your attitude toward sharks.	language arts, social studies, art

Time	Framework Links	Key Skills	Connections
one session	37, 40, 59, 60	interpreting (inferring, identifying cause and effect, reasoning, elaborating), evaluating (critiquing, identifying bias)	For other values-clarification activities, try "Sizing Up Shrimp" (pages 160-165), as well as "The Spice of Life" in *Biodiversity Basics* and "Perspectives" in *Wildlife for Sale.*
one to two sessions	1, 1.1, 2, 3, 17.1, 18, 21, 23.1	organizing (classifying, categorizing, arranging), interpreting (translating, relating, reasoning), applying (restructuring, composing)	Use "Sea for Yourself" (pages 72-81) to get students interested in learning about different marine species as well as the importance and uniqueness of their habitats.
two sessions, one night for homework	30.1, 40, 47.1, 50, 50.1	interpreting (drawing conclusions, inferring, defining problems, reasoning, elaborating), applying (predicting, hypothesizing)	"Where the Wild Sharks Are" (pages 216-228) can set the stage for helping students understand how certain fishing methods may affect some shark species more than others, based on where in the ocean the sharks live. For more on fishing methods that can affect shark populations, try "Catch of the Day (pages 166-175).
one to two sessions, depending on projects chosen	5, 37, 41, 42, 58	applying (creating, synthesizing, composing), presenting (writing, illustrating)	To further explore cultural connections with biodiversity, use "Salmon People" (online) and "The Culture/Nature Connection" in *Biodiversity Basics,* and "A Wild Pharmacy" in *Wildlife for Sale.*

Activities Overview Alien Species

Activities	At a Glance	Objectives	Subjects
1. America's "Most Wanted" (pages 260-269)	Track down information on one of several "most wanted" alien species, and find out how it is causing problems in the United States.	Define alien species. Describe some environmental problems associated with alien species. Give examples of several alien species that are causing problems in the United States.	science
2. Drawing Conclusions (pages 270-279)	Create a comic strip that depicts an alien species arriving in a new habitat. Then draw a conclusion (literally!) of what impacts this alien might have had.	Name some ways that introduced species behave in a new environment. Describe some characteristics of an alien species that are likely to cause problems in its new environment. Explain why scientists can't always predict how an alien species will affect its environment.	art, science
3. Aliens Among Us (pages 280-283)	Learn more about alien species in your area and their effects on local habitats.	Name several nonnative species in your area. Describe one problem caused by alien species that affects people in your community. Describe some of the different ways that alien species can affect people in other communities.	science, social studies
4. Decisions! Decisions! (pages 284-289)	Play the part of community decision-makers trying to take action on alien species.	Identify several methods for controlling alien species. Explain controversies and considerations surrounding the economic, ecological, ethical, social, and aesthetic issues involved in controlling alien invasions and restoring native species.	social studies, science, art (drama), language arts

Time	Framework Links	Key Skills	Connections
two to three sessions	3, 23, 44, 48, 49, 65	gathering (reading comprehension, researching, collecting, identifying main ideas), interpreting (summarizing, relating, inferring, drawing conclusions, identifying cause and effect), presenting (reporting, explaining, clarifying)	Start off with "Sea for Yourself" (pages 72-81) so students have the chance to explore local aquatic habitats and experience firsthand any invasive species that may be affecting your area. Follow up with "Decisions! Decisions!" (pages 284-289) to help students weigh pros and cons of programs designed to rid areas of harmful alien species.
three sessions	12, 22, 23, 44, 48, 49, 65	gathering (reading comprehension, researching), analyzing (identifying components and relationships among components, reasoning, confirming), presenting (illustrating, articulating, explaining)	Before conducting this activity, use "America's Most Wanted" (pages 260-269) to help students learn about the factors that may make alien species invasive. Afterward, encourage students to use "Aliens Among Us" (pages 280-283) to find out more about local nonnative species that may be causing trouble.
two to three sessions	23, 48, 49, 63	(Skills will vary depending on which activity idea is chosen.)	"America's Most Wanted" (pages 260-269) will provide students with clues of characteristics they should be looking for in nonnative plants and animals that will help assess whether those plants and animals might become invasive problem species. "Sea for Yourself" (pages 72-81) also complements this activity: It will allow students to further explore the local area and talk to resource-management professionals about their concerns with invasive species.
two to three sessions	37, 47, 48, 55, 59, 60, 62, 63, 69, 70, 71	interpreting (reasoning, identifying problems, drawing conclusions), evaluating (critiquing, assessing), presenting (acting), citizenship skills (working in a group, evaluating a position, evaluating the need for citizen action)	To learn about the problems associated with certain alien species, start with "America's Most Wanted" (pages 260-269). For other activities that deal with conservation-related conflict resolution, try "What's a Zoo to Do?" and "Thinking About Tomorrow," both in *Biodiversity Basics*.

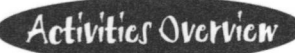 **Activities Overview**

Salmon
(On the Web at www.worldwildlife.org/windows/marine.)

Activities	At a Glance	Objectives	Subjects
1. A Fish Tale (pages 15-21)	Participate in a guided imagery exercise that explores the life cycle of salmon.	Describe the main stages in the life cycle of a salmon. Discuss some of the threats that salmon face at various points in their life cycle.	science, language arts (listening skills)
2. Salmon Scavenger Hunt (pages 23-29)	Go on a salmon scavenger hunt to find out about threats to salmon populations.	Describe several ways in which human activities impact salmon at various points in their life cycle.	science, social studies, language arts (research skills)
3. Much Ado About O$_2$ (pages 31-40)	Conduct experiments to measure the changes in dissolved oxygen (DO) when temperatures rise and fertilizer is added.	Define dissolved oxygen. Describe the importance of DO to salmon and other fish. Explain how rising temperatures and the addition of fertilizers affect DO. List some of the human activities that affect DO. Generate ideas for ways to reduce the effects of these activities.	science
4. Salmon People (pages 41-53)	Read and compare salmon myths from two different cultures. Compare the attitudes and beliefs presented in the myths to those in an article about modern-day Native Americans working for salmon recovery. Develop a presentation to convey the cultural importance of salmon.	Describe the cultural significance of salmon to some Native Americans and some of the ways in which those cultural traditions continue today.	social studies, language arts
5. Make Way For Salmon (pages 54-57)	Create a board game in which different salmon runs compete to make it to the ocean and then return to their home streams to spawn.	Demonstrate understanding of the life cycle of salmon, the threats salmon face, and some of the ways humans can help protect wild salmon.	science, social studies, language arts, art

Time	Framework Links	Key Skills	Connections
one session	12, 13, 21	gathering (listening, identifying main ideas), organizing (sequencing, arranging), analyzing (identifying components and relationships among components, discussing)	Use "Sea for Yourself" (pages 72-81) as an introduction to the incredible diversity of creatures that live in marine habitats. Link that activity to this one by stressing the importance of protecting not only marine but also freshwater habitats for anadromous species such as salmon. Follow up with "Salmon People" (pages 41-53 online) to encourage students to explore the way that people depend on and are inspired by salmon in different cultures.
research time plus one session	12, 13, 29, 30.1, 35, 38, 40, 47, 47.1, 48, 50, 52.1, 59, 60, 63, 72	gathering (researching, collecting), analyzing (identifying patterns, comparing and contrasting, discussing), interpreting (identifying cause and effect, drawing conclusions)	Students can further explore threats to salmon populations in "Much Ado About O$_2$" (pages 31-40 online) and "Make Way for Salmon" (pages 54-57 online).
two sessions plus observation time once a week for three to four weeks	10, 10.1, 11, 12, 13, 29, 33.3, 35, 46, 52, 52.1, 63, 72	gathering (observing, collecting, recording, measuring), analyzing (identifying patterns, questioning, calculating, discussing), interpreting (inferring, identifying cause and effect, reasoning), applying (experimenting, proposing solutions, problem solving)	To learn about other threats to salmon populations, try "Salmon Scavenger Hunt" (pages 23-29 online) and "Make Way for Salmon" (pages 54-57 online). For other biodiversity-related science experiments, see "Secret Services" in Biodiversity Basics.
two sessions	5, 10, 11, 12, 13, 26, 29, 35, 41, 46, 52.1, 63, 67.1, 72	gathering (listening, reading comprehension, identifying main ideas), analyzing (identifying components and relationships among components, comparing and contrasting, discussing), presenting (acting)	"A Wild Pharmacy" and "Perspectives" in Wildlife for Sale, as well as "The Culture/Nature Connection" in Biodiversity Basics, deal with the intersection of cultural beliefs and species conservation.
two or three sessions	12,13, 22, 23, 23.1, 29, 30.1, 35, 40, 47, 50, 52.1, 59, 60, 63, 72	applying (planning, designing, estimating, predicting, synthesizing, creating), presenting (writing, illustrating, explaining, clarifying), citizenship skills (working in a group)	To help students understand the threats facing wild salmon populations, have them complete the "Salmon Scavenger Hunt" (pages 23-29 online) and "Much Ado About O$_2$" (pages 31-40 online) before trying this activity.

Appendix G

Resources

The following resources can help you design and enhance a marine biodiversity program or unit. We've included organizations, books, curriculum guides, posters, audiotapes, CD-ROMs, videos, and Web sites. Keep in mind that this resource list includes materials we have found or used; however, there are many other resources available on biodiversity and marine biodiversity.

Please note that we accessed the Web sites in 2003, but Web addresses change often. We've included prices when available, but all costs are subject to change and don't include shipping or handling fees. Books that are out of print may be found by checking the Internet, libraries, and bookstores.

Professional Associations and Nonprofit Conservation Organizations

The following organizations are involved in activities that protect marine biodiversity and educate people about marine-related issues. These organizations can help you and your students find up-to-date information, curriculum guides, research projects, and other environmental education resources.

Adopt-a-Stream Foundation (AASF) is a nonprofit organization that targets watershed restoration through environmental education. AASF encourages and assists individuals, schools, and civic organizations in becoming "streamkeepers," or stewards of local streams. AASF works to ensure that streams continue to provide healthy spawning and rearing habitat for fish, clean drinking water for local residents, and a quality recreational environment. 600 128th Street, SE, Everett, WA 98208-6353. (425) 316-8592. **www.streamkeeper.org**

American Cetacean Society has developed materials that provide information on whales and dolphins. P.O. Box 2639, San Pedro, CA 90731. (310) 548-6279. **www.acsonline.org**

American Littoral Society encourages the study and conservation of marine life and its habitat in the coastal zone and develops scientific information to advocate pro-marine positions. The society offers publications and field trips. ALS funds the Baykeeper project in New York Harbor as well as the ReefKeeper coral reef monitoring projects in Florida and the Caribbean. Sandy Hook, Highlands, NJ 07732. (732) 291-0055. **www.alsnyc.org**

American Museum of Natural History promotes scientific literacy with exhibitions and educational materials on natural history. A newly renovated Ocean Life Hall opened in summer 2003. Central Park West, 79th St., New York, NY 10024. (212) 769-5000. **www.amnh.org/exhibitions/specials/ocean/index.html**

American Zoo and Aquarium Association promotes the advancement of zoos and aquariums. The association focuses on the conservation of wildlife and the preservation and propagation of endangered and rare species. 8403 Colesville Rd., Ste. 710, Silver Spring, MD 20910. (301) 562-0777. **www.aza.org**

Chesapeake Bay Foundation focuses on the management of the Chesapeake Bay and provides environmental education programs for students, teachers, and the general public. 162 Prince George St., Annapolis, MD 21401. (410) 268-8816. **www.savethebay.cbf.org**

The Coral Reef Alliance (CORAL) is a member-supported, nonprofit organization dedicated to keeping coral reefs alive around the world. 2014 Shattuck Ave., Berkeley, CA 94704-1117. (510) 848-0110. **www.coral.org**

The Cousteau Society, Inc. is dedicated to the protection of marine biodiversity and the improvement of the quality of life for present and future generations. It publishes *Dolphin Log* for young readers. 870 Greenbrier Circle, Ste. 402, Chesapeake, VA 23320. (757) 523-9335. **www.cousteau.org**

Ducks Unlimited, Inc. focuses on the conservation of waterfowl and wetlands. Children's memberships include *Puddler* magazine. One Waterfowl Way, Memphis, TN 38120. (800) 453-8257. **www.ducks.org**

Earthwater Stencils, Ltd.

engages community members in watershed protection through the use of storm drain stencils. Stencils on storm drains describe the flow of storm water runoff, raising public awareness about the connections among water pollution and the water cycle. Pre-made storm drain stencils and water-related educational guides can be ordered by contacting Earthwater Stencils, 4425 140th Ave., SW, Rochester, WA 98579. www.earthwater-stencils.com

Environmental Defense

addresses global issues such as ocean pollution, rain forest destruction, and global warming. 257 Park Ave. South, New York, NY 10010. (212) 505-2100. www.environmentaldefense.org

Global Rivers Environmental Education Network (GREEN)

is a project of Earth Force, a youth organization dedicated to creating lasting environmental solutions in local communities. GREEN, an international network of students, teachers, and institutions, conducts comprehensive watershed education programs designed to improve water quality around the world. 1908 Mount Vernon Ave., 2nd Floor, Alexandria, VA 22301. (703) 299-9400. www.green.org

The International Coral Reef Initiative (ICRI)

is an informal network of governments and international agencies working with scientific and conservation institutions to improve coral reef management practices, increase political support for coral reef protection, and share information on the health of coral reef ecosystems. World Bank, Rm. MC 4-4221 A, 1818 H St., NW, Washington DC 20433. www.icriforum.org

Izaak Walton League of America

promotes public education aimed at conserving and responsibly enjoying natural resources. 707 Conservation Ln., Gaithersburg, MD 20878. (800) 453-5463. www.iwla.org

Manomet Center for Conservation Sciences

is a New England-based conservation science and education organization whose mission is to conserve natural resources for the benefit of wildlife and human populations. Originally incorporated as a bird observatory, the center's programs now reach from the tip of South America to the North Slope of Alaska, and focus on biodiversity, habitat, and resource conservation in forests, wetlands, agricultural lands, and the marine environment. 81 Stage Point Rd., P.O. Box 1770, Manomet, MA 02345. (508) 224-6521. www.manomet.org

Marine Aquarium Council (MAC)

is an international nonprofit organization that brings marine aquarium animal collectors, exporters, importers, and retailers together with aquarium keepers, public aquariums, conservation organizations, and government agencies. MAC's mission is to conserve coral reefs and other marine ecosystems by creating standards to certify individuals and businesses that are involved in the collection and care of marine aquarium fishes. 923 Nu'uanu Ave., Honolulu, HI 96817. (808) 550-8217. www.aquariumcouncil.org

Marine Conservation Biology Institute

is a nonprofit scientific and conservation advocacy organization that works to protect and restore marine life by encouraging research and training in marine conservation biology; bringing scientists together to examine crucial marine conservation issues; doing policy research to frame the marine conservation agenda;

lecturing and producing books and other publications to educate scientists, the public, and decision makers on key issues; and building partnerships to solve problems affecting marine life and people. 15805 NE 47th Ct., Redmond, WA 98052. www.mcbi.org

Marine Stewardship Council (MSC)

promotes sustainable fishery practices and responsible consumerism through its certification program. A product labeled as "MSC certified" indicates to consumers that the seafood product was harvested from a well-managed fishery. The labeling practice also rewards those fisheries that follow environmentally sound practices. MSC's Sea into the Future campaign educates consumers about the dangers of overfishing, and encourages citizens to pledge to buy sustainably harvested seafood. 2110 N. Pacific St., Ste. 102, Seattle, WA 98103. www.msc.org

Monterey Bay Aquarium

seeks to stimulate interest, increase knowledge, and promote stewardship of Monterey Bay and the world's ocean environment. It hosts the Seafood Watch program, designed to raise consumer awareness about the importance of choosing sustainable seafood in markets and restaurants. Seafood Watch recommends which seafood to buy or avoid, helping consumers become active supporters of environmentally friendly seafood harvesting methods. Check out the Web site for a downloadable Seafood Watch wallet card. 886 Cannery Row, Monterey, CA 93940. (831) 648-4888. www.montereybayaquarium.org

The National Coalition for Marine Conservation

is the nation's oldest public advocacy group dedicated to conserving the world's ocean fish, habitat, and environment. Its mission is to build public awareness of the threats to marine fisheries, provide constructive solutions, and convince state, national, and international fishery managers to take appropriate action to reverse the depletion of marine fishery resources. 3 West Market St., Leesburg, VA 20176. (703) 777-0037. **www.savethefish.org**

The New England Aquarium

seeks to increase the understanding of aquatic life and environments, enable people to act to conserve the world of water, and provide leadership for the preservation and sustainable use of aquatic resources. The aquarium is home to the Pew Fellows Program in Marine Conservation. Central Wharf, Boston, MA 02110. (617) 973-5200. **www.neaq.org**

North American Association for Environmental Education

supports the work of professional environmental educators and students in North America and in more than 55 countries around the world. It publishes a bimonthly newsletter as well as many other environmental education resources, and it sponsors an annual conference. 410 Tarvin Rd., Rock Spring, GA 30739. (706) 764-2926. **www.naaee.org**

Oceana is a nonprofit international advocacy organization committed to protecting the world's oceans. Oceana brings together dedicated people from around the world, building an international movement to save the oceans through public policy advocacy, science and economics, legal action, grassroots mobilization, and public education. 2501 M St., NW, Ste. 300, Washington, DC 20037. (202) 833-3900. **www.oceana.org**

The Ocean Conservancy,

formerly known as the Center for Marine Conservation, focuses on protecting marine wildlife and habitats and conserving coastal and ocean resources. It publishes several types of educational materials for classrooms and young people. 1725 DeSales St., NW, Ste. 600, Washington, DC 20036. (202) 429-5609. **www.oceanconservancy.org**

Ocean Futures,

founded by Jean-Michel Cousteau (son of well-known oceanographer Jacques Cousteau), works to further public awareness of marine habitats and mammals through research and education efforts. The organization pursues a wide range of projects—from sustainable tourism to coastal protection initiatives—with a particular emphasis on using television and film to raise public awareness and support for ocean conservation. 325 Chapala St., Santa Barbara, CA 93101. (805) 899-8899. **www.oceanfutures.org**

The Ocean Project is an international network of aquariums, zoos, museums, and conservation organizations working to create an understanding of the significance of oceans and the role each person plays in conserving them for the future. 102 Waterman St., Ste. 16, Providence, RI 02906. (401) 272-8822. **www.theoceanproject.org**

The Ocean Wilderness Network

is a nonprofit organization made up of a coalition of ocean conservation organizations seeking to establish a network of marine protected areas along the Pacific coast of the United States. 2425 Porter St., Ste. 18, Soquel, CA 95073. (831) 462-2550. **www.ocean wildernessnetwork.org**

Pew Fellows Program in Marine Conservation,

a program of the Pew Charitable Trusts in partnership with the New England Aquarium, creates the bi-monthly electronic newsletter, *PFP SeaSpan*. New England Aquarium, Central Wharf, Boston, MA 02110. (617) 720-5100. **www.pewmarine.org**

Project AWARE, the educational branch of the Professional Association of Diving Instructors, provides information on marine biodiversity issues and gives grants to support marine education. 30151 Tomas St., Rancho Santa Margarita, CA 92688-2125. **www.projectaware.org**

Project Learning Tree,

a program of the American Forest Foundation, teaches environmental education to students in grades K through 12. Guides are distributed free to workshop participants. 1111 19th St., NW, Ste. 780, Washington, DC 20036. (202) 463-2462. **www.plt.org**

Project WET facilitates and promotes awareness, appreciation, knowledge, and stewardship of water resources through the development and dissemination of curriculum materials and the establishment of state and internationally sponsored Project WET programs. 201 Culbertson Hall, Montana State University, Bozeman, MT 59717. (406) 994-5392. **www.projectwet.org**

Project WILD is a national environmental education program that emphasizes wildlife conservation, ecology, and interdisciplinary teaching. The program offers an Aquatic K-12 Curriculum and Activity Guide, highlighting aquatic ecosystems and their inhabitants. Guides are distributed free to workshop participants. 5555 Morningside Dr., Ste. 212, Houston, TX 77005. (713) 520-1936. **www.projectwild.org**

SeaWeb is a nonprofit organization that works to raise awareness of the world's oceans and marine life. SeaWeb strives to make credible information about the ocean accessible to the public. 1731 Connecticut Ave., NW, 4th Fl., Washington, DC 20009. (202) 483-9570. **www.seaweb.org**

Shedd Aquarium promotes the enjoyment, appreciation, and conservation of aquatic life and environments through education, exhibits, and research. Shedd is the world's largest indoor aquarium and displays animals from all around the planet. 1200 South Lake Shore Dr., Chicago, IL 60605. (312) 939-2435. **www.sheddnet.org**

South Carolina Aquarium displays and interprets the aquatic environments of South Carolina. 100 Aquarium Wharf, Charleston, SC 29413. (843) 720-1990. **www.scaquarium.org**

The Surfrider Foundation USA is a nonprofit organization dedicated to protecting oceans, waves, and beaches. P.O. Box 6010, San Clemente, CA 92674. (949) 492-8170. **www.surfrider.org**

Windows on the Wild (WOW) is an international environmental education program of World Wildlife Fund. The program is designed to educate people of all ages about biodiversity issues and to stimulate critical thinking, discussion, and informed decision-making on behalf of the environment. Through educational materials (such as this marine module), training programs, and special events and programs, Windows on the Wild uses biodiversity as the organizing theme to help create an environmentally literate citizenry that has the knowledge, skills, and commitment to make responsible decisions about the environment. 1250 24th St., NW, Washington, DC 20037. (202) 293-4800. **www.worldwildlife.org/windows**

World Wildlife Fund (WWF) works worldwide in more than 100 countries to protect the diversity of life on Earth. To find out more about WWF's work on fisheries and marine issues, visit **www.worldwildlife.org/oceans**. WWF also supports a variety of education and community outreach programs, such as Windows on the Wild (see above). 1250 24th St., NW, Washington, DC 20037. (202) 293-4800. **www.worldwildlife.org**

Government Agencies and Organizations

National Marine Fisheries Service (NMFS) is the federal agency responsible for the stewardship of the nation's living marine resources and their habitats. 1315 East West Hwy., SSMC3, Silver Spring, MD 20910. (301) 713-2370. **www.nmfs.noaa.gov**

National Ocean Service (NOS) is NOAA's principal advocate for coastal and ocean stewardship through partnerships at all levels, including the National Marine Sanctuaries program. NOS developed Wavebreaking News, a multimedia report on activities within the service. 1305 East West Hwy., SSMC4, 13th Floor, Silver Spring, MD 20910. (301) 713-3070. **www.nos.noaa.gov**

The National Oceanic and Atmospheric Administration (NOAA), in the U.S. Department of Commerce, conducts research and gathers data about oceans, the atmosphere, outer space, and the sun, and applies this knowledge to science and services that affect the American public. NOAA includes the National Ocean Service, the National Sea Grant College Program, the National Marine Fisheries Service, and the National Weather Service, among other programs. NOAA's research, conducted through the Office of Oceanic and Atmospheric Research (OAR), is the driving force behind NOAA's environmental products and services. 14th St. and Constitution Ave., NW, Rm. 6013, Washington, DC 20230. (202) 482-6090. **www.noaa.gov**

The National Sea Grant College Program, a program of the National Oceanic and Atmospheric Administration in the Department of Commerce, is a partnership among government, academia, industry, scientists, and private citizens. The goal of the program is to help Americans understand and sustainably use the Great Lakes and ocean waters for long-term economic growth. 1315 East West Hwy., Silver Spring, MD 20910. (301) 713-2431. **www.nsgo.seagrant.org**

U.S. Environmental Protection Agency (EPA) sets and enforces environmental standards and conducts research on the causes, effects, and control of environmental problems. It supports environmental education programs and projects through the Office of Environmental Education. Ariel Rios Building, 1200 Pennsylvania Ave., NW, Washington, DC 20460. (202) 260-2090. **www.epa.gov/enviroed**

U.S. Fish and Wildlife Service is part of the Department of the Interior and was created to conserve fish and wildlife resources in the United States. The Web site provides information for educators on wildlife laws, environmental education, and the national wildlife refuge system. The site includes extensive database access for searches on wildlife information, including a current list of endangered and threatened species. Webb Bldg., 4040 North Fairfax Dr., Rm. 308, Arlington, VA 22203-1613. (202) 208-4131. **www.fws.gov**

Curriculum Resources

Biodiversity Basics (Middle School) explores the meaning of biological diversity, its significance, current status, and measures taken to protect it. Extensive background information is complemented by 34 interdisciplinary activities for teachers and nonformal educators. Appendices offer guidelines for action projects, a biodiversity education framework, language learning tips, current legislation, planning charts, glossary, and bibliography. (World Wildlife Fund, 1999). Available from Acorn Naturalists, 17821 East 17th St., #103, P.O. Box 2423, Tustin, CA 92781. (800) 422-8886. $39.95 www.acornnaturalists.com

Biodiversity: The Florida Story (Middle School) is a set of five teacher's guides that are linked to the Florida Sunshine State standards. Topics include urbanization and economics; agriculture and land management; water quality and its relationship to land use, management, and public/private partnerships; and nonnative (alien) species, indicator species, and human uses. Teachers may obtain a free set of the materials by attending a workshop sponsored by the Regional Service Project. (1998). Available from the Office of Environmental Education/Florida Gulf Coast University, 1311 Paul Russell Rd., Ste. 201A, Tallahassee, FL 32301-4880. (850) 487-7900. www.myfloridaedu cation.com/curriculum/environ/biodi versity.html

Bridge–Ocean Sciences Teacher Resource Center (All Levels) is a growing collection of the best marine education resources available online. It provides educators with a convenient source of accurate and useful information on global, national, and regional marine science topics and gives researchers a contact point for educational outreach. The Bridge is supported by the National Oceanographic Partnership Program and is sponsored by the National Marine Educators Association and the national network of NOAA/Sea Grant educators. Sea Grant Marine Advisory Services, Virginia Institute of Marine Science, College of William and Mary, Gloucester Point, VA 23062. www.vims.edu/bridge

Chesapeake Choices and Challenges (CCC), the Chesapeake Bay Foundation's educational curriculum, provides students with the skills necessary to investigate local bay issues, while allowing teachers to integrate bay-related activities with classroom instruction. Interdisciplinary activities in each unit are written for middle school students and correlate with state standards. Curriculum is available to teachers in Maryland, Pennsylvania, and Virginia. Teachers receive the curriculum and associated materials at regionally sponsored professional development workshops. Philip Merrill Environmental Center, 6 Herndon Ave., Annapolis, MD 21403. (410) 268-8816. www.cbf.org

Consortium for Oceanographic Activities for Students and Teachers (COAST) is a K-12 ocean and coastal science education initiative that provides links to such programs as Operation Pathfinder. Center for Educational and Training Technology, P.O. Box 9662, Mississippi State, MS 39762-9662. www.coast-nopp.org

FOR SEA (Middle School) is a marine science curriculum and teacher training program that recognizes the power and the fragility of the world's oceans as well as the need for individuals to understand and appreciate the sea. At workshops and residential institutes, teachers engage in FOR SEA hands-on activities, learn from leading scientists and science educators, practice successful marine science teaching strategies, and develop a plan for their own use of the FOR SEA materials. Institute of Marine Science, P.O. Box 188, Indianola, WA 98342. (360) 779-5122. www.forsea.org

Global Environmental Education Resource Guide (Middle School) is a selection of activities that help frame and clarify key issues associated with the global environment. Topics covered include acid rain, biodiversity, deforestation, and desertification, greenhouse gases, marine and estuarine pollution, overpopulation, ozone depletion, and sea level rise. (1996). Available from University of Southern Mississippi, Institute of Marine Sciences, and the J. L. Scott Marine Education Center and Aquarium, P.O. Box 7000, Ocean Springs, MS 39566-7000. (601) 374-5550. $10.00

The JASON Project (Middle School) is an educational project begun by Robert Ballard (discoverer of the wreck of the RMS Titanic). The JASON Project is dedicated to enabling teachers and students around the world to take part in global undersea explorations using advanced interactive telecommunications. Teacher's guides and links to technology resources for the classroom can also be found on this site. JASON Foundation for Education, 11 Second Ave., Needham Heights, MA 02494-2808. (781) 444-8858. www.jasonproject.org

Living in Water (Middle School)

is a classroom-based scientific study of water, aquatic environments, and the plants and animals that live in water. It integrates basic physical, life, and earth sciences; mathematics; and language arts and is a complete year-long science curriculum. Teachers or districts may select from among its 50 activities to create shorter curricula. Developed by the National Aquarium in Baltimore, *Living in Water* is available from Kendall/Hunt Publishing Company. (800) 228-0810. $23.95. www.kendallhunt.com

Marine Activities, Resources, and Education (MARE)

(Elementary-Middle School) in the Lawrence Hall of Science at the University of California, Berkeley, is a whole-school, interdisciplinary, marine science program that includes a series of eight curriculum guides. Each grade focuses on a different marine environment and integrates language arts, language development, social studies, and art with science and mathematics. University of California, Berkeley, Lawrence Hall of Science #5200, Berkeley, CA 94720-5200. (510) 642-5132. www.lhs.berkeley.edu/MARE

Marine Education Sites Directory (All Levels),

created by Maryland Sea Grant, provides links to sites containing materials for K-12, college, and community marine education. www.mid-atlantic.seagrant.org/links/education.html

Ocean News (Middle School-

Secondary) is a series of newsletters, teacher's guides, and computer disks that focus on five marine science issues: Exploring the Fluid Frontier, Marine Mammals, Seabirds, Marine Pollution, and Marine Biodiversity. (1996). Available from Bamfield Marine Station, Bamfield, British Columbia, VOR 1BO, Canada. (604) 728-3301. $73.00. www.oceanlink.island.net/onews/oceanews.html

Our Oceans, Ourselves: Marine Biodiversity for Educators (All Levels)

is a marine biodiversity framework that provides a template to help educators include marine biodiversity in their programs. It also includes background information on marine biodiversity and several activities. (1995). Available from Environment Canada, 351 St. Joseph Blvd., 5th Fl., Hull, Quebec, K1A OH3, Canada. (819) 953-4374. Free. naaee.org/npeee/biodiversity.php

Project Oceanography (Middle

School) is a live television program designed for science students. Each week during the school year, students can learn about a variety of ocean science topics right in the classroom. College of Marine Science, University of South Florida, St. Petersburg, FL 33701. (888) 51-OCEAN. www.marine.usf.edu/pjocean

Water Matters/El Aqua es Importante, Volume 3 (All

Levels) is a water education kit developed for the U.S. Geological Survey's Water Resources Education Initiative. Available in both English and Spanish, the curriculum emphasizes oceans and coastal hazards, watersheds, and hazardous waste through a set of colorful posters. Available from the National Science Teachers Association. P.O. Box 90214, Washington, DC 20090-0214. (800) 277-5300. $5.95. store.nsta.org. Free copies are available in Spanish from the EPA's National Center for Environmental Publications. P.O. Box 42419, Cincinnati, OH 45242-0419. (800) 490-9198. www.epa.gov/ncepihom/ordering.htm

WOW!: The Wonders of Wetlands (All Levels) is a

curriculum guide that includes background material for teachers and hands-on activities. Activities focus on general wetland concepts and definitions, wetland communities of plants and animals, the role of water in wetlands, the role of soils in

wetlands, and the interactions between humans and wetlands. (1995). Available from the Watercourse, 201 Culbertson Hall, Montana State University, Bozeman, MT 59717-0057. (406) 994-5392. $21.95. www.projectwet.org

Books and Articles for Adults

The following resources focus on various aspects of marine biodiversity. While some are appropriate for older students, most are written at an adult reading level.

The Blue Planet: Seas of Life

by Andrew Byatt reveals the diverse wildlife found in six distinct habitats that make up the watery parts of our world. With a forward by Sir David Attenborough and numerous full-color photographs, this book offers a comprehensive portrait of Earth's ocean ecosystems. (DK Publishing, Inc., 2001). $40.00

Education and Sustainability: Responding to the Global Challenge, edited by Daniella

Tilbury, Robert B. Stevenson, John Fien, and Danie Schreuder, discusses current issues in education for sustainability (EFS), particularly in light of the 2002 World Summit on Sustainable Development. Highlighting case studies from around the world, the book, and an accompanying publication/CD-ROM on the EFS debate, highlight issues relevant to teaching young people about the environment. (IUCN, 2002). $21.00. www.iucn.org/bookstore

Encyclopedia of Marine Mammals edited by William F. Perrin, Bernd Würsig, and J.G.M. Thewissen is a comprehensive work that covers a wide variety of aspects of marine mammals, including their physiology, evolution, behavior, reproduction, ecology, and diseases, as well as issues of exploitation, conservation, and management. (Academic Press, 2002). $139.95

Faces of Fishing: People, Food, and the Sea at the Beginning of the Twenty-First Century by Brad Matsen is a photo essay that highlights people around the world who are involved with industrial, small-scale, or recreational fishing. (Monterey Bay Aquarium Press, 1998). $19.95

From Abundance to Scarcity: A History of U.S. Marine Fisheries Policy by Michael Weber describes the historical development of the American government's fisheries policy and institutions. (Island Press, 2001). $27.50

Global Marine Biological Diversity: A Strategy for Building Conservation into Decision-Making by Elliot Norse focuses on threats to life in the sea and ways to save, study, and use that life sustainably. Developed by more than 100 experts, this book presents information and views on the challenge of conserving the living sea. (Island Press, 1993). $32.00

The Living Ocean: Understanding and Protecting Marine Biodiversity by Boyce Thorne-Miller, with a forward by Sylvia Earle, is an essential primer for anyone wishing to gain an understanding of marine biodiversity and how it can be protected. It provides an overview of basic concepts and principles as well as a review of relevant policy issues and existing laws. (Island Press, 1998). $17.95

National Geographic Atlas of the Ocean: The Deep Frontier by Sylvia Earle includes more than 150 detailed maps, dramatic photographs, and state-of-the-art satellite images. (National Geographic Society, 2001). $50.00

Ocean Planet: Writings and Images of the Sea by Peter Benchley and Judith Gradwohl explores the roles the sea plays in our lives and examines the impact of human endeavors on this vast yet vulnerable environment. (Harry N. Abrams Inc. Publishers and Times Mirror Magazines Inc., in association with the Smithsonian Institution, 1995). $39.95 hardback, $19.95 paperback

The Sea Around Us by Rachel Carson is a classic book that captures the mystery and allure of the ocean and provides a timely reminder of both the fragility and the importance of the ocean and the life that abounds within it. (Oxford University Press, 1991). $14.95

SeaLife: A Complete Guide to the Marine Environment by Geoffrey Waller includes detailed illustrations, essays by marine biologists, and descriptions of more than 600 species. (Smithsonian Institution Press, 1996). $55.00

Song for the Blue Ocean: Encounters Along the World's Coasts and Beneath the Seas by Carl Safina explores coasts, islands, reefs, and deep ocean environments around the world. (Henry Holt & Co., 1997). $30.00

"Thinking Like an Ocean" in Conservation in Practice (Fall 2002) by Scott Norris with Martin Hall, Edward Melvin, and Julia Parrish explores the dilemma of marine bycatch and its potential solutions. The article centers on the need for "ecological common sense,"

using the campaign to decrease dolphin bycatch as an example of a complicated case study with some unexpected ramifications. Available from the Society for Conservation Biology, 4245 North Fairfax Dr., Arlington, VA 22203. (703) 276-2384. $8.00. cbinpractice.org/InPractice/BackIssueForm.pdf

Wild Ocean: America's Parks Under the Sea by Sylvia Earle and Walcott Henry takes a tour of the deepest reaches of the ocean with famed marine biologist Sylvia Earle. Through stories, maps, and photos, Dr. Earle guides readers through America's national marine sanctuaries, exploring little-known creatures and plants, immense coral reefs, and the relics of shipwrecks. (National Geographic Society, 1999). $40.00

Books for Middle School Students

Beneath Blue Waters: Meetings with Remarkable Deep-Sea Creatures by Deborah Kovacs and Kate Madin explores the diversity of life in the deep regions of the ocean. (Viking, 1996). $16.99

Earth Kids by Jill Wheeler tells stories of individual children who have taken action to conserve the environment. Some of the children have helped to protect forests, defend animals, and clean up beaches, water, and air. (Abdo & Daughters, 1993). $22.83

The Endangered Animals Series by Dave Taylor includes books about animals from endangered oceans, wetlands, islands, grasslands, forests, mountains, deserts, and savannas. (Crabtree Publishing, 1992-93). $6.36 each

WOW!—A Biodiversity Primer
by World Wildlife Fund is a full-color, magazine-style primer that helps middle school students understand biodiversity. This award-winning primer includes fiction and nonfiction stories and articles. (World Wildlife Fund, 1994). Available from Acorn Naturalists, 17821 East 17th St., #103, P.O. Box 2423, Tustin, CA 92781. (800) 422-8886. $3.00. www.acornnaturalists.com

Young Explorer's Guide to Undersea Life by Pam Armstrong introduces children to sea lions, whales, sharks, moon jellies, and other sea life. (Monterey Bay Aquarium Press, 1996). $16.95

Zoobooks: Exploring Ocean Ecosystems contains eight student books of articles, photographs, illustrations, facts, games, and activities about animals, along with a curriculum guide. It takes a thematic approach to teaching science and the scientific process and encourages learning across the curriculum. Titles include *Dolphins and Porpoises*, *Penguins*, *Seals and Sea Lions*, *Sharks*, *Turtles*, and *Whales*. (1990-96). Available from Wildlife Education Ltd., 9820 Willow Creek Rd., Ste. 300, San Diego, CA 92131-1112. (800) 477-5034. $29.95

Multimedia Resources

Videos, Audiotapes, and CD-ROMs

Biodiversity! Exploring the Web of Life Education Kit
(Middle School) introduces young people to biodiversity. This half-hour video by World Wildlife Fund explores the meaning of biodiversity, its status, and what people can do to protect it. An educator's guide is included in the kit along with *WOW!—A Biodiversity Primer*, a full-color magazine for students. (1997). Available from Acorn Naturalists, 17821 East 17th St., #103, P.O. Box 2423, Tustin, CA 92781. (800) 422-8886. $24.95.
www.acornnaturalists.com

Footsteps in the Sea: Growing Up in the Fisheries Crisis features three teenagers in Gloucester, Massachusetts, who document the crisis in their fishing community. (1998). Video available from Bullfrog Films, P.O. Box 149, Oley, PA 19547. (610) 779-8226. $195.00. www.bullfrogfilms.com

The Great Ocean Rescue
(Middle School) introduces students to marine ecosystems and marine biodiversity. The kit includes a CD-ROM, four student reference booklets, and a teacher's guide with lesson plans, worksheets, and activities. Windows™ or Macintosh™ compatible. (1996). Available from Tom Snyder Productions, Inc., 80 Coolidge Hill Rd., Watertown, MA 02172-9718. (800) 342-0236. $328.00. www.teachtsp.com

Magnificent Fish: The Forgotten Giants focuses on three generally unknown large fish: sharks, tuna, and billfish. Hosted by Peter Benchley, the video demonstrates the behavior and physical feats of these creatures while explaining misconceptions about them. Available from WOW Films, Global Marine Programs, New England Aquarium, Central Wharf, Boston, MA 02110. (New England Aquarium, 1997). $19.95

Oceans Alive/Oceanos Vivos
(Middle School-Secondary) includes four videotapes, each containing multiple five-minute programs that illustrate the relationships among marine life. Filmed in the Red Sea, Caribbean, Sea of Cortez, Pacific, and other locations, "Oceans Alive!" encourages students to ask questions and to share their experiences about the sea. The first two videotapes of the program are also available in Spanish. (Environmental Media Corporation, 1996). $199.95. www.envmedia.com

Oceans of Life Kit
(Middle School) is one of the Radio Expeditions series produced by National Public Radio and the National Geographic Society. "Oceans of Life" (1995) focuses on marine biodiversity and includes a teacher's guide, a 30-minute audio cassette, and a map or poster. Hour-long CD recordings of Radio Expeditions shows are also available. To order kits, call (202) 414-2726. To order CDs, call (888) 677-3472. www.npr.org/programs/re/archivesdate/1995/oceans/index.html

The Shape of Life
(Secondary-Adult) An eight-part television and CD series that reveals the breakthroughs of scientific discovery that led to the understanding of the dramatic rise of the animal kingdom through evolution. The first six episodes emphasize marine animals and their evolutionary roles. "The Shape of Life" activity guide features explorations, activities, and lessons to accompany the series for informal science educators, classroom teachers, students, and families. (2002). Video: $79.98. www.pbs.org/kcet/shapeoflife/resources/

Unwanted Catch, produced specifically for zoo and aquarium audiences, analyzes the problem of bycatch, or non-targeted marine animals captured during fishing practices. The video examines five separate fisheries to illustrate the

issue and gain perspectives from various people. It describes which species are highly susceptible to being caught, and explains strategies for reducing bycatch. Available from WOW Films, Global Marine Programs, New England Aquarium, Central Wharf, Boston, MA 02110. (New England Aquarium, 1998). $19.95

The Video Project: Media for a Safe and Sustainable World is an eco-video collection for schools with themes related to marine biodiversity topics. Catalogues are available from P.O. Box 77188, San Francisco, CA 94107. (800) 4-PLANET. www.videoproject.net

Posters

Biodiversity—From Sea to Shining Sea (Middle School-Secondary) is a poster kit produced by World Wildlife Fund that includes two 22" x 34" double-sided posters (the front features 12 photographs highlighting the diversity of life on Earth; the back includes a map of the most threatened ecoregions in the United States). The kit also includes a 12-page educator's guide with background information about biodiversity and suggestions for how to use the poster. (1996). Available from Acorn Naturalists, 17821 East 17th St., #103, P.O. Box 2423, Tustin, CA 92781. (800) 422-8886. $5.00. www.acornnaturalists.com

Diversity Endangered (Middle School-Secondary) is a poster exhibition kit designed by the Smithsonian Institution Traveling Exhibition Service (SITES). The kit includes 15 posters that feature issues and challenges related to biodiversity, ranging from the variety and interrelation of species to the complex habitats of tropical rain forests, wetlands, and coral reefs. The kit is accompanied by a resource and programming guide. (1997). Available from SITES, 1100 Jefferson Dr., SW,

Quad 3146, Washington, DC 20560. (202) 357-3168. $100.00. www.sites.si.edu/education/search_pubs.asp

Joy to the Fishes and the Deep Blue Sea (All Levels), produced by World Wildlife Fund and Monterey Bay Aquarium, celebrates Earth's amazing diversity of marine life and the importance of protecting it. The front features a colorful painting by nature artist Larry Duke and includes species that live in five marine habitats—coral reefs, mangroves, estuaries, kelp forests, and the deep sea. The back of the poster describes the "Top Ten Things You Can Do to Save Marine Biodiversity" and identifies the animals, plants, and habitats depicted on the front of the poster. Available from Acorn Naturalists, 17821 East 17th St., #103, P.O. Box 2423, Tustin, CA 92781. (800) 422-8886. $5.00. www.acornnaturalists.com

Ocean Habitats by award-winning artist Larry Duke is a series of posters that feature intricately detailed and delicately colored paintings of the Kelp Forest, Rocky Shore, Sandy Beaches, Deep Sea, Coral Reefs, and Wetlands. Available from the Monterey Bay Aquarium, 886 Cannery Row, Monterey, CA 93940. (831) 648-4800. $15.00 each

Selected Web Sites

Educational Resources Information Center (ERIC) provides links to an extensive body of educational literature. The center allows access to the on-line ERIC Publications Catalogue, a Question-and-Answer Service, the ERIC Database, links to other education sites, and numerous products to help search and use the information in the ERIC system. The site is supported by the U.S. Department of Education

and the National Library of Education. www.eric.ed.gov

EE Link provides a variety of links to informative environmental sites and facilitates the investigation of current environmental issues. The site includes printable documents and links to environmental education-related organizations and materials on the Internet. www.eelink.net

Eisenhower National Clearinghouse (ENC) for Mathematics and Science Education contains information on the U.S. Department of Education's ENC initiative, which is dedicated to continuing efforts to reform K-12 math and science education. The clearinghouse includes activities, journal articles, a listing of conferences and special events, and links to other sites. www.enc.org

Marine Protected Areas of the United States, a Web site of the U.S. Departments of Commerce and the Interior, was created by the National MPA Initiative in 2000 as a resource on marine protected areas in the United States. The site offers basic facts on MPAs, such as a definition, a list of current U.S. MPAs, and a description of conservation challenges. For more in-depth information, a virtual library containing links to national and global MPA reference materials is provided. www.mpa.gov

Marinecareers.net, a Web site of NOAA's Sea Grant Program, introduces young adults to careers in marine-related professions such as oceanography, marine biology, and ocean engineering. The site highlights various occupations and includes a resource page with links to information on other aspects of working in marine-related fields. www.marinecareers.net

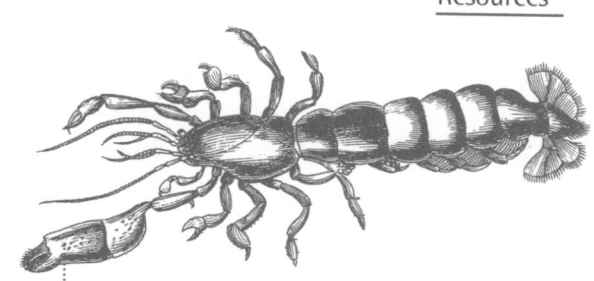

NOAA Ocean Explorations

follows ocean explorations in near real-time, accesses related lesson plans for grades 5 through 12, and promotes learning about ocean exploration technologies. A visitor can observe remote marine flora and fauna in the multimedia gallery, review NOAA's 200-year history of ocean exploration, and use additional NOAA resources in a virtual library. **www.oceanexplorer.noaa.gov/explorations/explorations.html**

OceanLink's Careers in Marine Science

describes a variety of marine-related career opportunities. Interviews with professionals from around the world provide firsthand descriptions of a range of marine science positions, and a "Career Links" page connects to Web sites of American and Canadian universities offering marine-related programs. **www.oceanlink.island.net/career/career2.html**

REVEL Project—Research and Education: Volcanoes, Exploration, and Life (REVEL)

is a National Science Foundation-sponsored program that sends teachers below the surface of the Pacific Ocean to participate in ongoing oceanographic research. Follow along with the current mission through the Web site's logbook. **www.ocean.washington.edu/outreach/revel**

Second Nature brings environmental education online with "Starfish." "Starfish" was developed to help educators infuse environmental and sustainability concepts into their teaching. Starfish provides database searches of university-level courses offered by faculty members in more than 35 disciplines, a bibliography with over 1,000 references, and comprehensive information on more than 20 innovative teaching and learning techniques. **www.secondnature.org/efs/efs.html**

The Sustainable Seas Expeditions,

sponsored by the National Geographic Society and the National Marine Sanctuary Program, takes students along as submersibles travel 2,000 feet below the surface to study the biology of 12 marine sanctuaries in U.S. waters. The project includes live Web chats and video uplinks with well-known scientists such as Sylvia Earle. **www.nationalgeographic.com/seas**

WWW Virtual Library: Environment

covers a wide range of ecology and biodiversity-related topics. The site also provides links to periodicals and journals for more advanced research. **www.earthsystems.org/virtuallibrary/vlhome.html**

Appendix H
Metric Conversions

When You Know	Multiply By	To Find
feet	.30	meters
yards	.91	meters
miles	1.61	kilometers
square feet	.09	square meters
square yards	.84	square meters
square miles	2.60	square kilometers
pounds	.45	kilograms
short tons (2,000 pounds)	.90	tonnes (metric ton)
gallons	3.79	liters

Boldfaced numbers indicate pages in the Salmon Case Study found at www.worldwildlife.org/windows/marine.
Italicized numbers indicate pages in the Framework found at www.worldwildlife.org/windows.

Windows on the Wild: Oceans of Life    **371**

Boldfaced numbers indicate pages in the Salmon Case Study found at www.worldwildlife.org/windows/marine.
Italicized numbers indicate pages in the Framework found at www.worldwildlife.org/windows.

Activities Index

Boldfaced type indicates pages in the Salmon Case Study found at **www.worldwildlife.org/windows/marine.**

Unit Plans Index

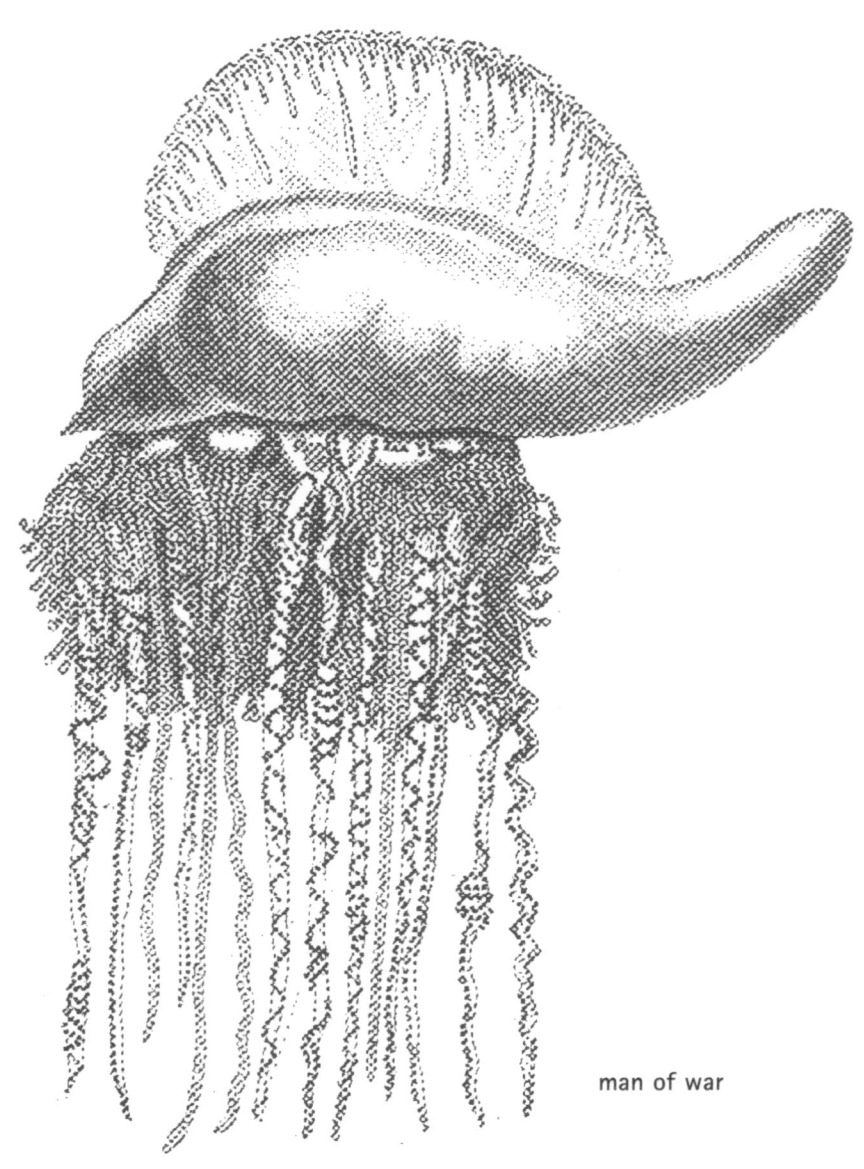

man of war

Notes:

Notes:

FEEDBACK FORM

WHAT DO YOU THINK?

Please take a few minutes to give us feedback so that we can improve *Oceans of Life* when we reprint. We will also use your comments to help us develop other materials in the *WOW* family. Thanks for taking the time to give us your ideas!

RATING THE SECTIONS

Please use the numbers below to rate the sections in *Oceans of Life*. Feel free to add any specific comments on content, design, and usefulness.

Ratings: 4 = Great!   3 = Good   2 = Average   1 = Poor

☐ **Background Information**   ☐ **Shark Case Study**   ☐ **Unit Plans**

☐ **Introducing Marine Biodiversity**   ☐ **Alien Species Case Study**

☐ **Coral Reefs Case Study**   ☐ **Salmon Case Study**   ☐ **Glossary & Resources**

☐ **Shrimp Case Study**   ☐ **Mini Case Studies**   ☐ **Overall Design**

GENERAL COMMENTS:

What do you like best about *Oceans of Life*?

What recommendations do you have to improve *Oceans of Life*? If you have comments about specific activities, please list the activity title(s) along with your suggestions.

Other comments or ideas:

The following errors appear in Oceans of Life:

Page Number _____

Error (please describe clearly) _____

_____

_____

_____

Page Number _____

Error (please describe clearly) _____

_____

_____

_____

Page Number _____

Error (please describe clearly) _____

_____

_____

_____

Please complete (optional):

Name _____

School/Organization _____

Address _____

_____

Phone _____   E-mail _____

Please send this form to:

World Wildlife Fund
Education Department
1250 24th Street, NW
Washington, DC 20037
Fax: (202) 887-5293 or
Fax: (202) 293-9211